Springer Series in Computational Physics

Springer Series in Computational Physics

A Computational Method in Plasma Physics
F. Bauer, O. Betancourt, P. Garabedian

Implementation of Finite Element Methods for Navier-Stokes Equations
F. Thomasset

Finite-Difference Techniques for Vectorized Fluid Dynamics Calculations
Edited by D. Book

Unsteady Viscous Flows. D. P. Telionis

Computational Methods for Fluid Flow. R. Peyret, T. D. Taylor

Computational Methods in Bifurcation Theory and Dissipative Structures
M. Kubicek, M. Marek

Optimal Shape Design for Elliptic Systems. O. Pironneau

The Method of Differential Approximation. Yu. I. Shokin

Computational Galerkin Methods. C. A. J. Fletcher

Numerical Methods for Nonlinear Variational Problems
R. Glowinski

Numerical Methods in Fluid Dynamics. Second Edition
M. Holt

Computer Studies of Phase Transitions and Critical Phenomena
O. G. Mouritsen

Finite Element Methods in Linear Ideal Magnetohydrodynamics
R. Gruber, J. Rappaz

Numerical Simulation of Plasmas. Y. N. Dnestrovskii, D. P. Kostomarov

Computational Methods for Kinetic Models of Magnetically Confined Plasmas
J. Killeen, G. D. Kerbel, M. C. McCoy, A. A. Mirin

Spectral Methods in Fluid Dynamics
C. Canuto, M. Y. Hussaini, A. Quarteroni, T. A. Zang

Computational Techniques for Fluid Dynamics 1. Second Edition
Fundamental and General Techniques. C. A. J. Fletcher

Computational Techniques for Fluid Dynamics 2. Second Edition
Specific Techniques for Different Flow Categories. C. A. J. Fletcher

Methods for the Localization of Singularities in Numerical Solutions of Gas Dynamics Problems. E. V. Vorozhtsov, N. N. Yanenko

Classical Orthogonal Polynomials of a Discrete Variable
A. F. Nikiforov, S. K. Suslov, V. B. Uvarov

Flux Coordinates and Magnetic Field Structure:
A Guide to a Fundamental Tool of Plasma Theory
W. D. D'haeseleer, W. N. G. Hitchon, J. D. Callen, J. L. Shohet

W.D. D'haeseleer W.N.G. Hitchon
J.D. Callen J.L. Shohet

Flux Coordinates and Magnetic Field Structure

A Guide to a Fundamental Tool of Plasma Theory

With 40 Figures

Springer-Verlag
Berlin Heidelberg New York London
Paris Tokyo Hong Kong Barcelona

Dr. William Denis D'haeseleer, Associate Professor (part time)
Katholieke Universiteit Leuven, School of Engineering,
Arenbergkasteel, Kard. Mercierlaan 94, B-3001 Leuven, Belgium
and
The NET Team, Max-Planck-Institut für Plasmaphysik,
Boltzmannstraße 2, D-8046 Garching bei München, Fed. Rep. of Germany

Dr. William Nicholas Guy Hitchon, Associate Professor
Dr. J. Leon Shohet, Professor
University of Wisconsin-Madison, Department of Electrical and Computer Engineering,
1415 Johnson Drive, Madison, WI 53706, USA

Dr. James D. Callen, D. W. Kerst Professor
University of Wisconsin-Madison,
Departments of Nuclear Engineering & Engineering Physics and Physics,
1500 Johnson Drive, Madison, WI 53706-1687, USA

ISBN-13: 978-3-642-75597-2 e-ISBN-13: 978-3-642-75595-8
DOI: 10.1007/978-3-642-75595-8

Library of Congress Cataloging-in-Publication Data. Flux coordinates and magnetic field structure : a guide to a fundamental tool of plasma theory / W. D. D'haeseleer ... [et al.]. p. cm. – (Springer series in computational physics) Includes bibliographical references and index. ISBN-13: 978-3-642-75597-2 : 1. Plasma (Ionized gases) 2. Magnetic fields. I. D'haeseleer, W. D. (William Denis), 1957 – . II. Series. QC718.5.M3F48 1990 530.4'41–dc20 90-10333

Softcover reprint of the hardcover 1st edition 1991

Typesetting: Asco Trade Typesetting Ltd., Hong Kong
56/3140-543210 – Printed on acid-free paper

Dedicated to the memory
of Michel Antoine D'haeseleer
and
ad majorem Dei gloriam

Preface

During the course of our research work on the theory of transport phenomena in tokamaks and stellarators, we learned to appreciate the convenience and utility of the special types of coordinate systems in which the magnetic field lines appear as straight lines. These so-called *flux coordinates* sweep, so to speak, the complexity of the magnetic-field structure "under the rug", and permit the user to concentrate on the (other) physical issues.

Although frequently used in theoretical calculations, at least in a formal way, it is striking that these coordinate systems are employed with little or no consistency, often in (over-)simplified form, and, regrettably, sometimes incorrectly. In addition, every researcher or group seems to have an individual preference concerning the type of flux-coordinate system used. It was rather surprising that there was no comprehensive review available that relates these flux-coordinate systems to each other in a rigorous but readable way.

Therefore, the idea grew to include a short survey of this subject as an appendix to one of our writings on transport. However, we soon discovered the unrealistic nature of our attempt and had to recognize that it was merely wishful thinking. In order to do a proper job, by which we mean producing a document that is comprehensive but still comprehensible for an audience of non-specialists, we realized that the quantity of material to be covered would by far surpass the volume of a reasonable appendix. This is the genesis of this book.

As we stated above, the target group of this monograph consists mainly of non-specialists. More specifically, we hope to reach an audience as varied as beginning graduate students in the field of plasmas and controlled fusion, as well as established researchers (especially experimentalists and those who undertake modeling in trying to interpret experimental results based on the available theories). Even for physicists with a more general interest in mathematical physics, or for applied mathematicians, this monograph might be useful in that it provides a self-contained treatment of curvilinear coordinates. The material on flux coordinates is then to be seen as an extensive example, illustrating the fundamental concepts.

In this spirit, we decided to concentrate our efforts on writing a document that is easily readable, but still quite rigorous. It aims to *teach* the reader. Therefore, we develop almost everything from first principles and cover the subject step by step, culminating in a fairly advanced level at the end of the treatise. In addition to the bare presentation of formulae and expressions, we give clarifications and provide comments and interpretations which help to put everything into context.

The book is divided into three parts.

In Part I we establish the foundation on which the remainder of the text is built. It comprises a first chapter on basics concerning vectors, differential geometry and curvilinear coordinates (Chap. 2), a second chapter deals with tensors (Chap. 3) and the last chapter of Part I (Chap. 4) deals with the basic concepts of the magnetic-field geometry in toroidal and open-ended devices. Chapters 2 and 4 in particular should be considered as prerequisites for the rest.

Part II treats the various distinct flux-coordinate systems. In Chaps. 5 through 9, we discuss Clebsch coordinates, the generic straight-field-line coordinate system, the Boozer toroidal flux coordinates, Hamada coordinates and the form of flux coordinates in axisymmetric tokamaks. In addition, we treat two coordinate systems that superficially resemble flux coordinates, but which are not flux coordinates: symmetry coordinates in a tokamak and so-called canonical coordinates.

In Part III of this monograph, we consider some topics that deserve attention, but which were tangential to the main issues in the previous chapters. Here, we treat "proper" toroidal coordinates (Chap. 10), the dynamic equilibrium of an ideal tokamak plasma (Chap. 11), the relationship between the proper length of a field line $\int dl/B$ and the specific volume of a flux surface $dV/d\Psi_{tor}$ (Chap. 12), the transformation properties of vector and tensor components (Chap. 13), and finally two alternative derivations for the divergence of a vector (Chap. 14).

There are a number of people who have been helpful in getting this monograph into its present shape.

We are most grateful to Mrs. Sarah Cohen, for whom it must have been a nightmare to type the various formulae with their variety of superscripts and subscripts. She retired soon after having finished this manuscript; we hope that we are not to blame for this decision.

We also would like to thank Springer-Verlag and its Physics Editor, Dr. H.-U. Daniel, for the professional handling of our request for the publication of this work. We also appreciate that we had the opportunity to update the manuscript with regard to recently (1988, 1989) published scientific literature. Also, we wish to express our gratitude to the referees for their useful comments.

One of us (W. D'H) wishes to thank Dr. J. A. Tataronis for numerous discussions on the subject of magnetic-field structure.

This work was supported by the US Department of Energy through Grants No. DE-FG02-87ER53216, No. DE-FG02-85ER53201 and No. DE-FG02-86ER53218, and in part by the Wisconsin Alumni Research Foundation.

We hope that the reader who works through all the steps given will reach a level where he/she can consider co- and contravariant representations to be quite simple. Whether we have succeeded in fulfilling this expectation must be judged by those patient and persistent enough to have worked through the book.

Garching and Madison,
November 1990

W. D. D'haeseleer W. N. G. Hitchon
J. D. Callen J. L. Shohet

Contents

Part II **Flux Coordinates**

Part III **Selected Topics**

1. Introduction and Overview

This treatise on magnetic-flux coordinates is intended to present a self-contained treatment of what are the natural coordinate systems for the description of magnetically confined plasmas. *Flux coordinates* are carefully chosen curvilinear coordinates with the property that the equation of a *magnetic-field line*, expressed in those coordinates, takes the form of a *straight line*. Sometimes we shall also refer to these coordinates systems as *straight-**B** systems*.

There does not appear to be a comprehensive review of the subject and so this document will attempt to provide a comprehensible and fairly complete treatment. This should help to make theoretical plasma calculations more accessible to non-specialists and researchers new to the field. To achieve this, we shall develop the mathematical tools such as tensor analysis in curvilinear coordinates from first principles as far as possible.

There is little or no consistency in definitions, nomenclature, interpretations and even conclusions in the area of tensors, dyadics, curvilinear coordinates and flux coordinates. The use of the words covariant and contravariant vectors is especially confusing, but ideas such as tensors, dyads and dyadics are no better. In this monograph we have tried to retain the most logical and straightforward formulations or the ones put forward most frequently. We are convinced that this important area deserves some clarification.

The above does not imply that no useful treatments of magnetic-flux coordinates exist. Useful contributions containing interesting material on magnetic-flux coordinates, magnetic geometries and/or equilibrium have been written by Morozov and Solov'ev (1966), Shafranov (1966), Solov'ev and Shafranov (1970), Solov'ev (1975), Greene and Johnson (1962), Boozer (1980, 1981, 1982, 1983), Kieras (1982), Hirshman (1982) and Hazeltine and Meiss (1985). Some basic ideas are presented in textbook treatments by Bateman (1978), Miyamoto (1980), Balescu (1988) and White (1989). However, although much of the plasma-physics literature uses these coordinates and some papers provide a discussion of them, all assume considerable familiarity with the subject (or its mathematical foundations). Many times, simplified forms are given for axisymmetric systems, or vacuum fields are used.

Since this monograph is to provide *practical* formulae and concepts, it is not appropriate to dwell on the theoretical framework behind the vector and tensor formalism, originating in an interplay between linear algebra and differential geometry. Rather, the focus is on the geometrical properties of vectors and tensors in curvilinear coordinates. Because magnetic-flux coordinates are located in the three-dimensional configuration space, we limit our discussion to three dimensions (3-D).

The book consists of three parts.

In Part I, containing Chapts. 2, 3 and 4, the mathematical groundwork is laid and the basic concepts concerning the magnetic geometry in magnetic-fusion devices are introduced. The chapter following this introduction, Chap. 2, will address the vector-related issues. Chapter 3 will generalize some ideas developed in Chap. 2 to tensors, because a vector can be considered as a very simple kind of tensor. In these two chapters, we shall treat vectors, dyads, triads (polyads) and tensors. The related, but for plasma physics unimportant, topic of quaternions will be skipped entirely here. (For a very elementary discussion on quaternions, see Kramer (1981), Chaps. 4 and 28; for more advanced treatments, see Morse and Feshbach (1953), Chap. 1, and Behnke (1974), Vol. 1, Chap. 8.) In Chap. 4, containing a preliminary discussion on magnetic geometries, the following subjects are considered: the equation of a magnetic-field line, the frozen-flux theorem, magnetic pressure and tension, field-line curvature, flux surfaces, types of coordinate systems in devices with "simple" flux surfaces, magnetic-surface labeling, the rotational transform, the flux-surface average and the magnetic differential equation.

Part II deals with the flux-coordinate systems. We treat each of the important flux-coordinate systems and, importantly, find the relationships between them. We start with the Clebsch system (Chap. 5), which is useful mainly in open-ended fusion devices and which has an intimate relationship with toroidal straight-$\boldsymbol{B}$ coordinates. A relative of the Clebsch system, the "Boozer-Grad" coordinate system, is also treated in considerable detail. The Boozer-Grad system has the property that one of its coordinates reduces to the magnetic scalar potential in vacuum.

In Chapt. 6, the flux coordinates for general toroidal systems are handled. After having introduced the generic form for toroidal straight-$\boldsymbol{B}$ coordinates in arbitrary toroidal confinement systems (with "good" magnetic surfaces), we specialize to the simplified case of an axisymmetric tokamak. In a next section, we discuss a special type of coordinate system that can always be constructed when an ignorable coordinate is present, but which is generally distinct from a flux-coordinate system. An important fact to recognize is that flux coordinates, in a general non-axisymmetric toroidal system, are not uniquely determined. Depending on how the extra degree of freedom is "consumed", the so-called toroidal Boozer flux-coordinate system or the Hamada coordinate system emerges. The Boozer system has the property that the periodic part of the scalar magnetic potential vanishes. There exists a simple linear relationship between the Boozer angle variables and a streaming function of the Boozer-Grad system. That streaming function reduces to the scalar magnetic potential in vacuum. The Hamada system has the convenience that both the magnetic-field lines and the current-density lines are straight. A further advantage is that the Jacobian is a constant.

In Chap. 7, we comment on the conversion from Clebsch-type coordinates to toroidal angle flux coordinates. Both the generic Clebsch system and the Boozer-Grad systems are treated.

Chapter 8 deals with the establishment of the flux-coordinate transformation. Besides being a recapitulation of the methods scattered through the text, it gives us the opportunity to say a few words on equilibrium calculations.

A second non-flux-coordinate system, appearing in Part II of this monograph, is developed in Chap. 9. The so-called "canonical coordinates" have the property that the magnetic-field expression in those coordinates resembles that written in flux coordinates. For special cases, these generalized coordinates reduce to flux coordinates. We shall stress the similarities and differences between the two types of coordinates.

Some important but less crucial topics have been deferred to Part III, Selected Topics, in order not to interrupt the flow of reading in the earlier parts. In Chap. 10, we treat "proper" toroidal coordinates, which are a three-dimensional version of bipolar coordinates. In Chap. 11, we examine the time-dependent tokamak dynamics of an ideal plasma. Here we show that both the plasma and the flux surfaces shrink radially inward under the influence of a perpendicular electric field, which is in turn induced by the change in transformer flux. In Chap. 12, we study the relationship between the specific flux volume $dV/d\Psi_{tor}$ and the integral $\int dl/B$. Chapter 13 deals with the transformation properties of vector and tensor components. It will explain where the names covariant and contravariant come from. Finally, in Chap. 14, we consider two alternatives to the derivation of the divergence formula given in the text.

A few comments on the notation used are appropriate. Vectors are represented by *boldface Roman italic* symbols, e.g., $\boldsymbol{A}$. Second-order tensors or dyadics are denoted by *boldface sanserif italic* characters, e.g., $\boldsymbol{\mathsf{A}}$.

These symbols are more elegant than the "arrow" or "over/under bar" notations: $\vec{A}$, $\bar{A}$ or $\underline{A}$ for vectors (which are first-order tensors) and $\overleftrightarrow{A}$, $\bar{\bar{A}}$ or $\underline{\underline{A}}$ for second-order tensors and dyadics. These notations have the advantage, though, that they reflect the tensorial order: n "arrow heads" or "over/under bars" for n-th-order tensors.

Because we ran out of reasonable typographical symbols, we shall use the "arrow" notation to denote the third-order tensors that we shall encounter in Chap. 3, e.g., $\overset{\rightleftarrows}{\boldsymbol{\mathsf{A}}}$.

Part I

Fundamental Concepts

Before one is ready to attack the different flux-coordinate systems studied in this book, the reader must possess a reasonable background in vector and tensor quantities in general curvilinear coordinate systems, on the one hand, and in magnetic-field-structure related concepts on the other.

Chapters 2 and 3 deal with mathematical preliminaries. Chapter 2, on vectors, is a prerequisite for the remainder of the book; it introduces basic ideas such as co- and contravariant vector components, the Jacobian, the metric coefficients and vector differentiation (including Christoffel symbols). Chapter 3, treating tensors, is not really required to understand what follows, but it completes the discussion of the previous chapter. Tensors, dyads, dyadics and the fundamental or metric tensor are considered.

Chapter 4, then again, is essential. Here we consider a variety of properties of the magnetic field and its structure in toroidal and open-ended devices. These "magnetic preliminaries" give us the vocabulary to deal with the rest of the text. This chapter deals with the following subjects. First, the equation of a magnetic-field line is derived. It will be clear that the $\boldsymbol{B}$-field components to be used are of the contravariant type. Next, we prove the "frozen-flux" theorem. We show that in a plasma with infinite conductivity the magnetic field is frozen into the plasma. Then, we discuss the concepts of magnetic pressure and tension. These seem to be very useful concepts for qualitative arguments in magnetically-confined plasmas. In the next section, we carefully define the magnetic-field-line curvature. The "normal" and "geodesic" curvature with respect to the flux surfaces are discussed. As a next topic, flux surfaces are rigorously defined. We say a few words on their "existence" in actual devices. Then the stage is set for a qualitative discussion of curvilinear coordinates in toroidal devices with "simple" magnetic surfaces. Here, we point out that a Clebsch system can be rather confusing in toroidal devices and that an angular system is preferable. In the following section, the different ways to label magnetic-flux surfaces are explained. We concentrate on the poloidal and toroidal magnetic fluxes and the volume inside a flux surface. The section before last of Chap. 4 defines the concept of rotational transform and gives an expression in terms of the ratio of (the change of) poloidal and toroidal magnetic fluxes. Finally, we consider an important type of differential equation, known as the magnetic differential equation because it contains the operator $\boldsymbol{B} \cdot \nabla$. We sketch the solution procedure and discuss its solubility conditions.

2. Vector Algebra and Analysis in Curvilinear Coordinates

2.1 Introduction

In this chapter we derive and compile some fundamental relations concerning the 3-D vector algebra and calculus in curvilinear coordinates. More detailed treatments (containing additional material which is less important for our needs) can be found in standard works on vectors and tensors such as Borisenko and Tarapov (1968), Wrede (1963), and Spiegel (1959). Classic textbooks on mathematical physics such as Morse and Feshbach (1953), Margenau and Murphy (1976), Mathews and Walker (1970) and Butkov (1968), contain some instructive chapters on the subject matter. Also worth mentioning are the treatments in well-known physics and engineering books such as Symon (1971, Mechanics), Stratton (1941, Electromagnetic Theory), Milne-Thomson (1968, Hydrodynamics) and Bird, Stewart and Lightfoot (1960, Transport Phenomena).

As we try to develop a self-consistent formalism, we must be most careful with the *notation.* Only by consistent and straightforward usage of a correct notation can these mathematical tools prove their power.

In this Chapter, we take a pragmatic approach without spending too much space on abstract foundations on the one hand or elementary concepts on the other. The treatment will be quite rigorous but will assume that the reader knows what a vector, a basis, a curve, a surface, etc. are.

In Sect. 2.2, we introduce the important concept of reciprocal sets of basis vectors. This idea is then used in Sect. 2.3. to define two kinds of basis vectors in a general curvilinear-coordinate system, leading to the concept of covariant and contravariant components of a vector. In Sect. 2.4, co- and contravariant "vectors" are discussed. In the Euclidean space we work in, no such things exist; only *vectors* exist. Section 2.5 treats some properties of curvilinear-coordinate systems: the metric coefficients, the Jacobian, the form of the dot and cross products, and the differential arc length, area and volume. In Sect. 2.6, vector differentiation in curvilinear systems is put in perspective. We explain why "covariant derivative" might not be a good name and show how the Christoffel symbols arise naturally from differentiation of the basis vectors. These symbols are discussed here because they are occasionally used in plasma physics (see e.g., Stacey, 1981). In Sect. 2.6, we also write the proper forms of the gradient, divergence and curl in curvilinear coordinates. Section 2.7 consists of a few words on the perpendicular and parallel components of one vector with respect to another. To conclude the chapter, Sect. 2.8 summarizes the most important vector identities.

2.2 Reciprocal Sets of Vectors

The concept of reciprocal sets of vectors is a crucial building block for the vector description of vectors in curvilinear coordinates. Below we shall see that the usual basis vectors along which a vector is expanded are reciprocal vectors in the sense to be explained.

The sets of vectors $\boldsymbol{A}, \boldsymbol{B}, \boldsymbol{C}$ and $\boldsymbol{a}, \boldsymbol{b}, \boldsymbol{c}$ are called *reciprocal sets of vectors* if:

$$\boldsymbol{A}\cdot\boldsymbol{a} = \boldsymbol{B}\cdot\boldsymbol{b} = \boldsymbol{C}\cdot\boldsymbol{c} = 1$$

$$\boldsymbol{A}\cdot\boldsymbol{b} = \boldsymbol{A}\cdot\boldsymbol{c} = \boldsymbol{B}\cdot\boldsymbol{a} = \boldsymbol{B}\cdot\boldsymbol{c} = \boldsymbol{C}\cdot\boldsymbol{a} = \boldsymbol{C}\cdot\boldsymbol{b} = 0 \ . \tag{2.2.1}$$

Each vector of one set is orthogonal to two vectors of the other set, and produces a dot product of unity with the remaining vector of that other set. These relations can only be satisfied if the sets $\boldsymbol{a}, \boldsymbol{b}, \boldsymbol{c}$ and $\boldsymbol{A}, \boldsymbol{B}, \boldsymbol{C}$ are each comprised of linearly independent vectors (non-coplanar, nor parallel). This implies that their respective triple products $\boldsymbol{a}\cdot(\boldsymbol{b}\times\boldsymbol{c})$ and $\boldsymbol{A}\cdot(\boldsymbol{B}\times\boldsymbol{C})$ are non-zero, meaning that the parallelepiped formed by their arrow heads and a common origin has a finite volume.

To find the vectors of one set $\boldsymbol{a}, \boldsymbol{b}, \boldsymbol{c}$ in terms of its reciprocal set $\boldsymbol{A}, \boldsymbol{B}, \boldsymbol{C}$, we have the inter-relationship:

$$\boldsymbol{a} = \frac{\boldsymbol{B}\times\boldsymbol{C}}{\boldsymbol{A}\cdot(\boldsymbol{B}\times\boldsymbol{C})} \ , \qquad \boldsymbol{b} = \frac{\boldsymbol{C}\times\boldsymbol{A}}{\boldsymbol{B}\cdot(\boldsymbol{C}\times\boldsymbol{A})} \ , \qquad \boldsymbol{c} = \frac{\boldsymbol{A}\times\boldsymbol{B}}{\boldsymbol{C}\cdot(\boldsymbol{A}\times\boldsymbol{B})} \ , \tag{2.2.2}$$

where $\boldsymbol{A}\cdot\boldsymbol{B}\times\boldsymbol{C} = \boldsymbol{B}\cdot\boldsymbol{C}\times\boldsymbol{A} = \boldsymbol{C}\cdot\boldsymbol{A}\times\boldsymbol{B} \neq 0$. Actually, (2.2.1) are necessary and sufficient conditions for $\boldsymbol{a}, \boldsymbol{b}, \boldsymbol{c}$ and $\boldsymbol{A}, \boldsymbol{B}, \boldsymbol{C}$ to be reciprocal sets. Equation (2.2.2) also holds with the pairs $\boldsymbol{a}, \boldsymbol{A}$; $\boldsymbol{b}, \boldsymbol{B}$; and $\boldsymbol{c}, \boldsymbol{C}$ interchanged. That (2.2.1) and (2.2.2) are equivalent can be shown as follows. From the properties $\boldsymbol{a}\cdot\boldsymbol{B} = \boldsymbol{a}\cdot\boldsymbol{C} = 0$ one concludes that $\boldsymbol{a}$ must be perpendicular to $\boldsymbol{B}$ and $\boldsymbol{C}$. For $\boldsymbol{B}$ and $\boldsymbol{C}$ being non collinear vectors, this means that $\boldsymbol{a}$ is along $\pm(\boldsymbol{B}\times\boldsymbol{C})$, or $\boldsymbol{a} = K(\boldsymbol{B}\times\boldsymbol{C})$, where K is a constant. After this expression has been dot-multiplied by the vector $\boldsymbol{A}$, one finds that the constant K equals $(\boldsymbol{A}\cdot(\boldsymbol{B}\times\boldsymbol{C}))^{-1}$ because $\boldsymbol{a}\cdot\boldsymbol{A} = 1$. This gives the first expression in (2.2.2). The formulae for $\boldsymbol{b}$ and $\boldsymbol{c}$ can be proved similarly or obtained by cyclic permutation.

The fact that each set $\boldsymbol{a}, \boldsymbol{b}, \boldsymbol{c}$ and $\boldsymbol{A}, \boldsymbol{B}, \boldsymbol{C}$ consists of vectors forming a trihedron with non-zero volume suggests that they can be utilized to express any 3-D vector as a linear combination of them. In other words, the sets $\boldsymbol{a}, \boldsymbol{b}, \boldsymbol{c}$ and $\boldsymbol{A}, \boldsymbol{B}, \boldsymbol{C}$ qualify as legitimate sets of basis vectors in 3-D. Indeed, any vector $\boldsymbol{W}$ can be written as

$$\boldsymbol{W} = (\boldsymbol{W}\cdot\boldsymbol{a})\boldsymbol{A} + (\boldsymbol{W}\cdot\boldsymbol{b})\boldsymbol{B} + (\boldsymbol{W}\cdot\boldsymbol{c})\boldsymbol{C} \tag{2.2.3a}$$

or

$$\boldsymbol{W} = (\boldsymbol{W}\cdot\boldsymbol{A})\boldsymbol{a} + (\boldsymbol{W}\cdot\boldsymbol{B})\boldsymbol{b} + (\boldsymbol{W}\cdot\boldsymbol{C})\boldsymbol{c} \ . \tag{2.2.3b}$$

Application of the so-called "*bac-cab*" rule to the expression $\boldsymbol{x} \times (\boldsymbol{C} \times \boldsymbol{W}) \equiv (\boldsymbol{A} \times \boldsymbol{B}) \times (\boldsymbol{C} \times \boldsymbol{W})$, i.e., with $\boldsymbol{A} \times \boldsymbol{B}$ considered as one vector $\boldsymbol{x}$, leads to

$$\begin{aligned}(\boldsymbol{A} \times \boldsymbol{B}) \times (\boldsymbol{C} \times \boldsymbol{W}) &= \boldsymbol{C}(\boldsymbol{A} \times \boldsymbol{B} \cdot \boldsymbol{W}) - \boldsymbol{W}(\boldsymbol{A} \times \boldsymbol{B} \cdot \boldsymbol{C}) \\ &= \boldsymbol{C}(\boldsymbol{A} \cdot \boldsymbol{B} \times \boldsymbol{W}) - \boldsymbol{W}(\boldsymbol{A} \cdot \boldsymbol{B} \times \boldsymbol{C}) \ . \end{aligned} \tag{2.2.4}$$

The same rule performed on $(\boldsymbol{C} \times \boldsymbol{W}) \times (\boldsymbol{A} \times \boldsymbol{B}) \equiv \boldsymbol{y} \times (\boldsymbol{A} \times \boldsymbol{B})$ gives

$$\begin{aligned}(\boldsymbol{A} \times \boldsymbol{B}) \times (\boldsymbol{C} \times \boldsymbol{W}) &= -(\boldsymbol{C} \times \boldsymbol{W}) \times (\boldsymbol{A} \times \boldsymbol{B}) \\ &= -\boldsymbol{A}(\boldsymbol{C} \times \boldsymbol{W} \cdot \boldsymbol{B}) + \boldsymbol{B}(\boldsymbol{C} \times \boldsymbol{W} \cdot \boldsymbol{A}) \\ &= -\boldsymbol{A}(\boldsymbol{B} \cdot \boldsymbol{C} \times \boldsymbol{W}) + \boldsymbol{B}(\boldsymbol{A} \cdot \boldsymbol{C} \times \boldsymbol{W}) \ . \end{aligned} \tag{2.2.5}$$

Equating the right-hand sides of both equations, and solving for $\boldsymbol{W}$ in the second term of (2.2.4), we obtain

$$\boldsymbol{W} = \frac{\boldsymbol{A}(\boldsymbol{B} \cdot \boldsymbol{C} \times \boldsymbol{W})}{\boldsymbol{A} \cdot (\boldsymbol{B} \times \boldsymbol{C})} - \frac{\boldsymbol{B}(\boldsymbol{A} \cdot \boldsymbol{C} \times \boldsymbol{W})}{\boldsymbol{A} \cdot (\boldsymbol{B} \times \boldsymbol{C})} + \frac{\boldsymbol{C}(\boldsymbol{A} \cdot \boldsymbol{B} \times \boldsymbol{W})}{\boldsymbol{A} \cdot (\boldsymbol{B} \times \boldsymbol{C})} \tag{2.2.6}$$

or

$$\boldsymbol{W} = \boldsymbol{A}\boldsymbol{W} \cdot \frac{\boldsymbol{B} \times \boldsymbol{C}}{\boldsymbol{A} \cdot \boldsymbol{B} \times \boldsymbol{C}} + \boldsymbol{B}\boldsymbol{W} \cdot \frac{\boldsymbol{C} \times \boldsymbol{A}}{\boldsymbol{A} \cdot \boldsymbol{B} \times \boldsymbol{C}} + \boldsymbol{C}\boldsymbol{W} \cdot \frac{\boldsymbol{A} \times \boldsymbol{B}}{\boldsymbol{A} \cdot \boldsymbol{B} \times \boldsymbol{C}} \ . \tag{2.2.7}$$

With (2.2.2), this becomes

$$\boldsymbol{W} = \boldsymbol{A}(\boldsymbol{W} \cdot \boldsymbol{a}) + \boldsymbol{B}(\boldsymbol{W} \cdot \boldsymbol{b}) + \boldsymbol{C}(\boldsymbol{W} \cdot \boldsymbol{c}) \ ,$$

proving relation (2.2.3a). Thus, the expansion coefficients of a vector $\boldsymbol{W}$ expressed along the basis set $\boldsymbol{A}, \boldsymbol{B}, \boldsymbol{C}$ can be found from the dot product of $\boldsymbol{W}$ with the "complementary" vector of the set $\boldsymbol{a}, \boldsymbol{b}, \boldsymbol{c}$ which is reciprocal to the original basis set. As can be seen from (2.2.3b), we could also expand along the set $\boldsymbol{a}, \boldsymbol{b}, \boldsymbol{c}$. The expressions in (2.2.3) are of crucial importance for our upcoming development of non-orthogonal coordinate systems.

Specializing to the Cartesian coordinate system, the set of unit basis vectors $\hat{\boldsymbol{i}}, \hat{\boldsymbol{j}}$, and $\hat{\boldsymbol{k}}$, is reciprocal to itself. (Unit vectors are denoted by " ^ ".) This implies that in such a system, a vector $\boldsymbol{W}$ can be written as

$$\boldsymbol{W} = (\boldsymbol{W} \cdot \hat{\boldsymbol{i}})\hat{\boldsymbol{i}} + (\boldsymbol{W} \cdot \hat{\boldsymbol{j}})\hat{\boldsymbol{j}} + (\boldsymbol{W} \cdot \hat{\boldsymbol{k}})\hat{\boldsymbol{k}} \tag{2.2.8}$$

or equivalently,

$$\boldsymbol{W} = W(1)\hat{\boldsymbol{i}} + W(2)\hat{\boldsymbol{j}} + W(3)\hat{\boldsymbol{k}} \ . \tag{2.2.9}$$

We used a bracketed notation for the indices rather than subscripts or superscripts. The notation will be adjusted after we have established the difference between subscripts and superscripts. Equation (2.2.9) is the "usual" expansion of a vector in its components in the Cartesian coordinate system.

2.3 Curvilinear Coordinates

This section on curvilinear coordinates contains the fundamental ideas on which the development of flux coordinates is based. At the end of this section we arrive at the important conclusion that a vector can be conveniently expanded along two reciprocal sets of basis vectors having a most natural relationship to the chosen coordinate system. This leads to the concepts of covariant and contravariant components.

We first establish the curvilinear transformation and define coordinate curves and coordinate surfaces. Then, a first set of basis vectors, namely the tangent-basis vectors, is introduced. The second set of basis vectors will be comprised of the gradients to the coordinate surfaces.

We assume a knowledge of differential geometry, as described by Stoker (1969), Goetz (1970), Guggenheimer (1977), Spivak (1970), Buck (1965), Mathews and Walker (1970), or Spiegel (1959), among others.

2.3.1 Transformation to Curvilinear Coordinates

Consider a transformation $\boldsymbol{R}(u^1, u^2, u^3)$ by means of which any point, determined by the position vector $\boldsymbol{R}$ of 3-D space, is expressed as a function of three parameters u^1, u^2 and u^3. When we expand $\boldsymbol{R}$ in its components x, y, z in a Cartesian coordinate system, we can rewrite this transformation as

$$\boldsymbol{R}(u^1,u^2,u^3)\colon\quad \begin{aligned} x &= x(u^1,u^2,u^3) \\ y &= y(u^1,u^2,u^3) \\ z &= z(u^1,u^2,u^3)\ . \end{aligned} \tag{2.3.1}$$

If the transformation is one-to-one, it can be inverted:

$$\begin{aligned} u^1 &= u^1(x,y,z) \\ u^2 &= u^2(x,y,z) \\ u^3 &= u^3(x,y,z)\ . \end{aligned} \tag{2.3.2}$$

The requirements for a transformation to be invertible are that the functions x, y, z of (2.3.1) must have continuous partial derivatives with respect to u^1, u^2 and u^3 respectively, and that the determinant made up by the nine partial derivatives $\partial x/\partial u^1, \partial x/\partial u^2, \partial x/\partial u^3, \partial y/\partial u^1, \ldots$ is not zero (at least in the domain under consideration). Below, we shall call this determinant the Jacobian of the transformation. (For a detailed discussion, see Buck (1965).) Then the correspondence between (x, y, z) and (u^1, u^2, u^3) is unique. In some practical cases the single-valuedness does not hold, e.g., for periodic functions on "closed" surfaces such as spheres, cylinders and tori; this problem can usually be circumvented.

Equation (2.3.2) shows that a certain point (determined by) $\boldsymbol{R}$ with Cartesian coordinates (x, y, z) can also be described uniquely by the independent parameters u^1, u^2 and u^3, so they are *coordinates* as well. (The (u^1, u^2, u^3) coordinates will be called *curvilinear* as a generalization of "rectilinear".) The reason for using a *superscript notation* (which could at times be confused with powers) will become clear below. The position of the indices is important.

In the curvilinear system there are three naturally occurring families of *coordinate surfaces*, obtained when one coordinate u^i is held fixed while the other two are varied continuously. The equations for these surfaces are:

$$u^1 = c^1 \qquad (u^2, u^3 \text{ variable})$$

$$u^2 = c^2 \qquad (u^1, u^3 \text{ variable})$$

$$u^3 = c^3 \qquad (u^1, u^2 \text{ variable}) \ .$$

c^1, c^2 and c^3 are arbitrary constants. It will be assumed that a unique surface of each family goes through any given point determined by the position vector $\boldsymbol{R}$ in keeping with the hypothesis that the endpoint of $\boldsymbol{R}$ is uniquely determined by its coordinates u^1, u^2, and u^3.

Analogously, three families of *coordinate curves* are produced when one coordinate u^i is allowed to vary while the other two, u^j and u^k, are held fixed. The equations are:

$$u^2 = c^2 \ , \qquad u^3 = c^3 \qquad (u^1 \text{ variable})$$

$$u^3 = c^3 \ , \qquad u^1 = c^1 \qquad (u^2 \text{ variable})$$

$$u^1 = c^1 \ , \qquad u^2 = c^2 \qquad (u^3 \text{ variable}) \ .$$

The direction in which a variable point of the curve moves as u^i is increased is taken to be the positive direction along that coordinate curve. The general coordinate system is summarized in Fig. 2.1.

If the coordinate curves intersect at right angles, then the curvilinear system is called *orthogonal*.

Examples of simple curvilinear coordinate systems are the familiar cylindrical and spherical systems. These are orthogonal.

2.3.2 Tangent-Basis Vectors

Consider the system of generalized coordinates u^1, u^2 and u^3 as shown in Fig. 2.2. We define a *basis* at a point P, determined by the position vector $\boldsymbol{R}$, as any set of vectors $\boldsymbol{e}_1$, $\boldsymbol{e}_2$ and $\boldsymbol{e}_3$ of fixed length pointing in the positive direction of the coordinate curves. In other words, the *basis vectors* $\boldsymbol{e}_1$, $\boldsymbol{e}_2$ and $\boldsymbol{e}_3$ are tangent vectors of the coordinate curves at the point P.

The curvilinear coordinates have been defined by the transformation $\boldsymbol{R}(u^1, u^2, u^3)$. Simple differential geometry considerations show that $\partial\boldsymbol{R}/\partial u^i$ is a

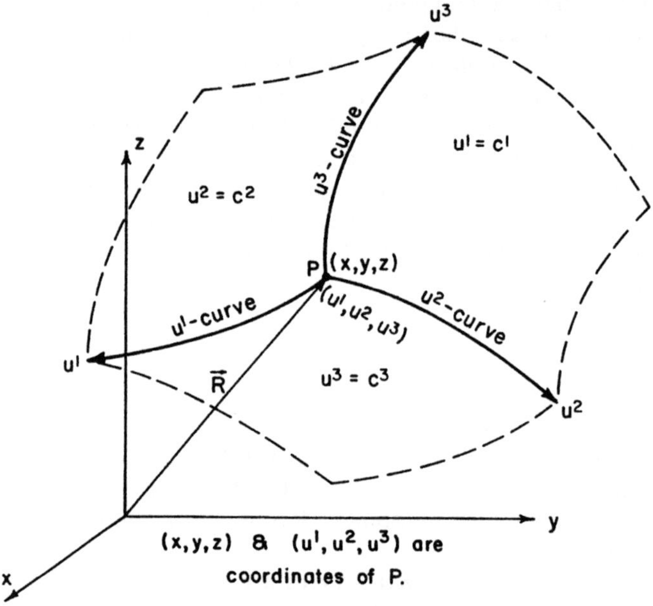

Fig. 2.1. General curvilinear coordinate system with coordinates (u^1, u^2, u^3). The coordinate curves and coordinate surfaces are represented with respect to a Cartesian system with coordinates (x, y, z)

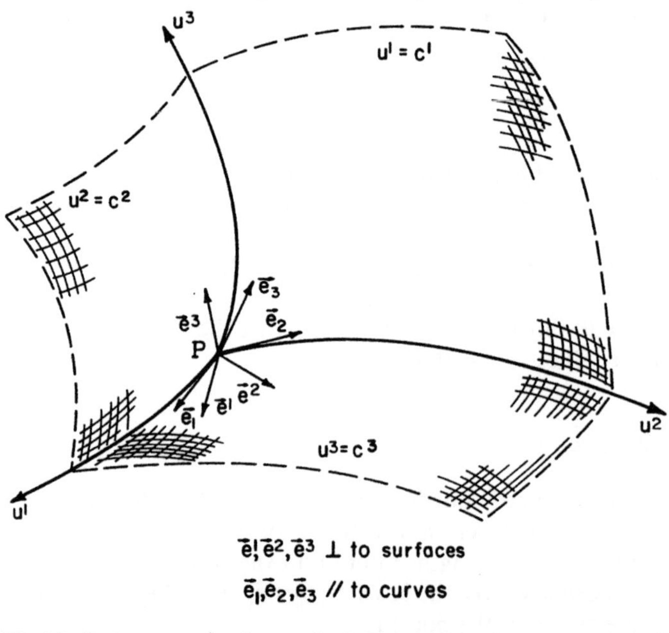

Fig. 2.2. Basis vectors $\boldsymbol{e}^i$ and $\boldsymbol{e}_j$ at point P. The contravariant basis vectors $\boldsymbol{e}^i = \nabla u^i$ are perpendicular to the constant coordinate surfaces; the covariant basis vectors $\boldsymbol{e}_j = \partial \boldsymbol{R}/\partial u^j$ are parallel to the coordinate curves

tangent vector of the u^i-coordinate curve. (A tangent vector to a one parameter curve $\boldsymbol{R} = \boldsymbol{R}(u)$ is $d\boldsymbol{R}/du$). It seems logical to choose these vectors as the *tangent-basis vectors*:

$$\boldsymbol{e}_1 = \frac{\partial \boldsymbol{R}}{\partial u^1} \; ; \qquad \boldsymbol{e}_2 = \frac{\partial \boldsymbol{R}}{\partial u^2} \; ; \qquad \boldsymbol{e}_3 = \frac{\partial \boldsymbol{R}}{\partial u^3} \; . \tag{2.3.3}$$

The definition in (2.3.3) is consistent with the requirement that a set of basis vectors must be linearly independent; the triple product $\boldsymbol{e}_1 \cdot \boldsymbol{e}_2 \times \boldsymbol{e}_3$ is non-zero.

This basis made up by the basis vectors $\boldsymbol{e}_1$, $\boldsymbol{e}_2$ and $\boldsymbol{e}_3$ is said to be *local*, since, in general, it varies from point to point. (By "fixed length" in the definition of a basis, we mean that once the length of a basis vector at a point P_1 has been chosen, we always keep that length at P_1; at a point P_2, the basis vector might have a different length.) Also, in general, these basis vectors are neither orthogonal nor of unit length, nor even dimensionless, since e.g., $\partial \boldsymbol{R}/\partial \theta$ has units of length. Only in coordinate systems with straight lines as coordinate "curves", i.e., with rectangular and oblique coordinate lines, do the basis vectors not vary from point to point.

If the variable u^i is taken to be the length along the coordinate curve, then the vector $\boldsymbol{e}_i$ is a *unit tangent vector*. In the more general case where u^i is just a parameter, one could define a unit tangent vector according to the obvious prescription:

$$\hat{\boldsymbol{e}}_i \equiv \frac{\boldsymbol{e}_i}{|\boldsymbol{e}_i|} = \frac{\partial \boldsymbol{R}}{\partial u^i} \Big/ \left| \frac{\partial \boldsymbol{R}}{\partial u^i} \right| = \frac{\boldsymbol{e}_i}{h_i} \qquad \text{(no summation)} \; . \tag{2.3.4}$$

The magnitude $|\boldsymbol{e}_i| = |\partial \boldsymbol{R}/\partial u^i|$ is conventionally represented by the symbol h_i and is called a *scale factor* (or a metric coefficient).[1] Thus,

$$h_i = \left| \frac{\partial \boldsymbol{R}}{\partial u^i} \right| = |\boldsymbol{e}_i| \; . \tag{2.3.5}$$

Although the h_i's are usually mostly used for orthogonal systems, the above definition is completely general and valid for any coordinate system.

Note that the tangent-basis vector $\boldsymbol{e}_i$ has a *subscript* (i.e., the index is underneath). Although the coordinates u^i have superscripts, the fact that $\boldsymbol{e}_i$ has a subscript in (2.3.3) is suggestive because the index in

$$\frac{\partial \boldsymbol{R}}{\partial u^i}$$

is "underneath", i.e., below the "fraction-line".

[1] Sometimes these scale factors are called "Lamé factors" (especially in the Russian literature; see, e.g., Kovrizhnykh 1984). However, the name "Lamé factor" is also used in the context of elasticity theory, where it appears as the proportionality constant between the elasticity tensor and the strain tensor (Butkov 1968).

2.3.3 Reciprocal-Basis Vectors

The gradient, $\nabla\Phi$, of a function Φ is defined so that the differential $d\Phi$ is given by $d\Phi = \nabla\Phi\cdot d\boldsymbol{R}$. With $\Phi \equiv u^i$, this becomes

$$du^i = \nabla u^i \cdot d\boldsymbol{R} \ . \tag{2.3.6}$$

If we consider the position vector $\boldsymbol{R}$ as a function of u^i, i.e., $\boldsymbol{R} = \boldsymbol{R}(u^1, u^2, u^3)$, we find from the chain rule that

$$d\boldsymbol{R} = \sum_{j=1}^{3} \frac{\partial \boldsymbol{R}}{\partial u^j} du^j = \frac{\partial \boldsymbol{R}}{\partial u^j} du^j = \boldsymbol{e}_j\, du^j \ . \tag{2.3.7}$$

Here we have used the definition of $\boldsymbol{e}_i$ given in (2.3.3) and in the last two forms of (2.3.7) the *summation convention* has been applied. The summation convention says that if a letter used as an index appears once as a superscript and once as a subscript on the same side of an equation, then that side of the equation should be considered to be summed over all values of that index. The summation sign is then superfluous. A repeated (or dummy) index implies summation and may be replaced by another letter; e.g. $\boldsymbol{e}_j du^j = \boldsymbol{e}_1 du^1 + \boldsymbol{e}_2 du^2 + \boldsymbol{e}_3 du^3 = \boldsymbol{e}_k du^k$.

Substitution of $d\boldsymbol{R}$ as given by (2.3.7) into (2.3.6) leads to

$$du^i = \nabla u^i \cdot \boldsymbol{e}_j du^j \ , \tag{2.3.8}$$

which can only hold if and only if

$$\nabla u^i \cdot \boldsymbol{e}_j = \delta^i_j \ , \tag{2.3.9}$$

where δ^i_j is the Kronecker delta (which is unity if the indices coincide and zero otherwise).

The relationship in (2.3.9) states that the sets of vectors ∇u^i and $\boldsymbol{e}_j$ form reciprocal sets of vectors as defined in (2.2.1). Therefore, we define the set ∇u^i as the *reciprocal-basis vectors*:

$$\boldsymbol{e}^i \equiv \nabla u^i \tag{2.3.10}$$

in which case (2.3.9) reads:

$$\boldsymbol{e}^i \cdot \boldsymbol{e}_j = \delta^i_j \ . \tag{2.3.11}$$

As was pointed out with respect to the $\boldsymbol{e}_j$, the $\boldsymbol{e}^i$ are normally not of unit length nor are they perpendicular. While the basis vectors $\boldsymbol{e}_j$ are tangent to the u^j coordinate curves, the reciprocal basis vectors $\boldsymbol{e}^i$ are *perpendicular to the coordinate surfaces* $u^i = c^i$ (see Fig. 2.2).

The two sets of basis vectors $\boldsymbol{e}_j$ and $\boldsymbol{e}^i$ have been recognized as reciprocal sets of vectors. Consequently they satisfy the relationships (2.2.2), which allow any vector of one set to be calculated if the three vectors of the other set are known:

$$\nabla u^i = \frac{\dfrac{\partial \boldsymbol{R}}{\partial u^j} \times \dfrac{\partial \boldsymbol{R}}{\partial u^k}}{\dfrac{\partial \boldsymbol{R}}{\partial u^i} \cdot \left(\dfrac{\partial \boldsymbol{R}}{\partial u^j} \times \dfrac{\partial \boldsymbol{R}}{\partial u^k}\right)} = \frac{\boldsymbol{e}_j \times \boldsymbol{e}_k}{\boldsymbol{e}_i \cdot (\boldsymbol{e}_j \times \boldsymbol{e}_k)} = \boldsymbol{e}^i \tag{2.3.12}$$

(and similarly for ∇u^j and ∇u^k), and

$$\frac{\partial \boldsymbol{R}}{\partial u^i} = \frac{\nabla u^j \times \nabla u^k}{\nabla u^i \cdot (\nabla u^j \times \nabla u^k)} = \frac{\boldsymbol{e}^j \times \boldsymbol{e}^k}{\boldsymbol{e}^i \cdot (\boldsymbol{e}^j \times \boldsymbol{e}^k)} = \boldsymbol{e}_i \tag{2.3.13}$$

(and analogously for $\partial \boldsymbol{R}/\partial u^j$ and $\partial \boldsymbol{R}/\partial u^k$). Here i, j and k must be chosen such that (i, j, k) forms a cyclic permutation of $(1, 2, 3)$. This sentence will henceforth be abbreviated as "i, j, k cyc 1, 2, 3". The word "cyclic" is equivalent to the requirement that only "even" permutations of 1, 2, 3 are acceptable: 1, 2, 3 or 2, 3, 1, or 3, 1, 2. An ordering of numbers $1, 2, \ldots, n$ is understood to be an even (odd) permutation of $1, 2, \ldots, n$ if it results from the standard order by an even (odd) number of interchanges of pairs of numbers.

Alternatively, the reciprocal-basis vectors $\boldsymbol{e}^i$ are sometimes called *dual* basis vectors (Stoker 1969; Davis and Snider 1975). The reciprocal-basis vectors $\boldsymbol{e}^i$ are chosen to have a *superscript*, because ∇u^i does, consistently with $\boldsymbol{e}_i$ having a subscript.

When we use the term "*basis vector*" with no adjective, we mean the *tangent-basis vectors* $\boldsymbol{e}_i$; when the dual-basis vectors $\boldsymbol{e}^j$ are meant, we shall use the adjective "*reciprocal*". Below, we will introduce other names for $\boldsymbol{e}_i$ and $\boldsymbol{e}^j$, being related to the position of their indices i or j.

2.3.4 Covariant and Contravariant Components of a Vector

According to our earlier discussions, a vector can be written as a linear combination of the vectors from either of the (reciprocal) sets of basis vectors. From (2.2.3) a vector $\boldsymbol{D}$ can be written as

$$\begin{aligned}\boldsymbol{D} &= \left(\boldsymbol{D} \cdot \frac{\partial \boldsymbol{R}}{\partial u^1}\right) \nabla u^1 + \left(\boldsymbol{D} \cdot \frac{\partial \boldsymbol{R}}{\partial u^2}\right) \nabla u^2 + \left(\boldsymbol{D} \cdot \frac{\partial \boldsymbol{R}}{\partial u^3}\right) \nabla u^3 \\ &= (\boldsymbol{D} \cdot \boldsymbol{e}_1)\boldsymbol{e}^1 + (\boldsymbol{D} \cdot \boldsymbol{e}_2)\boldsymbol{e}^2 + (\boldsymbol{D} \cdot \boldsymbol{e}_3)\boldsymbol{e}^3 \end{aligned} \tag{2.3.14a}$$

or

$$\boldsymbol{D} \equiv D_1 \boldsymbol{e}^1 + D_2 \boldsymbol{e}^2 + D_3 \boldsymbol{e}^3 \ , \tag{2.3.15a}$$

which is an expansion along the set of reciprocal-basis vectors. Analogously, $\boldsymbol{D}$ can be written as a linear combination of the tangent-basis vectors:

$$\begin{aligned}\boldsymbol{D} &= (\boldsymbol{D} \cdot \nabla u^1) \frac{\partial \boldsymbol{R}}{\partial u^1} + (\boldsymbol{D} \cdot \nabla u^2) \frac{\partial \boldsymbol{R}}{\partial u^2} + (\boldsymbol{D} \cdot \nabla u^3) \frac{\partial \boldsymbol{R}}{\partial u^3} \\ &= (\boldsymbol{D} \cdot \boldsymbol{e}^1)\boldsymbol{e}_1 + (\boldsymbol{D} \cdot \boldsymbol{e}^2)\boldsymbol{e}_2 + (\boldsymbol{D} \cdot \boldsymbol{e}^3)\boldsymbol{e}_3 \end{aligned} \tag{2.3.14b}$$

or,

$$D \equiv D^1 e_1 + D^2 e_2 + D^3 e_3 \ . \tag{2.3.15b}$$

In summary, we have

$$D = D_i e^i \qquad \text{with} \qquad D_i \equiv D \cdot e_i \tag{2.3.16a}$$

$$D = D^i e_i \qquad \text{with} \qquad D^i \equiv D \cdot e^i \ . \tag{2.3.16b}$$

The summation convention is implicit here.

One can look upon this expansion of (2.3.14) as follows: $\boldsymbol{R}$ is the position vector pointing from the origin of a Cartesian coordinate system to the initial point of the vector we wish to describe, in this case the vector $\boldsymbol{D}$. The endpoint of $\boldsymbol{R}$ is then the place where the two kinds of bases e_i and e^i are attached, and these allow us to find the components of $\boldsymbol{D}$ by means of the dot products (2.3.16) (see Figs. 2.1 and 2.2).

The sub- or superscript notation replaces the bracket notation used in (2.2.9). There is more than one kind of component of a vector, depending on the choice of basis vectors.

The coefficients D_i appearing in (2.3.16a) are called the *covariant components* of the vector $\boldsymbol{D}$ while the D^j in (2.3.16b) are called *contravariant components.* Equation (2.3.16a) shows that our notation is consistent; to find $\boldsymbol{D}$, one must sum over the sub- and superscript i; the covariant component D_i results from the dot product of $\boldsymbol{D}$ with the basis vector e_i (the index is underneath on both sides of the equation $D_i = D \cdot e_i$).

From now on we adopt the following convention to name the components of a vector:

$$\text{covariant} \equiv \text{subscripts} \equiv (\ )_i$$

$$\text{contravariant} \equiv \text{superscripts} \equiv (\ )^j \ . \tag{2.3.17}$$

Thus an index *underneath* implies a *covariant* object, an index *above* denotes *contravariant* object. The reason for the words co- and contravariant has to do with the transformation properties of the components of a vector. This is discussed in Chap. 13. A simple mnemonic aid to distinguish these terms is as follows. Since there is an "*R*" in *super*script, we call D^i a contravariant component. There is no "*R*" in *sub*script, hence the name covariant.

The functions $u^i(x, y, z)$ can have physical dimensions (units), as can the basis vectors. If the u^i are dimensionless, the e_i and e^j have dimensions of L and L^{-1}, respectively. Consequently, the *components of a vector do not necessarily have the same dimensions as the vector itself.*

For the special case of a Cartesian coordinate system it is clear from the definition formulae, (2.3.3), that the $\hat{\imath}, \hat{\jmath}, \hat{k}$ basis vectors are the tangent-basis vectors. Moreover, they have a unit length since the coordinates are the "arc-lengths" along the coordinate axes. $\hat{\imath}, \hat{\jmath}, \hat{k}$ thus form an orthonormal basis. On the other hand, since the relationship in (2.3.11) is exactly that obeyed by the $\hat{\imath}$,

$\hat{j}$, $\hat{k}$ basis vectors, it follows that they are reciprocal basis vectors as well. (This is in agreement with our earlier observation that the set $\hat{i}$, $\hat{j}$, $\hat{k}$ is reciprocal to itself; see (2.2.8).) As a consequence, there is no distinction between co- and contravariant components of a vector in a Cartesian system: $\boldsymbol{W}\cdot\hat{\boldsymbol{e}}_1 = \boldsymbol{W}\cdot\hat{\boldsymbol{e}}^1 = \boldsymbol{W}\cdot\hat{\boldsymbol{i}} = W(1)$, nor is there a need to make a distinction between sub- or superscripts in the component notation. Without any ambiguity, we can now rewrite (2.2.9) as

$$\boldsymbol{W} = W_1\hat{\boldsymbol{i}} + W_2\hat{\boldsymbol{j}} + W_3\hat{\boldsymbol{k}}$$

or

$$\boldsymbol{W} = W_x\hat{\boldsymbol{i}} + W_y\hat{\boldsymbol{j}} + W_z\hat{\boldsymbol{k}} \ .$$

If we deal with a *position vector* $\boldsymbol{R}$, where $R_x = x$, $R_y = y$ and $R_z = z$, we obtain $\boldsymbol{R} = x\hat{\boldsymbol{i}} + y\hat{\boldsymbol{j}} + z\hat{\boldsymbol{k}}$.

2.4 Covariant and Contravariant "Vectors"

A lot of confusion about the terminology co- or contravariant vectors and/or components arises from the fact that both co- and contravariant vectors, and co- and contravariant components do exist, albeit in different contexts. Moreover, the wording "co- or contravariant vector" is often used to mean "co- or contravariant components of a vector". Here, we shall assert that when dealing with only one (Euclidean) vector space such as 3-D space, there exist only *vectors* (see e.g., Borisenko and Tarapov 1968). There are *no* contravariant and covariant vectors *as such* in the three-dimensional space: a vector is just a vector. If one digresses into abstract algebra, one can define contravariant vectors in one vector space and covariant vectors in *another* vector space, which is the dual of the first. References to this development are given at the end of Chap. 3.

A vector will always be identified by its *contravariant components* or by its *covariant components.* It must be stressed that, in principle, both representations contain all the information needed and are equally valid.

It is often said that the vector $d\boldsymbol{R}$ (the infinitesimal vector connecting the two position vectors $\boldsymbol{R}$ and $\boldsymbol{R} + d\boldsymbol{R}$) is a contravariant vector. More specifically, the (misleading) statement would be that the coordinate differentials du^i form a contravariant vector. This statement should be modified to say that the "natural" and easiest way of writing a vector $d\boldsymbol{R}$ in components is by means of its contravariant components. These components are obtained by a simple chain-rule application, as can be seen from (2.3.7). Comparison of (2.3.7) with (2.3.16b) shows that the du^i do indeed represent the contravariant components of $d\boldsymbol{R}$. Since $d\boldsymbol{R}$ is the vector between two position vectors $\boldsymbol{R}$ and $\boldsymbol{R} + d\boldsymbol{R}$ with coordinates (u^1, u^2, u^3) and $(u^1 + du^1, u^2 + du^2, u^3 + du^3)$, respectively., it is obvious why we used *superscripts to label the coordinates* u^i: contravariant implies above, and the

du^i are contravariant.

However, with equal validity, one could have chosen to write $\boldsymbol{dR}$ using its covariant components according to (2.3.16a)

$$\boldsymbol{dR} = (\boldsymbol{dR})_i \boldsymbol{e}^i = (\boldsymbol{dR})_i \nabla u^i \ . \tag{2.4.1}$$

Here, $(\boldsymbol{dR})_i$ is given by

$$\begin{aligned}(\boldsymbol{dR})_i &= \boldsymbol{dR}\cdot \boldsymbol{e}_i = \boldsymbol{dR}\cdot\frac{\partial \boldsymbol{R}}{\partial u^i} = du^j\frac{\partial \boldsymbol{R}}{\partial u^j}\cdot\frac{\partial \boldsymbol{R}}{\partial u^i}\\ &= du^j \boldsymbol{e}_j\cdot \boldsymbol{e}_i = \boldsymbol{e}_i\cdot \boldsymbol{e}_j du^j\\ &= g_{ij}du^j \ ,\end{aligned} \tag{2.4.2}$$

where we have replaced the dot product of the basis vectors $\boldsymbol{e}_i$ and $\boldsymbol{e}_j$ by the symbol g_{ij}. (At this point, the symbol g_{ij} is a shorthand notation for $\boldsymbol{e}_i\cdot \boldsymbol{e}_j$; below, we shall give it a real meaning and stress its importance.) Thus, $\boldsymbol{dR}$ is just a vector, which can be written in terms of two sorts of components.

The confusion we alluded to above follows from the fact that some branches of science *define* a vector only by its components, without introducing a symbol for the vector itself. As a consequence, one (misleadingly) calls the quantity du^i a contravariant vector. However, here we employ the more correct term of contravariant *components*.

Similarly, it is sometimes said that the gradient of a function is a covariant vector. Once again this is an ambiguous statement and follows from the "normal" way of "decomposing" the gradient by means of chain-rule differentiation:

$$\nabla\Phi(u^1,u^2,u^3) = \nabla\Phi = \frac{\partial\Phi}{\partial u^i}\nabla u^i = \frac{\partial\Phi}{\partial u^i}\boldsymbol{e}^i \ . \tag{2.4.3}$$

Therefore, we have that,

$$(\nabla\Phi)_i = \frac{\partial\Phi}{\partial u^i} \ , \tag{2.4.4}$$

showing that $(\partial\Phi/\partial u^i)$ are the covariant components of the gradient vector $\nabla\Phi$.

$\nabla\Phi$ can be written in terms of its contravariant components as:

$$\nabla\Phi = (\nabla\Phi)^i \boldsymbol{e}_i$$

with

$$\begin{aligned}(\nabla\Phi)^i &= \nabla\Phi\cdot \boldsymbol{e}^i = \nabla\Phi\cdot\nabla u^i = \frac{\partial\Phi}{\partial u^j}\nabla u^j\cdot\nabla u^i\\ &= \frac{\partial\Phi}{\partial u^j}\boldsymbol{e}^j\cdot \boldsymbol{e}^i = \boldsymbol{e}^i\cdot \boldsymbol{e}^j\frac{\partial\Phi}{\partial u^j}\\ &= g^{ij}\frac{\partial\Phi}{\partial u^j} \ .\end{aligned} \tag{2.4.5}$$

This time, we have replaced $(\boldsymbol{e}^i \cdot \boldsymbol{e}^j)$ by the symbol g^{ij}.

Calling $\partial\Phi/\partial u^i$ "*the* covariant vector" makes little sense; all one can say is that $\partial\Phi/\partial u^i$ are the covariant components, in this case with respect to the commonly chosen sets of bases $\partial\boldsymbol{R}/\partial u^i$ and ∇u^i. The safest way (in the sense that the symbol is valid with respect to any kind of basis vectors) to denote covariant and contravariant components is by $(\nabla\Phi)_i$ and $(\nabla\Phi)^i$.

We shall, nevertheless, use the terms co- and contravariant *vectors* for certain types of vectors: *the tangent and reciprocal basis vectors.* It is common practice to call the tangent-basis vectors $\boldsymbol{e}_i$ "*the covariant-basis vectors*" (because they are labeled by a *subscript*) (Boozer 1980, 1981, 1982, 1983; Bateman 1978; Wrede 1963). Similarly the reciprocal basis vectors $\boldsymbol{e}^i$ are called "*the contravariant-basis vectors*" (because of the superscript). One should be aware of the fact that this is *only for convenience*, since the basis vectors $\boldsymbol{e}_i$ and $\boldsymbol{e}^i$ themselves can actually be written with covariant and contravariant *components* as well. Indeed, applying (2.3.16) we obtain for $\boldsymbol{e}_i$ the following: $\boldsymbol{e}_i = (\boldsymbol{e}_i)^j\boldsymbol{e}_j = \boldsymbol{e}_i \cdot \boldsymbol{e}^j\boldsymbol{e}_j = \delta_i^j\boldsymbol{e}_j = 1\boldsymbol{e}_i$ and $\boldsymbol{e}_i = (\boldsymbol{e}_i)_j\boldsymbol{e}^j = \boldsymbol{e}_i \cdot \boldsymbol{e}_j\boldsymbol{e}^j = g_{ij}\boldsymbol{e}^j$; analogously, for $\boldsymbol{e}^i$, we would obtain $\boldsymbol{e}^i = 1\boldsymbol{e}^i$ and $\boldsymbol{e}^i = g^{ij}\boldsymbol{e}_j$. Here we have again used the symbols g_{ij} and g^{ij} that we introduced above. Now, the surprising thing is that the "trivial" components (being equal to 1) for the $\boldsymbol{e}_i$ are the contravariant components and for the $\boldsymbol{e}^i$ are the covariant components! Nevertheless, one calls $\boldsymbol{e}_i$ the covariant and $\boldsymbol{e}^i$ the contravariant-basis vectors. This is thus even worse than for du^i and $d\boldsymbol{R}$, or $\partial\Phi/\partial u^i$ and $\nabla\Phi$ discussed above; we are now advocating a "reverse logic"![2] Since basis vectors are a privileged kind of vectors, one concentrates on the position of the indices and provides the basis vectors themselves with the names co- and contravariant.[3] This is why we have been deliberately ambiguous below (2.3.17). We consider (2.3.17) as valid for *everything* that is labeled with subscripts and superscripts. Actually, there is a more profound reason for this policy. From the transformation laws discussed in Chap. 13, one can conclude that the description of $\boldsymbol{e}_i$ as the "covariant" basis vectors seems reasonable in the sense that they transform in the same way as the tangent-basis vectors, i.e., similarly to themselves. There it is explained that objects that transform similarly to the $\boldsymbol{e}_i$ are called covariant.

---

[2] For the same reason, our proposed nomenclature is *not* in agreement with the abstract-algebraic development of vectors in dual spaces. The tangent-basis vectors $\boldsymbol{e}_i$ are members of the "vector space of contravariant vectors" such that they are contravariant vectors viewed from that standpoint. On the other hand, the dual vector space (which is different from the first vector space), with the covariant vectors as elements, does contain the dual- or reciprocal-basis vectors. Hence, a reciprocal-basis vector is a covariant vector. However, when only one (Euclidean) vector space is involved, the "abstract algebraic" distinction between co- and contravariant vectors as such disappears. Only the distinction between co- and contravariant components of a vector can now be made. The fact that the basis vectors $\boldsymbol{e}_j$ and $\boldsymbol{e}^i$ are then merely vectors grants us the liberty to use the words co- and contravariant basis vectors according to the position of their indices.

[3] There appears to be a typographical error in Bateman (1978), page 126, regarding this commonly used nomenclature in plasma physics.

2.5 Vector Relationships in Curvilinear Coordinates

2.5.1 The Metric Coefficients g_{ij} and g^{ij}

We now formally introduce the important metric coefficients g_{ij} and g^{ij}, which form the backbone of general curvilinear coordinate systems. They determine the differential arc length along a curve, they allow us to switch back and forth between the covariant and contravariant components of a vector, and they will provide us with a means to calculate the dot and cross products of two vectors.

The metric coefficients g_{ij} are defined as the dot product of the tangent-basis vectors:

$$g_{ij} = \boldsymbol{e}_i \cdot \boldsymbol{e}_j = \frac{\partial \boldsymbol{R}}{\partial u^i} \cdot \frac{\partial \boldsymbol{R}}{\partial u^j} . \tag{2.5.1}$$

Similarly, the reciprocal or conjugate metric coefficients g^{ij} (also called simply metric coefficients) result from the dot product of the reciprocal-basis vectors:

$$g^{ij} = \boldsymbol{e}^i \cdot \boldsymbol{e}^j = \nabla u^i \cdot \nabla u^j . \tag{2.5.2}$$

From their definition, it follows that these coefficients are *symmetric*: $g_{ij} = g_{ji}$ and $g^{ij} = g^{ji}$.

When we deal with an orthogonal but not necessarily orthonormal coordinate system (such as a spherical, a cylindrical or a toroidal one) we find that $g_{ij} = 0$ if $i \neq j$ and similarly for g^{ij}. Thus *off-diagonal metric coefficients which are all zero imply an orthogonal coordinate system.*

Recall that the so called scale factors h_i were defined as $h_i = |\boldsymbol{e}_i|$. Therefore, (2.5.1) becomes

$$g_{ij} = h_i h_j \hat{\boldsymbol{e}}_i \cdot \hat{\boldsymbol{e}}_j = h_i h_j \cos \measuredangle (\boldsymbol{e}_i, \boldsymbol{e}_j) , \tag{2.5.3}$$

leading to

$$g_{ii} = h_i^2 \qquad \text{or} \qquad h_i = \sqrt{g_{ii}} . \tag{2.5.4}$$

This is the reason why the h_i are also often called metric coefficients.

The differential arc length along a curve is defined as:

$$dl = |d\boldsymbol{R}| = \sqrt{d\boldsymbol{R} \cdot d\boldsymbol{R}} . \tag{2.5.5}$$

The differential vector $d\boldsymbol{R}$ equals

$$d\boldsymbol{R} = \frac{\partial \boldsymbol{R}}{\partial u^1} du^1 + \frac{\partial \boldsymbol{R}}{\partial u^2} du^2 + \frac{\partial \boldsymbol{R}}{\partial u^3} du^3 = \frac{\partial \boldsymbol{R}}{\partial u^j} du^j = \boldsymbol{e}_j du^j . \tag{2.5.6}$$

so that the square of the arc length is given by

$$(dl)^2 = \sum_{i,j} du^i du^j \boldsymbol{e}_i \cdot \boldsymbol{e}_j . \tag{2.5.7}$$

Because of (2.5.1) this can be written as

$$(dl)^2 = g_{ij} du^i du^j \tag{2.5.8a}$$

in terms of the g_{ij}, or as

$$(dl)^2 = h_i h_j \cos \sphericalangle (\hat{e}_i, \hat{e}_j)\, du^i\, du^j \ , \tag{2.5.8b}$$

expressed in terms of the h_k. These equations are often used to define the metric coefficients. Instead, we preferred to define the g_{ij} through (2.5.1), which seems to provide more "physical" insight.

Once the metric coefficients g_{ij} and g^{ij} are known, it turns out to be a trivial exercise to change covariant vector components into contravariant components, and vice versa. From the definitions in (2.3.16) we obtain:

$$D_i = \boldsymbol{D} \cdot \boldsymbol{e}_i = D^j \boldsymbol{e}_j \cdot \boldsymbol{e}_i = D^j g_{ji}$$

or, utilizing the symmetry of g_{ij}

$$D_i = g_{ij} D^j \ . \tag{2.5.9}$$

Similarly,

$$D^k = \boldsymbol{D} \cdot \boldsymbol{e}^k = D_i \boldsymbol{e}^i \cdot \boldsymbol{e}^k = D_i g^{ik}$$

or

$$D^k = g^{ki} D_i \ . \tag{2.5.10}$$

Recall that (2.5.9) and (2.5.10) must be summed over j and i respectively.

Analogous expressions hold for the covariant and contravariant basis vectors as well. Following (2.3.14a) but with $\boldsymbol{D} \equiv \boldsymbol{e}_i$ leads to

$$\boldsymbol{e}_i = (\boldsymbol{e}_i \cdot \boldsymbol{e}_j) \boldsymbol{e}^j$$

or

$$\boldsymbol{e}_i = g_{ij} \boldsymbol{e}^j \ . \tag{2.5.11}$$

By the same token, (2.3.14b) gives then, for $\boldsymbol{D} \equiv \boldsymbol{e}^i$,

$$\boldsymbol{e}^i = (\boldsymbol{e}^i \cdot \boldsymbol{e}^j) \boldsymbol{e}_j$$

or

$$\boldsymbol{e}^i = g^{ij} \boldsymbol{e}_j \ . \tag{2.5.12}$$

Whenever covariant quantities are to be converted to contravariant components (and vice-versa), one multiplies them (allowing for the implicit summation) by the metric coefficients having the other index position.

A final interesting relationship concerns the metric coefficients g_{ij}, g^{ij} and the Kronecker delta δ^i_j. Consider (2.5.11).

$$e_i = g_{ij} e^j \ .$$

If we dot-multiply this equation with the reciprocal-basis vector e^k, we obtain

$$e_i \cdot e^k = g_{ij} e^j \cdot e^k$$

or

$$\delta_i^k = g_{ij} g^{jk} \ . \tag{2.5.13}$$

Upon interchanging i and k, (2.5.12) leads to an equivalent expression (since g_{ij} and g^{ij} are symmetric, this observation is trivial):

$$\delta_i^k = g^{kj} g_{ji} \ . \tag{2.5.14}$$

When we form a matrix with elements g_{ij} and denote it by $[g_{ij}]$, and similarly for g^{jk}, and recall the definition of the product of two matrices, then we can interpret (2.5.13) and (2.5.14) by saying that the product of the matrices $[g_{ij}]$ and $[g^{jk}]$ equals the identity matrix. (If $[C] = [A]\,[B]$ then $C_{ij} = \sum_m A_{im} B_{mj}$.) Consequently, the coefficients g^{ij} (or g_{ij}) can be computed from the g_{ij} (or g^{ij}) by calculation of the inverse matrix. Thus, $[g^{ij}] = [g_{ij}]^{-1}$ and $[g_{ij}] = [g^{ij}]^{-1}$, or $g^{ij} = G^{ij}/g$ and $g_{ij} = G_{ij}/g^{-1} = G_{ij} g$. Here G_{ij} (or G^{ij}) represents the cofactor of the matrix element g^{ij} (or g_{ij}) while g is the determinant of the matrix $[g_{ij}]$. Thus

$$g = \det[g_{ij}] \qquad \text{and} \qquad g^{-1} = \det[g^{ij}] \ . \tag{2.5.15}$$

The expressions in (2.5.11–14), which imply summations, can be recalled by noting that the summation over an index, appearing above and below on the same side of the equation, can be thought of as a "saturation" of that index. As a consequence, that index disappears from the result. The indices in the result are the ones which we did not sum over, and are simply carried over to the other side of the equality sign, at the same position (above or below). This rule of thumb suggests that we can write (2.5.13) and (2.5.14) as follows:

$$g_i^k = g_{ij} g^{jk} = g^{kj} g_{ji} \ . \tag{2.5.16}$$

Indeed, as we shall discuss later in detail in Chap. 3. on tensor-related issues, the symbol g_i^k has the same values, 0 or 1, as the Kronecker delta δ_i^k:

$$g_i^k = \delta_i^k \ . \tag{2.5.17}$$

In the chapter on tensors we shall point out that in principle the only legitimate Kronecker delta must be written as δ^i_j and that (at least as long as we deal with covariant and contravariant quantities, i.e., with sub- and superscripts) one should be cautious in employing symbols like δ_{ij} or δ^{ij}.

2.5.2 The Jacobian

a) The Jacobian of the Curvilinear Coordinate System

Consider the curvilinear coordinate transformation introduced in (2.3.1) and (2.3.2). The Jacobian of the transformation (2.3.1) is by definition the determinant of the matrix produced by the nine partial derivatives $\partial x/\partial u^i$, $\partial y/\partial u^i$, $\partial z/\partial u^i$; $i = 1, 2, 3$. Its proper name is "the Jacobian of (x, y, z) with respect to the u^i". Thus we have

$$J \equiv J(u^1, u^2, u^3) \equiv J\left(\frac{x, y, z}{u^1, u^2, u^3}\right) \equiv \frac{\partial(x, y, z)}{\partial(u^1, u^2, u^3)}$$

$$\equiv \begin{vmatrix} \dfrac{\partial x}{\partial u^1} & \dfrac{\partial x}{\partial u^2} & \dfrac{\partial x}{\partial u^3} \\ \dfrac{\partial y}{\partial u^1} & \dfrac{\partial y}{\partial u^2} & \dfrac{\partial y}{\partial u^3} \\ \dfrac{\partial z}{\partial u^1} & \dfrac{\partial z}{\partial u^2} & \dfrac{\partial z}{\partial u^3} \end{vmatrix} \equiv \begin{vmatrix} \dfrac{\partial x}{\partial u^1} & \dfrac{\partial y}{\partial u^1} & \dfrac{\partial z}{\partial u^1} \\ \dfrac{\partial x}{\partial u^2} & \dfrac{\partial y}{\partial u^2} & \dfrac{\partial z}{\partial u^2} \\ \dfrac{\partial x}{\partial u^3} & \dfrac{\partial y}{\partial u^3} & \dfrac{\partial z}{\partial u^3} \end{vmatrix} . \tag{2.5.18}$$

The last equality is a consequence of equal determinants of a matrix and its transpose. The notation $J(u^1, u^2, u^3)$ stresses that J is a function of the u^i.

If we recall that the triple product $\boldsymbol{A} \cdot \boldsymbol{B} \times \boldsymbol{C}$ can be written as a determinant of the matrix with elements A_x, A_y, A_z, B_x, B_y, etc. in a Cartesian system and think of the transformation in (2.3.1) as $\boldsymbol{R}[x(u^1, u^2, u^3), y(u^1, u^2, u^3), z(u^1, u^2, u^3)]$, we can rewrite (2.5.18) in a more compact form:

$$J = \frac{\partial \boldsymbol{R}}{\partial u^1} \cdot \frac{\partial \boldsymbol{R}}{\partial u^2} \times \frac{\partial \boldsymbol{R}}{\partial u^3} . \tag{2.5.19}$$

All the even permutations of 1, 2, 3 are equally valid. The inverse transformation given by (2.3.2) has a Jacobian as well:

$$\mathscr{J} \equiv \mathscr{J}(x, y, z) \equiv \mathscr{J}\left(\frac{u^1, u^2, u^3}{x, y, z}\right) \equiv \frac{\partial(u^1, u^2, u^3)}{\partial(x, y, z)}$$

$$= \begin{vmatrix} \dfrac{\partial u^1}{\partial x} & \dfrac{\partial u^1}{\partial y} & \dfrac{\partial u^1}{\partial z} \\ \dfrac{\partial u^2}{\partial x} & \dfrac{\partial u^2}{\partial y} & \dfrac{\partial u^2}{\partial z} \\ \dfrac{\partial u^3}{\partial x} & \dfrac{\partial u^3}{\partial y} & \dfrac{\partial u^3}{\partial z} \end{vmatrix} = \begin{vmatrix} \dfrac{\partial u^1}{\partial x} & \dfrac{\partial u^2}{\partial x} & \dfrac{\partial u^3}{\partial x} \\ \dfrac{\partial u^1}{\partial y} & \dfrac{\partial u^2}{\partial y} & \dfrac{\partial u^3}{\partial y} \\ \dfrac{\partial u^1}{\partial z} & \dfrac{\partial u^2}{\partial z} & \dfrac{\partial u^3}{\partial z} \end{vmatrix} . \tag{2.5.20}$$

Now $\partial u^i/\partial x$, $\partial u^i/\partial y$, $\partial u^i/\partial z$ are considered to be the Cartesian components of $\boldsymbol{\nabla} u^i$, and so we have equivalently:

$$\mathscr{J} = \nabla u^1 \cdot \nabla u^2 \times \nabla u^3 \ . \tag{2.5.21}$$

We recognize the vectors in (2.5.19) and (2.5.21) as the members of our bases: the tangent-basis vectors $\boldsymbol{e}_i \equiv \partial \boldsymbol{R}/\partial u^i$, and the reciprocal-basis vectors $\boldsymbol{e}^i \equiv \nabla u^i$. Hence, the Jacobians are identical to

$$J = \boldsymbol{e}_1 \cdot \boldsymbol{e}_2 \times \boldsymbol{e}_3 \ ; \qquad \mathscr{J} = \boldsymbol{e}^1 \cdot \boldsymbol{e}^2 \times \boldsymbol{e}^3 \ . \tag{2.5.22a, b}$$

Since the $\boldsymbol{e}_i$ and $\boldsymbol{e}^i$ form reciprocal sets, we can prove the expected relationship between the two Jacobians $\mathscr{J}$ and J, namely, that they are the inverses of each other:

$$\mathscr{J} = J^{-1} \ . \tag{2.5.23}$$

Starting from $\mathscr{J} = \boldsymbol{e}^1 \cdot \boldsymbol{e}^2 \times \boldsymbol{e}^3$, and substituting the expression for the $\boldsymbol{e}^i$ as functions of the $\boldsymbol{e}_i$, as given by (2.3.12), we obtain

$$\begin{aligned} \mathscr{J} &= \frac{(\boldsymbol{e}_2 \times \boldsymbol{e}_3) \cdot [(\boldsymbol{e}_3 \times \boldsymbol{e}_1) \times (\boldsymbol{e}_1 \times \boldsymbol{e}_2)]}{[\boldsymbol{e}_1 \cdot \boldsymbol{e}_2 \times \boldsymbol{e}_3]^3} \\ &= \frac{(\boldsymbol{e}_2 \times \boldsymbol{e}_3)}{J^3} \cdot (\boldsymbol{e}_1(\boldsymbol{e}_2 \cdot \boldsymbol{e}_3 \times \boldsymbol{e}_1) - \boldsymbol{e}_2(\boldsymbol{e}_1 \cdot \boldsymbol{e}_3 \times \boldsymbol{e}_1)] \\ &= \frac{[\boldsymbol{e}_1 \cdot (\boldsymbol{e}_2 \times \boldsymbol{e}_3)][\boldsymbol{e}_2 \cdot \boldsymbol{e}_3 \times \boldsymbol{e}_1]}{J^3} = \frac{J^2}{J^3} = J^{-1} \ . \end{aligned}$$

(The distinction between superscripts and powers should be clear from the context.)

From now on, we shall always employ the symbol J for the Jacobian. The Jacobian of (2.5.21) or (2.5.22b) will be referred to as J^{-1}. It is customary to call J simply the Jacobian of the curvilinear coordinate system. Having said this, we can rewrite the "$\boldsymbol{e}_i - \boldsymbol{e}^i$ back-and-forth" expressions of (2.3.12) and (2.3.13) more compactly:

$$\boldsymbol{e}^i \equiv \nabla u^i = \frac{1}{J}\left(\frac{\partial \boldsymbol{R}}{\partial u^j} \times \frac{\partial \boldsymbol{R}}{\partial u^k}\right) = \frac{1}{J}(\boldsymbol{e}_j \times \boldsymbol{e}_k) \qquad i, j, k \text{ cyc } 1, 2, 3 \tag{2.5.24}$$

and

$$\boldsymbol{e}_i \equiv \frac{\partial \boldsymbol{R}}{\partial u^i} = J(\nabla u^j \times \nabla u^k) = J(\boldsymbol{e}^j \times \boldsymbol{e}^k) \qquad i, j, k \text{ cyc } 1, 2, 3 \tag{2.5.25}$$

where

$$J = \frac{1}{\nabla u^1 \cdot \nabla u^2 \times \nabla u^3} = \frac{\partial \boldsymbol{R}}{\partial u^1} \cdot \frac{\partial \boldsymbol{R}}{\partial u^2} \times \frac{\partial \boldsymbol{R}}{\partial u^3} \ . \tag{2.5.26}$$

b) The Relationship Between J and g

In a previous section we defined the determinant g of the matrix of metric coefficients: $g = \det[g_{ij}]$. For $\boldsymbol{R} = \boldsymbol{R}(x, y, z)$, where (x, y, z) are Cartesian coordinates, the dot product $\boldsymbol{e}_i \cdot \boldsymbol{e}_j$ is readily computed:

$$g_{ij} = \frac{\partial \boldsymbol{R}}{\partial u^i} \cdot \frac{\partial \boldsymbol{R}}{\partial u^j} = \frac{\partial x}{\partial u^i}\frac{\partial x}{\partial u^j} + \frac{\partial y}{\partial u^i}\frac{\partial y}{\partial u^j} + \frac{\partial z}{\partial u^i}\frac{\partial z}{\partial u^j} \,. \tag{2.5.27}$$

It is a theorem of matrix algebra that the determinant of the product of two matrices $[A]$ and $[B]$ equals the product of the determinants of the matrices: $(\det[A])(\det[B]) = \det([A][B])$. Recall that if $[C] = [A][B]$, then

$$C_{ij} = \sum_{q=1}^{3} A_{iq}B_{qj} \,.$$

If we now, for a moment, represent x by $R(1)$, y by $R(2)$, and z by $R(3)$, then (2.5.27) can be written as

$$g_{ij} = \sum_{q=1}^{3} \frac{\partial R(q)}{\partial u^i}\frac{\partial R(q)}{\partial u^j} \,. \tag{2.5.28}$$

Thus the g_{ij} of (2.5.28) can be thought of as produced by the product of two matrices with elements $\partial R(q)/\partial u^i$ and $\partial R(q)/\partial u^j$,

$$g = \det[g_{ij}] = \left(\det\left[\frac{\partial R(q)}{\partial u^i}\right]\right)\left(\det\left[\frac{\partial R(q)}{\partial u^j}\right]\right)$$

$$= \begin{vmatrix} \frac{\partial x}{\partial u^1} & \frac{\partial x}{\partial u^2} & \frac{\partial x}{\partial u^3} \\ \frac{\partial y}{\partial u^1} & \frac{\partial y}{\partial u^2} & \frac{\partial y}{\partial u^3} \\ \frac{\partial z}{\partial u^1} & \frac{\partial z}{\partial u^2} & \frac{\partial z}{\partial u^3} \end{vmatrix} \cdot \begin{vmatrix} \frac{\partial x}{\partial u^1} & \frac{\partial y}{\partial u^1} & \frac{\partial z}{\partial u^1} \\ \frac{\partial x}{\partial u^2} & \frac{\partial y}{\partial u^2} & \frac{\partial z}{\partial u^2} \\ \frac{\partial x}{\partial u^3} & \frac{\partial y}{\partial u^3} & \frac{\partial z}{\partial u^3} \end{vmatrix} \,.$$

$$= \left(\frac{\partial \boldsymbol{R}}{\partial u^1} \cdot \frac{\partial \boldsymbol{R}}{\partial u^2} \times \frac{\partial \boldsymbol{R}}{\partial u^3}\right)^2 = (J)^2 \,. \tag{2.5.29}$$

Hence we arrive at the important result that the determinant of the "metric matrix", g, is equal to the square of the Jacobian J:

$$g = (J)^2 \qquad \text{or} \qquad J = \sqrt{g} \,. \tag{2.5.30}$$

In an *orthogonal* coordinate system, (2.5.26) becomes

$$J = \sqrt{g} = h_1\hat{\boldsymbol{e}}_1 \cdot h_2\hat{\boldsymbol{e}}_2 \times h_3\hat{\boldsymbol{e}}_3 = h_1h_2h_3 \tag{2.5.31}$$

since $\hat{\boldsymbol{e}}_1 \cdot \hat{\boldsymbol{e}}_2 \times \hat{\boldsymbol{e}}_3 = 1$. (Recall that $\hat{\boldsymbol{e}}_i$ is a unit vector.)

2.5.3 The Dot and Cross-Products in Curvilinear Coordinates

Now we turn to the computation of the dot and cross-products between two vectors. The existence of two convenient bases along which to expand a vector into its components has implications for the computation of the dot and cross products.

a) Dot Products

Two vectors $\boldsymbol{A}$ and $\boldsymbol{B}$ can be written in terms of their contravariant or covariant components:

$$\boldsymbol{A} = A^i \boldsymbol{e}_i = A_i \boldsymbol{e}^i \tag{2.5.32a}$$

$$\boldsymbol{B} = B^j \boldsymbol{e}_j = B_j \boldsymbol{e}^j \ . \tag{2.5.32b}$$

The dot product between two vectors can be "shifted" to their basis-vectors, giving rise to four possible combinations:

$$\boldsymbol{A} \cdot \boldsymbol{B} = A^i \boldsymbol{e}_i \cdot B_j \boldsymbol{e}^j = A^i B_j \boldsymbol{e}_i \cdot \boldsymbol{e}^j = A^i B_j \delta_i^j \tag{2.5.33a}$$

$$= A_i \boldsymbol{e}^i \cdot B^j \boldsymbol{e}_j = A_i B^j \boldsymbol{e}^i \cdot \boldsymbol{e}_j = A_i B^j \delta_j^i \tag{2.5.33b}$$

$$= A^i \boldsymbol{e}_i \cdot B^j \boldsymbol{e}_j = A^i B^j \boldsymbol{e}_i \cdot \boldsymbol{e}_j = A^i B^j g_{ij} \tag{2.5.33c}$$

$$= A_i \boldsymbol{e}^i \cdot B_j \boldsymbol{e}^j = A_i B_j \boldsymbol{e}^i \cdot \boldsymbol{e}^j = A_i B_j g^{ij} \ . \tag{2.5.33d}$$

In summary, the dot product of two vectors in general curvilinear coordinates is

$$\boldsymbol{A} \cdot \boldsymbol{B} = A^i B_i = A_i B^i = g_{ij} A^i B^j = g^{ij} A_i B_j \ . \tag{2.5.34}$$

We should note that the last two forms of (2.5.34) also follow from the first two (and vice versa) when we recall the $D_i \leftrightarrow D^j$ formulae previously derived in (2.5.9) and (2.5.10). In *orthogonal* coordinate systems $g_{ij} = g^{ij} = 0$ for $i \neq j$ and (2.5.34) reduces to

$$\boldsymbol{A} \cdot \boldsymbol{B} = A^i B_i = A_i B^i = (h_i)^2 A^i B^i = \frac{A_i B_i}{(h_i)^2} \ . \tag{2.5.35}$$

A summation over i is implied in all formulae in (2.5.35). If we restrict our attention further to an *orthonormal* coordinate system, where $h_i = 1$ and $A_i = A^i$, we retrieve the trivial expression $\Sigma A(i)\, B(i) \equiv \Sigma A_i B_i$.

The magnitude of a vector equals the square root of the dot product with itself:

$$A = |\boldsymbol{A}| = \sqrt{\boldsymbol{A} \cdot \boldsymbol{A}} = \sqrt{A^i A_i} = \sqrt{A_i A^i} = \sqrt{g_{ij} A^i A^j}$$

$$= \sqrt{g^{ij} A_i A_j} \ . \tag{2.5.36}$$

We have essentially already applied the expressions derived above, when we determined the differential arc length along a curve in (2.5.8).

b) Cross Products

For the vector product, there are only two convenient combinations of basis vectors:

$$\boldsymbol{A} \times \boldsymbol{B} = A^i B^j \boldsymbol{e}_i \times \boldsymbol{e}_j = A_i B_j \boldsymbol{e}^i \times \boldsymbol{e}^j \ . \tag{2.5.37}$$

Here, as always, a summation over the indices i and j is understood (where we note that the terms involving $i = j$ give zero). The "$\boldsymbol{e}_i - \boldsymbol{e}^j$ back-and-forth" formulae of (2.5.24) and (2.5.25) give

$$\boldsymbol{A} \times \boldsymbol{B} = J \sum_{i,j}{}' (\pm) A^i B^j \boldsymbol{e}^k = J^{-1} \sum_{i,j}{}' (\pm) A_i B_j \boldsymbol{e}_k \ . \tag{2.5.38}$$

The index k must be chosen such that i, j, k form a permutation of 1, 2, 3; an even permutation is given a "+" sign, an odd one is assigned a "−" sign. The prime on the summation sign indicates that the values $i = j$ must be omitted from the summation. This means that we can rewrite (2.5.38) as follows:

$$\begin{aligned} \boldsymbol{A} \times \boldsymbol{B} &= J \sum_k (A^i B^j - A^j B^i) \boldsymbol{e}^k \qquad i, j, k \ \text{cyc} \ 1, 2, 3 \\ &= \frac{1}{J} \sum_k (A_i B_j - A_j B_i) \boldsymbol{e}_k \qquad i, j, k \ \text{cyc} \ 1, 2, 3 \ . \end{aligned} \tag{2.5.39}$$

The change in summation index is made possible by the sentence "i, j, k cyc 1, 2, 3." This expression can be written more compactly with the help of the Levi-Civita symbols, ε^{ijk} and ε_{ijk}, which are defined as

$$\varepsilon^{ijk} = \begin{cases} +1 & \text{if } i, j, k \text{ form an even permutation of } 1, 2, 3 \\ -1 & \text{if } i, j, k \text{ form an odd permutation of } 1, 2, 3 \\ 0 & \text{otherwise} \ , \end{cases} \tag{2.5.40}$$

and similarly for ε_{ijk}.
This leads to:

$$(\boldsymbol{A} \times \boldsymbol{B})_k = \varepsilon_{ijk} J A^i B^j = \varepsilon_{ijk} \sqrt{g} A^i B^j \qquad i, j, k \ \text{cyc} \ 1, 2, 3 \tag{2.5.41}$$

for the first expression in (2.5.38) and

$$(\boldsymbol{A} \times \boldsymbol{B})^k = \frac{\varepsilon^{ijk}}{J} A_i B_j = \frac{\varepsilon^{ijk}}{\sqrt{g}} A_i B_j \qquad i, j, k \ \text{cyc} \ 1, 2, 3 \tag{2.5.42}$$

for the second expression. Note that in these expressions on the left- and right-hand side of the equality, components of the other kind appear, and there is an additional factor of $J = \sqrt{g}$ or $J^{-1} = (\sqrt{g})^{-1}$.

Expressions for the vector product in terms of components of the same sort are given by the expressions given in (2.5.9) and (2.5.10), i.e.,

$$(\boldsymbol{A} \times \boldsymbol{B})_i = g_{ik} (\boldsymbol{A} \times \boldsymbol{B})^k \quad \text{and} \quad (\boldsymbol{A} \times \boldsymbol{B})^i = g^{ik} (\boldsymbol{A} \times \boldsymbol{B})_k \ . \tag{2.5.43}$$

2.5.4 The Differential Elements $dl(i)$, $dS(i)$, d^3R

a) The Differential Arc Length $dl(i)$

The differential arc length along a coordinate curve u^i, which we denote by $dl(i)$ is, according to (2.5.5)

$$dl(\text{along } u^i) \equiv dl(i) = |d\boldsymbol{R}(i)| = \sqrt{d\boldsymbol{R}(i)\cdot d\boldsymbol{R}(i)} \ . \tag{2.5.44}$$

The general expression for the vector $d\boldsymbol{R}$ is given in (2.5.6). However, since the coordinate curve u^i is defined by the intersection of two surfaces $u^j = \text{constant}$ and $u^k = \text{constant}$, we must require that $du^j = du^k = 0$ for $j, k \neq i$. Therefore, we obtain

$$\begin{aligned} d\boldsymbol{R}(i) &= \frac{\partial \boldsymbol{R}}{\partial u^i} du^i = \boldsymbol{e}_i du^i \qquad \text{no summation} \\ &= h_i du^i \hat{\boldsymbol{e}}_i = \sqrt{g_{ii}}\, du^i \hat{\boldsymbol{e}}_i \qquad \text{no summation} \ . \end{aligned} \tag{2.5.45}$$

Consequently, (2.5.44) leads to

$$dl(i) = h_i du^i = \sqrt{g_{ii}}\, du^i \ , \qquad \text{no summation} \ . \tag{2.5.46}$$

This is the usual definition of h_i. Note that the form $h_i du^i$ is also valid for non-orthogonal systems.

From the definition of g_{ij} in (2.5.1) and the $\boldsymbol{e}_i \leftrightarrow \boldsymbol{e}^i$ formulae in (2.5.25) we can derive an alternative expression for $dl(i)$:

$$dl(i) = J|\nabla u^j \times \nabla u^k|\, du^i \qquad i, j, k \text{ cyc } 1, 2, 3 \ . \tag{2.5.47}$$

b) The Differential Area Element $dS(i)$

The differential element of area in the coordinate surface $u^i = \text{constant}$ (at a point P determined by the position vector $\boldsymbol{R}$) is easily computed if we recall that the magnitude of the vector product equals the area of the parallelogram made by the two vectors:

$$\begin{aligned} dS(\text{in } u^i = \text{constant}) &\equiv dS(i) = |d\boldsymbol{R}(j) \times d\boldsymbol{R}(k)| \\ &= \left|\frac{\partial \boldsymbol{R}}{\partial u^j} \times \frac{\partial \boldsymbol{R}}{\partial u^k}\right| du^j du^k \ , \qquad \text{no summation over } j \text{ or } k \ . \end{aligned} \tag{2.5.48a}$$

Or

$$dS(i) = |\boldsymbol{e}_j \times \boldsymbol{e}_k|\, du^j du^k \ , \qquad \text{no summation over } j \text{ or } k \ . \tag{2.5.48b}$$

Here again, i, j, k form a cyclic permutation of 1, 2, 3.

With the expression for $\boldsymbol{e}^i \equiv \nabla u^i$ as a function of $\boldsymbol{e}_j$ and $\boldsymbol{e}_k$ as given in (2.5.24) we can rewrite $dS(i)$ as

$$dS(i) = J|\nabla u^i|\,du^j\,du^k \quad . \tag{2.5.49a}$$

This expression is useful when we need the differential area vector $d\boldsymbol{S}(i)$. Since we know that the gradient is perpendicular to the surface, we obtain

$$d\boldsymbol{S}(i) \equiv \frac{\nabla u^i}{|\nabla u^i|}dS = \pm J\,du^j du^k \nabla u^i \ , \qquad \text{no summation} \ . \tag{2.5.49b}$$

The sign must be chosen according to the convention adopted for the normal to the surface.

We now rearrange the expression given in (2.5.48b). Making use of the vector identity $(\boldsymbol{A} \times \boldsymbol{B})\cdot(\boldsymbol{C} \times \boldsymbol{D}) = (\boldsymbol{A}\cdot\boldsymbol{C})(\boldsymbol{B}\cdot\boldsymbol{D}) - (\boldsymbol{A}\cdot\boldsymbol{D})(\boldsymbol{B}\cdot\boldsymbol{C})$, to obtain

$$\begin{aligned} dS(i) &= \sqrt{(\boldsymbol{e}_j \times \boldsymbol{e}_k)\cdot(\boldsymbol{e}_j \times \boldsymbol{e}_k)}\,du^j\,du^k \\ &= \sqrt{(\boldsymbol{e}_j\cdot\boldsymbol{e}_j)(\boldsymbol{e}_k\cdot\boldsymbol{e}_k) - (\boldsymbol{e}_j\cdot\boldsymbol{e}_k)(\boldsymbol{e}_j\cdot\boldsymbol{e}_k)}\,du^j\,du^k \\ &\quad \text{no summation over } j \text{ or } k \ . \end{aligned} \tag{2.5.50}$$

Subsequently we employ the definition of g_{ij}, and find

$$dS(i) = \sqrt{g_{jj}g_{kk} - g_{jk}{}^2}\,du^j\,du^k \ , \qquad \text{no summation over } j \text{ or } k \ ; \quad i, j, k \text{ cyc } 1, 2, 3 \ . \tag{2.5.51}$$

The formulae for *orthogonal* systems are quite simple since the cross-metric coefficients g_{jk} are identically zero. Using (2.5.4), we have

$$dS(i) = h_j h_k du^j du^k \ , \qquad \text{no summation over } j \text{ or } k \ ; \quad i, j, k \text{ cyc } 1, 2, 3 \ . \tag{2.5.52}$$

c) The Differential Volume Element d^3R

d^3R represents the differential-volume element at a point P determined by the position vector $\boldsymbol{R}$. It is "infinitesimal" in all three coordinates. The symbol dV is kept for a volume element that is infinitesimal in only one direction (e.g. in the spherical system d^3R would be $r^2 dr \sin\theta\, d\theta\, d\phi$, while dV is for instance $4\pi\, r^2\, dr$). The volume element d^3R is the volume constructed by the differential vectors along each of the coordinate curves:

$$d^3R = d\boldsymbol{R}(1)\cdot d\boldsymbol{R}(2) \times d\boldsymbol{R}(3) = du^1 du^2 du^3 \boldsymbol{e}_1\cdot(\boldsymbol{e}_2 \times \boldsymbol{e}_3) \tag{2.5.53}$$

or

$$d^3R = J\,du^1 du^2 du^3 = \sqrt{g}\,du^1 du^2 du^3 \ . \tag{2.5.54}$$

This result is positive because by definition $\boldsymbol{e}_1, \boldsymbol{e}_2, \boldsymbol{e}_3$ form a right-handed system.

2.6 Vector Differentiation in Curvilinear Coordinates

2.6.1 The Covariant Derivative

a) Differentiation of Vector Components and Basis Vectors

Vector differentiation is a straightforward operation, as long as we keep in mind that, in curvilinear coordinates, not only the components of the vector but also the basis vectors (tangent and reciprocal) are position-dependent.

Consider a vector field $\boldsymbol{W} = \boldsymbol{W}(\boldsymbol{R}(u^1, u^2, u^3))$, i.e. the vector $\boldsymbol{W}$ is a function of position as indicated by the position vector $\boldsymbol{R}$. $\boldsymbol{W}$ can be expanded as usual:

$$\boldsymbol{W} = W^i \boldsymbol{e}_i \tag{2.6.1}$$

or

$$\boldsymbol{W} = W_i \boldsymbol{e}^i \ . \tag{2.6.2}$$

The W^i, W_i, $\boldsymbol{e}_i$ and $\boldsymbol{e}^i$ are all dependent upon (u^1, u^2, u^3). The differential vector $d\boldsymbol{W}$ is given by

$$d\boldsymbol{W} = \frac{\partial \boldsymbol{W}}{\partial u^k} du^k \ . \tag{2.6.3}$$

The partial derivatives $\partial \boldsymbol{W}/\partial u^k$ follow from (2.6.1) or (2.6.2):

$$\frac{\partial \boldsymbol{W}}{\partial u^k} = \frac{\partial W^i}{\partial u^k} \boldsymbol{e}_i + W^i \frac{\partial \boldsymbol{e}_i}{\partial u^k} \tag{2.6.4}$$

or

$$\frac{\partial \boldsymbol{W}}{\partial u^k} = \frac{\partial W_i}{\partial u^k} \boldsymbol{e}^i + W_i \frac{\partial \boldsymbol{e}^i}{\partial u^k} \ . \tag{2.6.5}$$

For a single value of k, these expressions represent vectors, which can in turn be written with covariant or contravariant components:

$$\frac{\partial \boldsymbol{W}}{\partial u^k} = \left(\frac{\partial \boldsymbol{W}}{\partial u^k}\right)^j \boldsymbol{e}_j \equiv \left(\frac{\partial \boldsymbol{W}}{\partial u^k} \cdot \boldsymbol{e}^j\right) \boldsymbol{e}_j \tag{2.6.6}$$

or

$$\frac{\partial \boldsymbol{W}}{\partial u^k} = \left(\frac{\partial \boldsymbol{W}}{\partial u^k}\right)_j \boldsymbol{e}^j \equiv \left(\frac{\partial \boldsymbol{W}}{\partial u^k} \cdot \boldsymbol{e}_j\right) \boldsymbol{e}^j \tag{2.6.7}$$

where $(\partial \boldsymbol{W}/\partial u^k)^j$ and $(\partial \boldsymbol{W}/\partial u^k)_j$ represent the contravariant and covariant components of the vector $\partial \boldsymbol{W}/\partial u^k$, respectively. By combining (2.6.4) through (2.6.7), we can write the components of the partial derivative of $\boldsymbol{W}$ with respect to u^k in four different ways:

$$\left(\frac{\partial \boldsymbol{W}}{\partial u^k}\right)^j = \frac{\partial W^i}{\partial u^k}\boldsymbol{e}_i\cdot\boldsymbol{e}^j + W^i\frac{\partial \boldsymbol{e}_i}{\partial u^k}\cdot\boldsymbol{e}^j \tag{2.6.8}$$

$$\left(\frac{\partial \boldsymbol{W}}{\partial u^k}\right)^j = \frac{\partial W_i}{\partial u^k}\boldsymbol{e}^i\cdot\boldsymbol{e}^j + W_i\frac{\partial \boldsymbol{e}^i}{\partial u^k}\cdot\boldsymbol{e}^j \tag{2.6.9}$$

for the contravariant components, and

$$\left(\frac{\partial \boldsymbol{W}}{\partial u^k}\right)_j = \frac{\partial W^i}{\partial u^k}\boldsymbol{e}_i\cdot\boldsymbol{e}_j + W^i\frac{\partial \boldsymbol{e}_i}{\partial u^k}\cdot\boldsymbol{e}_j \tag{2.6.10}$$

$$\left(\frac{\partial \boldsymbol{W}}{\partial u^k}\right)_j = \frac{\partial W_i}{\partial u^k}\boldsymbol{e}^i\cdot\boldsymbol{e}_j + W_i\frac{\partial \boldsymbol{e}^i}{\partial u^k}\cdot\boldsymbol{e}_j \tag{2.6.11}$$

for the covariant components.

Two of the four expressions above contain a Kronecker delta, and thus reduce to a simpler form. For this reason, only (2.6.8) and (2.6.11) are normally retained for the components of the partial derivative of $\boldsymbol{W}$:

$$\left(\frac{\partial \boldsymbol{W}}{\partial u^k}\right)^j = \frac{\partial W^j}{\partial u^k} + W^i\left\{\frac{\partial \boldsymbol{e}_i}{\partial u^k}\cdot\boldsymbol{e}^j\right\} \tag{2.6.12}$$

and

$$\left(\frac{\partial \boldsymbol{W}}{\partial u^k}\right)_j = \frac{\partial W_j}{\partial u^k} + W_i\left\{\frac{\partial \boldsymbol{e}^i}{\partial u^k}\cdot\boldsymbol{e}_j\right\} . \tag{2.6.13}$$

If we wish to employ the contravariant components of $\boldsymbol{W}$, we use the formula leading to the contravariant components of $\partial \boldsymbol{W}/\partial u^k$ as given in (2.6.12) and analogously for the covariant components, where we use (2.6.13)

An important observation is that the components of the derivative of $\boldsymbol{W}$ are *not* equal to the derivatives of the corresponding components, since terms involving the change of the basis vectors must be added. The expressions in (2.6.12) and (2.6.13) are called the components of the *covariant derivative*. $d\boldsymbol{W}$ of (2.6.3), with (2.6.12) and (2.6.13) substituted is called the *intrinsic* or *absolute differential.*

The reason for the adjectives must be sought in abstract-vector analysis. There one defines a vector by its components without using a symbol for the vector itself; one suppresses explicit reference to a basis. With that convention, the derivative of the "vector" would "at first sight" be the derivative of the component. One then has to construct a "correction term" – the second term of (2.6.12) and (2.6.13) – in a formal and artificial way such that "the derivative" has the right transformation properties.[4] The adjective indicates that one is after the proper derivative.

[4] After reading Chap. 13, the reader will see that one needs the second term in (2.6.12) and (2.6.13) to make sure that the LHS of these equations are the components of a second-order tensor.

In our notation, a component of a intrinsic or absolute differential is represented by δW^j (or δW_j) and is equal to

$$\delta W^j \equiv (d\boldsymbol{W})^j = \left(\frac{\partial \boldsymbol{W}}{\partial u^k}\right)^j du^k = \frac{\partial W^j}{\partial u^k} du^k + W^i \boldsymbol{e}^j \cdot \frac{\partial \boldsymbol{e}_i}{\partial u^k} du^k$$

$$= dW^j + \left\{\boldsymbol{e}^j \cdot \frac{\partial \boldsymbol{e}_i}{\partial u^k}\right\} W^i du^k \ . \tag{2.6.14}$$

In the "component ≡ vector" convention one would say that δW^j (or δW_j) is the intrinsic or absolute derivative itself. It is clear that by always referring explicitly to a basis we do not require any patchwork. Everything follows neatly from straightforward use of the common rules of calculus. It is only when one wishes to use the convention "component ≡ vector" that one must be careful to construct all the terms.

The components of the covariant-derivative vector $(\partial \boldsymbol{W}/\partial u^k)$ – we still think of k as a fixed index – are granted new symbols:

$$W^j_{,k} \equiv \left(\frac{\partial \boldsymbol{W}}{\partial u^k}\right)^j = \frac{\partial W^j}{\partial u^k} + W^i \left\{\frac{\partial \boldsymbol{e}_i}{\partial u^k} \cdot \boldsymbol{e}^j\right\} \tag{2.6.15}$$

and

$$W_{j,k} \equiv \left(\frac{\partial \boldsymbol{W}}{\partial u^k}\right)_j = \frac{\partial W_j}{\partial u^k} + W_i \left\{\frac{\partial \boldsymbol{e}^i}{\partial u^k} \cdot \boldsymbol{e}_j\right\} \ . \tag{2.6.16}$$

The comma in front of k on the LHS indicates that the differentiation is with respect to u^k. (In some texts, (Mathews and Walker 1970; Butkov 1968), a semi-colon is used instead of a comma for the components of the covariant derivative; the comma is then reserved for simple partial derivatives, i.e., the first terms in (2.6.15) and (2.6.16).)

b) The Covariant Derivative in Terms of the Metric Coefficients

The information about the chosen basis is largely contained in the metric coefficients g_{ij}. It can be expected that the "correction terms" in (2.6.15) and (2.6.16) are expressible in terms of these metric coefficients. Consider the correction term in (2.6.15): $W^i\{(\partial \boldsymbol{e}_i/\partial u^k) \cdot \boldsymbol{e}^j\}$. Keeping k still fixed, the vector $\partial \boldsymbol{e}_i/\partial u^k$ can be expanded in two ways, along $\boldsymbol{e}_j$ or $\boldsymbol{e}^j$ as follows:

$$\frac{\partial \boldsymbol{e}_i}{\partial u^k} = \left\{\frac{\partial \boldsymbol{e}_i}{\partial u^k} \cdot \boldsymbol{e}^j\right\} \boldsymbol{e}_j \equiv \left(\frac{\partial \boldsymbol{e}_i}{\partial u^k}\right)^j \boldsymbol{e}_j \tag{2.6.17}$$

or

$$\frac{\partial \boldsymbol{e}_i}{\partial u^k} = \left[\frac{\partial \boldsymbol{e}_i}{\partial u^k} \cdot \boldsymbol{e}_j\right] \boldsymbol{e}^j \equiv \left(\frac{\partial \boldsymbol{e}_i}{\partial u^k}\right)_j \boldsymbol{e}^j \ . \tag{2.6.18}$$

The expression in { } in (2.6.17) is what we need in our correction term. Since $e^j = g^{jn}e_n$, we see from equating (2.6.17) and (2.6.18) that

$$\left\{\frac{\partial e_i}{\partial u^k}\cdot e^j\right\} = g^{jn}\left[\frac{\partial e_i}{\partial u^k}\cdot e_n\right] \tag{2.6.19a}$$

$$= g^{jn}\frac{1}{2}\left[e_n\cdot\frac{\partial e_i}{\partial u^k} + e_n\cdot\frac{\partial e_k}{\partial u^i}\right] . \tag{2.6.19b}$$

Here, we have used the property

$$\frac{\partial e_i}{\partial u^k} \equiv \frac{\partial}{\partial u^k}\frac{\partial \boldsymbol{R}}{\partial u^i} = \frac{\partial}{\partial u^i}\frac{\partial \boldsymbol{R}}{\partial u^k} \equiv \frac{\partial e_k}{\partial u^i} . \tag{2.6.20}$$

The right-hand side of (2.6.19) can be manipulated as follows:

$$g^{jn}\frac{1}{2}\left[\frac{\partial}{\partial u^k}(e_n\cdot e_i) + \frac{\partial}{\partial u^i}(e_n\cdot e_k) - e_i\cdot\frac{\partial e_n}{\partial u^k} - e_k\cdot\frac{\partial e_n}{\partial u^i}\right]$$

$$= g^{jn}\frac{1}{2}\left[\frac{\partial g_{ni}}{\partial u^k} + \frac{\partial g_{nk}}{\partial u^i} - e_i\cdot\frac{\partial e_k}{\partial u^n} - e_k\cdot\frac{\partial e_i}{\partial u^n}\right]$$

$$= g^{jn}\frac{1}{2}\left[\frac{\partial g_{ni}}{\partial u^k} + \frac{\partial g_{nk}}{\partial u^i} - \frac{\partial}{\partial u^n}(e_i\cdot e_k)\right] .$$

Hence, we obtain

$$\left\{\frac{\partial e_i}{\partial u^k}\cdot e^j\right\} = \frac{1}{2}g^{jn}\left[\frac{\partial g_{ni}}{\partial u^k} + \frac{\partial g_{nk}}{\partial u^i} - \frac{\partial g_{ik}}{\partial u^n}\right] . \tag{2.6.21}$$

To find a similar correction term in (2.6.16) we observe that $e_j\cdot e^i = \delta^i_j$, so

$$\frac{\partial}{\partial u^k}(e_j\cdot e^i) = 0$$

and

$$\left\{e_j\cdot\frac{\partial e^i}{\partial u^k}\right\} = -\left\{e^i\cdot\frac{\partial e_j}{\partial u^k}\right\} . \tag{2.6.22}$$

The part of the correction term of (2.6.16) inside the brackets is almost the same as in (2.6.21) but with a minus sign, and the indices i and j interchanged. In summary,

$$W^j_{,k} = \frac{\partial W^j}{\partial u^k} + \frac{1}{2}g^{jn}\left[\frac{\partial g_{ni}}{\partial u^k} + \frac{\partial g_{nk}}{\partial u^i} - \frac{\partial g_{ik}}{\partial u^n}\right]W^i \tag{2.6.23}$$

and

$$W_{j,k} = \frac{\partial W_j}{\partial u^k} - \frac{1}{2} g^{in} \left[\frac{\partial g_{nj}}{\partial u^k} + \frac{\partial g_{nk}}{\partial u^j} - \frac{\partial g_{jk}}{\partial u^n} \right] W_i \ . \tag{2.6.24}$$

c) The Christoffel Symbols

The *Christoffel symbol of the second kind* is defined as

$$\left\{ \begin{matrix} j \\ i \ k \end{matrix} \right\} \equiv \left\{ e^j \cdot \frac{\partial e_i}{\partial u^k} \right\} = \frac{1}{2} g^{jn} \left[\frac{\partial g_{ni}}{\partial u^k} + \frac{\partial g_{nk}}{\partial u^i} - \frac{\partial g_{ik}}{\partial u^n} \right] . \tag{2.6.25}$$

This second Christoffel symbol is the jth contravariant component of the vector $\partial e_i/\partial u^k$ (for k fixed). Analogously, the *Christoffel symbol of the first kind* is the jth covariant component of the same vector $\partial e_i/\partial u^k$:

$$[j, ik] \equiv \left[e_j \cdot \frac{\partial e_i}{\partial u^k} \right] = \frac{1}{2} \left[\frac{\partial g_{ji}}{\partial u^k} + \frac{\partial g_{jk}}{\partial u^i} - \frac{\partial g_{ik}}{\partial u^j} \right] . \tag{2.6.26}$$

Some authors use $[ik,j]$. The second part of (2.6.26) follows from (2.6.19a) recognizing that 1/2 $[\ldots]$ of (2.6.25), without the g^{jn}, equals $[n, ik]$. The comma in $[j, ik]$ does not refer to the differentiating index here as can be seen from its definition. A better notation would perhaps have been $[j/ik]$.

The positions of the indices in the Christoffel symbols are very important. They are a shorthand for the { } and [ ] in (2.6.17) and (2.6.18): delete the e's but maintain the index positions. As usual, an upper index refers to a contravariant quantity, etc. Because of (2.6.20) we observe that the Christoffel symbols are symmetric in the indices appearing in $\partial e_i/\partial u^k$. Thus,

$$\left\{ \begin{matrix} j \\ i \ k \end{matrix} \right\} = \left\{ \begin{matrix} j \\ k \ i \end{matrix} \right\} \tag{2.6.27}$$

and

$$[j, ik] = [j, ki] \ . \tag{2.6.28}$$

The relationship between the two kinds of Christoffel symbols follows from (2.6.19a)

$$[i, jk] = g_{in} \left\{ \begin{matrix} n \\ j \ k \end{matrix} \right\} \tag{2.6.29}$$

$$\left\{ \begin{matrix} i \\ j \ k \end{matrix} \right\} = g^{in} [n, jk] \ . \tag{2.6.30}$$

Alternative symbols for the Christoffel symbols are $\Gamma^i_{jk} \equiv \left\{ \begin{matrix} i \\ j \ k \end{matrix} \right\}$ and $\Gamma_{i,jk} \equiv [i, jk]$. These Γ-symbols suggest a tensor character, which they do not have (see Chap. 3 or 13). Note that the so-called "Christoffel symbols" as defined by Stacey (1981) differ from those defined here and by other authors (Morse and Feshbach 1953, Spiegel 1959).

The introduction of the Christoffel symbols allows us to rewrite the components of the covariant derivative, given in (2.6.23) and (2.6.24) in a compact form:

$$W^j_{,k} = \frac{\partial W^j}{\partial u^k} + \begin{Bmatrix} j \\ i\ k \end{Bmatrix} W^i \tag{2.6.31}$$

and

$$W_{j,k} = \frac{\partial W_j}{\partial u^k} - \begin{Bmatrix} i \\ j\ k \end{Bmatrix} W_i \ . \tag{2.6.32}$$

In all the foregoing, the summation convention was used.

The covariant derivative was obtained using the ordinary rules of differentiation. Thus the derivative of a sum is the sum of the derivatives and the derivative of a product is the sum of similar products where one of the factors, in turn, is replaced by its derivative.

By definition, the Christoffel symbols originate from the dependence on position of the basis vectors, so in a rectilinear coordinate system, the Christoffel symbols are zero.

2.6.2 The Del Operator

a) Definition of the Del Operator

The del operator, $\boldsymbol{\nabla}$, is usually defined via the relationship $d\Phi = \boldsymbol{\nabla}\Phi \cdot d\boldsymbol{R}$. Expansion of $d\Phi$ via the chain rule leads to $d\Phi = (\partial\Phi/\partial u^i)\, du^i$, which can be written as a dot product using $\delta^i_j = \boldsymbol{e}^i \cdot \boldsymbol{e}_j$:

$$\frac{\partial \Phi}{\partial u^i} du^i = \frac{\partial \Phi}{\partial u^i} \boldsymbol{e}^i \cdot \boldsymbol{e}_j du^j = \frac{\partial \Phi}{\partial u^i} \boldsymbol{\nabla} u^i \cdot \boldsymbol{e}_j du^j \ . \tag{2.6.33}$$

Identification of this expression with $\boldsymbol{\nabla}\Phi \cdot d\boldsymbol{R}$ shows that we can formally introduce *the del operator* as

$$\boldsymbol{\nabla} \equiv \boldsymbol{\nabla} u^i \frac{\partial}{\partial u^i} \equiv \boldsymbol{e}^i \frac{\partial}{\partial u^i} \ . \tag{2.6.34}$$

We now derive the formulae for the gradient of a scalar field, the divergence and the curl of a vector field, using the material of the previous sections.

b) Gradient

We operate with $\boldsymbol{\nabla}$ on a scalar function $\Phi(u^1, u^2, u^3)$ and simply juxtapose the two symbols with *no operator sign* in between them. We obtain:

$$\operatorname{grad} \Phi \equiv \boldsymbol{\nabla}\Phi = \boldsymbol{\nabla} u^i \frac{\partial \Phi}{\partial u^i} = \frac{\partial \Phi}{\partial u^i} \boldsymbol{e}^i \tag{2.6.35}$$

or

$$(\boldsymbol{\nabla}\Phi)_i = \frac{\partial \Phi}{\partial u^i} . \tag{2.6.36}$$

Equation (2.6.36) is the covariant component of the gradient vector and does not necessarily have the same units as $\boldsymbol{\nabla}\Phi$, since the reciprocal-basis vectors can have units themselves.

c) Divergence

The simplest derivation of the divergence makes use of the properties $\boldsymbol{\nabla} \times \boldsymbol{\nabla}\Phi \equiv 0$ and $\boldsymbol{\nabla}\cdot(\boldsymbol{\nabla} \times \boldsymbol{A}) \equiv 0$ (S.P. Hirshman 1982). We first note that the $\boldsymbol{e}_i \leftrightarrow \boldsymbol{e}^j$ transformation formulae of (2.5.25) can be written in a curl form:

$$\frac{\boldsymbol{e}_i}{J} = (\boldsymbol{\nabla} u^j \times \boldsymbol{\nabla} u^k) = \boldsymbol{\nabla} \times (u^j \boldsymbol{\nabla} u^k) \qquad i, j, k \text{ cyc } 1, 2, 3 . \tag{2.6.37}$$

Because $\boldsymbol{e}_i/J$ is the curl of a vector, its divergence is zero:

$$\boldsymbol{\nabla}\cdot\left(\frac{\boldsymbol{e}_i}{J}\right) = \boldsymbol{\nabla}\cdot[\boldsymbol{\nabla} \times (u^j \boldsymbol{\nabla} u^k)] \equiv 0 . \tag{2.6.38}$$

Therefore, we have for the *divergence* of a vector $\boldsymbol{A}$:

$$\begin{aligned}\boldsymbol{\nabla}\cdot \boldsymbol{A} &= \boldsymbol{\nabla}\cdot(A^j \boldsymbol{e}_j) = \boldsymbol{\nabla}\cdot\left(JA^j\frac{\boldsymbol{e}_j}{J}\right) = JA^j\boldsymbol{\nabla}\cdot\left(\frac{\boldsymbol{e}_j}{J}\right) + \frac{\boldsymbol{e}_j}{J}\cdot \boldsymbol{\nabla}(JA^j) \\ &= \frac{\boldsymbol{e}_j}{J}\cdot \boldsymbol{\nabla}(JA^j) ,\end{aligned}$$

which upon using the definition of the $\boldsymbol{\nabla}$ operator, (2.6.34), becomes

$$\boldsymbol{\nabla}\cdot \boldsymbol{A} = \frac{1}{J}\frac{\partial}{\partial u^i}(JA^i) = \frac{1}{\sqrt{g}}\frac{\partial}{\partial u^i}(\sqrt{g}\, A^i) . \tag{2.6.39}$$

For completeness, we discuss two alternative derivations of the formula in (2.6.39) in Chap. 14. It concerns those that are encountered most frequently in the literature but quite often with unsatisfactory notations.

d) Curl

The curl of a vector field is obtained as the *cross*-product of the $\boldsymbol{\nabla}$-operator with the vector $\boldsymbol{A}$, $\boldsymbol{\nabla} \times \boldsymbol{A}$ or sometimes represented by rot $\boldsymbol{A}$.

First, we rewrite (2.5.25), with the constraint "i, j, k, cyc 1, 2, 3" absorbed into the formula:

$$\boldsymbol{e}^i \times \boldsymbol{e}^j = \varepsilon^{ijk}\frac{\boldsymbol{e}_k}{J} . \tag{2.6.40}$$

The curl of A is then

$$\nabla \times A = \nabla \times (A_j e^j) = A_j(\nabla \times e^j) + \nabla A_j \times e^j = \frac{\partial A_j}{\partial u^i} e^i \times e^j \ . \tag{2.6.41}$$

Here we have used $\nabla \times \nabla u^j \equiv 0$ and the definition of the del operator. Upon using (2.6.40) we obtain

$$\nabla \times A = \frac{\varepsilon^{ijk}}{J} \frac{\partial A_j}{\partial u^i} e_k = \frac{\varepsilon^{ijk}}{\sqrt{g}} \frac{\partial A_j}{\partial u^i} e_k \ . \tag{2.6.42}$$

More explicitly,

$$\begin{aligned} \nabla \times A &= \frac{1}{J} \sum_k \left(\frac{\partial A_j}{\partial u^i} - \frac{\partial A_i}{\partial u^j} \right) e_k \qquad i, j, k \text{ cyc } 1, 2, 3 \\ &= \frac{1}{\sqrt{g}} \sum_k \left(\frac{\partial A_j}{\partial u^i} - \frac{\partial A_i}{\partial u^j} \right) e_k \qquad i, j, k \text{ cyc } 1, 2, 3 \ . \end{aligned} \tag{2.6.43}$$

This derivation may also be performed via the covariant derivative:

$$\begin{aligned} \nabla \times A &= e^i \frac{\partial}{\partial u^i} \times A = e^i \times \frac{\partial A}{\partial u^i} = e^i \times e^j \left(\frac{\partial A}{\partial u^i} \right)_j \\ &= e^i \times e^j A_{j,i} = \frac{1}{J} \sum_{ij}{}' (\pm) A_{j,i} e_k \qquad i, j, k \text{ cyc } 1, 2, 3 \ . \end{aligned} \tag{2.6.44}$$

The prime on $\sum$ again implies $i \neq j$. We used (2.5.37) and (2.5.38) for the cross product. This implies:

$$\nabla \times A = J^{-1} \sum_k (A_{j,i} - A_{i,j}) e_k \qquad i, j, k \text{ cyc } 1, 2, 3 \ . \tag{2.6.45}$$

After substitution of the expressions for $A_{j,i}$ and $A_{i,j}$ according to (2.6.32) and recognition that the Christoffel symbols cancel each other because of their symmetry properties, we retrieve (2.6.43).

2.7 The Parallel and Perpendicular Components of a Vector

Sometimes one wants to find the parallel or perpendicular components of a vector V with respect to a certain given vector, say B (e.g., the magnetic field).

The dot product is obtained via the orthogonal projection of the one vector onto the other. Consequently, it should be clear that the *parallel* component of V (denoted by $V_\parallel$) with respect to (w.r.t) a vector B is simply:

$$V_{\parallel} = \frac{\boldsymbol{V}\cdot\boldsymbol{B}}{B} = \boldsymbol{V}\cdot\hat{\boldsymbol{B}} \;, \tag{2.7.1}$$

$\hat{\boldsymbol{B}}$ being the unit vector along $\boldsymbol{B}$.

On the other hand, the component of $\boldsymbol{V}$ *perpendicular* to $\boldsymbol{B}$, can be derived from the definition of the vector product:

$$\boldsymbol{V}_{\perp} = -\hat{\boldsymbol{B}} \times (\hat{\boldsymbol{B}} \times \boldsymbol{V}) \;. \tag{2.7.2}$$

Indeed, $|\hat{\boldsymbol{B}} \times \boldsymbol{V}|$ is the magnitude of the perpendicular component (w.r.t. $\hat{\boldsymbol{B}}$) and $-\hat{\boldsymbol{B}} \times (\hat{\boldsymbol{B}} \times \boldsymbol{V})$ is the vector $\perp$ to $\hat{\boldsymbol{B}}$, lying in the plane formed by the vectors $\hat{\boldsymbol{B}}$ and $\boldsymbol{V}$, which means that it must be in the direction of $\pm \boldsymbol{V}_{\perp}$.

Thus, any vector $\boldsymbol{V}$ can be written in terms of its parallel and perpendicular components with respect to another vector $\boldsymbol{B}$ as:

$$\boldsymbol{V} = V_{\parallel}\hat{\boldsymbol{B}} + \boldsymbol{V}_{\perp} = \hat{\boldsymbol{B}}(\hat{\boldsymbol{B}}\cdot \boldsymbol{V}) - \hat{\boldsymbol{B}} \times (\hat{\boldsymbol{B}} \times \boldsymbol{V}) \;, \tag{2.7.3}$$

where $\hat{\boldsymbol{B}} = \boldsymbol{B}/B$.

It is useful to know how the $\parallel$ and $\perp$ labels can "shift" in dot products ($\parallel$ and $\perp$ are with respect to an arbitrary vector $\boldsymbol{B}$, which can coincide with $\boldsymbol{A}$ or $\boldsymbol{C}$). We obtain:

$$\boldsymbol{A}_{\perp}\cdot\boldsymbol{C} = \boldsymbol{A}_{\perp}\cdot\boldsymbol{C}_{\perp} = \boldsymbol{A}\cdot\boldsymbol{C}_{\perp} \tag{2.7.4}$$

$$\boldsymbol{A}_{\parallel}\cdot\boldsymbol{C} = A_{\parallel}C_{\parallel} = \boldsymbol{A}\cdot\boldsymbol{C}_{\parallel} \;. \tag{2.7.5}$$

2.8 A Summary of Vector Related Identities

Here we summarize the most important vector identities used in plasma physics, of the NRL Plasma Formulary (Book 1983). More formulae involving vectors and tensors can be found at the end of Chap. 3.

$$\begin{aligned}\boldsymbol{A}\cdot\boldsymbol{B}\times\boldsymbol{C} &= \boldsymbol{A}\times\boldsymbol{B}\cdot\boldsymbol{C} = \boldsymbol{B}\cdot\boldsymbol{C}\times\boldsymbol{A} = \boldsymbol{B}\times\boldsymbol{C}\cdot\boldsymbol{A}\\ &= \boldsymbol{C}\cdot\boldsymbol{A}\times\boldsymbol{B} = \boldsymbol{C}\times\boldsymbol{A}\cdot\boldsymbol{B}\end{aligned} \tag{V.1}$$

$$\boldsymbol{A}\times(\boldsymbol{B}\times\boldsymbol{C}) = (\boldsymbol{C}\times\boldsymbol{B})\times\boldsymbol{A} = \boldsymbol{B}(\boldsymbol{A}\cdot\boldsymbol{C}) - \boldsymbol{C}(\boldsymbol{A}\cdot\boldsymbol{B}) \tag{V.2}$$

$$\boldsymbol{A}\times(\boldsymbol{B}\times\boldsymbol{C}) + \boldsymbol{B}\times(\boldsymbol{C}\times\boldsymbol{A}) + \boldsymbol{C}\times(\boldsymbol{A}\times\boldsymbol{B}) = 0 \tag{V.3}$$

$$(\boldsymbol{A}\times\boldsymbol{B})\cdot(\boldsymbol{C}\times\boldsymbol{D}) = (\boldsymbol{A}\cdot\boldsymbol{C})(\boldsymbol{B}\cdot\boldsymbol{D}) - (\boldsymbol{A}\cdot\boldsymbol{D})(\boldsymbol{B}\cdot\boldsymbol{C}) \tag{V.4}$$

$$(\boldsymbol{A}\times\boldsymbol{B})\times(\boldsymbol{C}\times\boldsymbol{D}) = (\boldsymbol{A}\times\boldsymbol{B}\cdot\boldsymbol{D})\boldsymbol{C} - (\boldsymbol{A}\times\boldsymbol{B}\cdot\boldsymbol{C})\boldsymbol{D} \tag{V.5}$$

$$\nabla(fg) = \nabla(gf) = f\nabla g + g\nabla f \tag{V.6}$$

$$\nabla\cdot(f\boldsymbol{A}) = f\nabla\cdot\boldsymbol{A} + \boldsymbol{A}\cdot\nabla f \tag{V.7}$$

$$\nabla \times (f\boldsymbol{A}) = f\nabla \times \boldsymbol{A} + \nabla f \times \boldsymbol{A} \tag{V.8}$$

$$\nabla \cdot (\boldsymbol{A} \times \boldsymbol{B}) = \boldsymbol{B} \cdot \nabla \times \boldsymbol{A} - \boldsymbol{A} \cdot \nabla \times \boldsymbol{B} \tag{V.9}$$

$$\nabla \times (\boldsymbol{A} \times \boldsymbol{B}) = \boldsymbol{A}(\nabla \cdot \boldsymbol{B}) - \boldsymbol{B}(\nabla \cdot \boldsymbol{A}) + (\boldsymbol{B} \cdot \nabla)\boldsymbol{A} - (\boldsymbol{A} \cdot \nabla)\boldsymbol{B} \tag{V.10}$$

$$\boldsymbol{A} \times (\nabla \times \boldsymbol{B}) = (\nabla \boldsymbol{B}) \cdot \boldsymbol{A} - (\boldsymbol{A} \cdot \nabla)\boldsymbol{B} \tag{V.11}$$

$$\nabla(\boldsymbol{A} \cdot \boldsymbol{B}) = \boldsymbol{A} \times (\nabla \times \boldsymbol{B}) + \boldsymbol{B} \times (\nabla \times \boldsymbol{A}) + (\boldsymbol{A} \cdot \nabla)\boldsymbol{B} + (\boldsymbol{B} \cdot \nabla)\boldsymbol{A} \tag{V.12}$$

$$\nabla^2 f = \nabla \cdot \nabla f \tag{V.13}$$

$$\nabla^2 \boldsymbol{A} = \nabla(\nabla \cdot \boldsymbol{A}) - \nabla \times \nabla \times \boldsymbol{A} \tag{V.14}$$

$$\nabla \times \nabla f = 0 \tag{V.15}$$

$$\nabla \cdot \nabla \times \boldsymbol{A} = 0 \tag{V.16}$$

$$\nabla \cdot \boldsymbol{R} = 3 \;, \quad \boldsymbol{R} = \text{position vector} \;. \tag{V.17}$$

3. Tensorial Objects

3.1 Introduction

A brief discussion of second-order tensorial objects such as dyads and dyadics is necessary in a comprehensive treatment of flux coordinates. The reason is twofold. Firstly, we regularly deal with dyadic objects in studying plasma equilibrium, even in scalar pressure cases. We frequently come across expressions such as $\boldsymbol{A} \cdot \boldsymbol{BC}$ or $\boldsymbol{A} \cdot \nabla \boldsymbol{B}$, where $\boldsymbol{BC}$ and $\nabla \boldsymbol{B}$ are dyads. An important example is the curvature along a magnetic-field line: $\hat{\boldsymbol{B}} \cdot \nabla \hat{\boldsymbol{B}}$. Secondly, tensor analysis in curvilinear coordinates is merely an extension of vector analysis; more accurately, vectors are a special case of tensors: vectors are first-order tensors.

Tensors are of considerable importance in general plasma physics. Examples of tensors used are the above mentioned dyads of the form $\boldsymbol{AB}$ or $\nabla \boldsymbol{B}$ needed for a description of the curvature and fluid inertial terms $\boldsymbol{v} \cdot \nabla \boldsymbol{v}$ among others, the pressure and viscous-stress tensors, the electromagnetic stress tensor, the dyadics utilized in the derivation of the Fokker-Planck equation, etc.

This is not the place to give a detailed exposition on tensors, but we can explain the main ideas. We shall take a pragmatic viewpoint and stress the similarities with and generalizations of vectors.

3.2 The Concept of Tensors; A Pragmatic Approach

The simplest way to introduce tensors is via a dot-product operation between vectors. A *dyad* or *second-order "simple" tensor*, $\boldsymbol{AB}$, is defined by

$$\boldsymbol{X} \cdot \boldsymbol{AB} \equiv (\boldsymbol{X} \cdot \boldsymbol{A}) \boldsymbol{B} \tag{3.2.1}$$

or

$$\boldsymbol{AB} \cdot \boldsymbol{Y} \equiv \boldsymbol{A}(\boldsymbol{B} \cdot \boldsymbol{Y}) \ . \tag{3.2.2}$$

$\boldsymbol{X}$ and $\boldsymbol{Y}$ are two arbitrary vectors.

The result of (3.2.1) and (3.2.2) is a vector as is clear from their RHS. The LHS states that this vector can be considered as being produced by the dot product of a vector and a dyad. The notation for a dyad is a simple juxtaposition of the two vectors making up the dyad, without a symbol in between. In more abstract mathematical treatments, the symbol $\otimes$ is used and is called the tensor product or the outer product (sometimes it is called the exterior product, although this name is mostly reserved for yet another product, namely, that related to Grassmann Algebra). Thus, $\boldsymbol{AB} \equiv \boldsymbol{A} \otimes \boldsymbol{B}$ in that notation. Note that the Russian notations for dot and cross products are also juxtapositions, distinguished by different brackets: $(\boldsymbol{AB}) \equiv$ what we write as $\boldsymbol{A} \cdot \boldsymbol{B}$ and $[\boldsymbol{AB}] \equiv$ what we write as $\boldsymbol{A} \times \boldsymbol{B}$.

From the definition, (3.2.1) and (3.2.2), it follows that the dot product of a dyad with a vector is (in general) *not commutative*. The result of $\boldsymbol{X} \cdot \boldsymbol{AB} = (\boldsymbol{X} \cdot \boldsymbol{A})\boldsymbol{B}$ is a vector along $\boldsymbol{B}$, whereas $\boldsymbol{AB} \cdot \boldsymbol{X} = \boldsymbol{A}(\boldsymbol{B} \cdot \boldsymbol{X})$ is a vector proportional to $\boldsymbol{A}$. These equations also explain why *the order* of the vectors making up a dyad is important i.e.,

$$\boldsymbol{AB} \neq \boldsymbol{BA}. \tag{3.2.3}$$

The reason is self-explanatory; $\boldsymbol{X} \cdot \boldsymbol{AB} = (\boldsymbol{X} \cdot \boldsymbol{A})\boldsymbol{B} \propto \boldsymbol{B}$, whereas $\boldsymbol{X} \cdot \boldsymbol{BA} = (\boldsymbol{X} \cdot \boldsymbol{B})\boldsymbol{A} \propto \boldsymbol{A}$.

We now expand the vectors $\boldsymbol{A}$ and $\boldsymbol{B}$ along the basis vectors $\boldsymbol{e}_i \equiv \partial \boldsymbol{R}/\partial u^i$ and $\boldsymbol{e}^j = \nabla u^j$; $\boldsymbol{A} = A^i \boldsymbol{e}_i = A_j \boldsymbol{e}^j$ with $A^i = \boldsymbol{A} \cdot \boldsymbol{e}^i$ and $A_j = \boldsymbol{A} \cdot \boldsymbol{e}_j$, and $\boldsymbol{B} = B^k \boldsymbol{e}_k = B_l \boldsymbol{e}^l$ with $B^k = \boldsymbol{B} \cdot \boldsymbol{e}^k$ and $B_l = \boldsymbol{B} \cdot \boldsymbol{e}_l$. The summation convention must be used over repeated indices. This leads to four possible forms of the dyad:

$$\boldsymbol{AB} = A^i \boldsymbol{e}_i B^k \boldsymbol{e}_k = A^i B^k \boldsymbol{e}_i \boldsymbol{e}_k \tag{3.2.4a}$$

$$\boldsymbol{AB} = A^i \boldsymbol{e}_i B_l \boldsymbol{e}^l = A^i B_l \boldsymbol{e}_i \boldsymbol{e}^l \tag{3.2.4b}$$

$$\boldsymbol{AB} = A_j \boldsymbol{e}^j B^k \boldsymbol{e}_k = A_j B^k \boldsymbol{e}^j \boldsymbol{e}_k \tag{3.2.4c}$$

$$\boldsymbol{AB} = A_j \boldsymbol{e}^j B_l \boldsymbol{e}^l = A_j B_l \boldsymbol{e}^j \boldsymbol{e}^l. \tag{3.2.4d}$$

The second form of (3.2.4a–d) follows trivially from the definition, (3.2.1) and (3.2.2).

Sometimes, we use a general symbol $A(m)\,B(n)$ to denote the four variants of A's and B's when we make statements that are valid for all of them. In three dimensions, there are nine possible combinations $A(m)\,B(n)$ in each of the four cases above. The coefficients $A(m)\,B(n)$, i.e., A^iB^k, A^iB_l, A_jB^k, A_jB_l are called the components of the dyad for the basis vector sets $\boldsymbol{e}_i$ and $\boldsymbol{e}^j$. The dyads $\boldsymbol{e}_i\boldsymbol{e}_k$ are called "basis-dyads" or "basis-tensors" in analogy with basis vectors. There are nine "basis-dyads" of each sort (in the sense of (3.2.4a–d) in a three-dimensional space).

We introduce first a single symbol for the dyad $\boldsymbol{AB}$

$$\boldsymbol{D} \equiv \boldsymbol{AB}. \tag{3.2.5}$$

Second-order tensorial quantities are represented by a bold face, sanserif, italic notation. (Other notations instead of $\boldsymbol{D}$ are $\overleftrightarrow{D}$, $\vec{\vec{D}}$ or $\underline{\underline{D}}$; the two "arrow heads" or the two bars refer to the second-order character.) Depending on the index positions, we have the following definitions:

$$D^{ik} \equiv A^i B^k \equiv \text{contravariant components of } \boldsymbol{AB}\,(\equiv \boldsymbol{D}) \tag{3.2.6a}$$

$$D_{jl} \equiv A_j B_l \equiv \text{covariant components of } \boldsymbol{AB}\,(\equiv \boldsymbol{D}) \tag{3.2.6b}$$

$$\left.\begin{array}{l} D^{i}_{\cdot l} \equiv A^i B_l \\ D_j^{\cdot k} \equiv A_j B^k \end{array}\right\} \equiv \text{mixed components of } \boldsymbol{AB}\,(\equiv \boldsymbol{D}). \tag{3.2.6c}$$

The dots in the symbols $D^{i}_{\cdot l}$ and $D_j^{\cdot k}$ remind us that i and j are the first indices in each case. The dot is necessary because otherwise, D^i_l could be interpreted as either the components of $\boldsymbol{D} = D^i_l \boldsymbol{e}_i \boldsymbol{e}^l$ or as the components of $\boldsymbol{D} = D^i_l \boldsymbol{e}^l \boldsymbol{e}_i$. But since, in general, $\boldsymbol{e}_i \boldsymbol{e}^l$ is different from $\boldsymbol{e}^l \boldsymbol{e}_i$, the D^i_l are also different, creating the need to indicate which basis vector must come first.

The components $D(m, n)$ can be ordered as a square array. They are then said to form a 3×3 matrix. In our co- and contravariant formalism, there are four different matrices belonging to the same dyad, at least for the choice $\boldsymbol{e}_i = \partial \boldsymbol{R}/\partial u^i$ and $\boldsymbol{e}^j = \nabla u^j$. When we change basis vectors, the components of the dyad will change as will the matrices. As is the case for vectors, tensors (or dyads) in Euclidean space are no longer covariant, contravariant or mixed. All we have are dyads or tensors which can be expressed in covariant, contravariant or mixed *components*, depending on the choice of basis vectors.

We now examine $\boldsymbol{D}$ from another viewpoint. The second-order tensorial object $\boldsymbol{D}$ that we referred to as a dyad or "simple" second-order tensor has a particular set of components which, in a certain sense, is special. The nine components, say D_{jl}, can be written as the mixed product of two groups of three numbers: (A_1, A_2, A_3) and (B_1, B_2, B_3), such that $D_{jl} = A_j B_l$, $j = 1, 2, 3$ and $l = 1$, 2, 3. An analogous statement can be made for the nine contravariant, and 18 mixed components of that *same* second-order tensor $\boldsymbol{D}$, as shown in (3.2.6). This means that $\boldsymbol{D}$ is a tensor which is produced by two vectors, $\boldsymbol{AB}$.

We now treat more general second-order tensors. Consider a second-order tensor $\boldsymbol{F}$

$$\boldsymbol{F} = F^{ik} \boldsymbol{e}_i \boldsymbol{e}_k = F_{jl} \boldsymbol{e}^j \boldsymbol{e}^l = F^{i}_{\cdot l} \boldsymbol{e}_i \boldsymbol{e}^l = F_j^{\cdot k} \boldsymbol{e}^j \boldsymbol{e}_k. \tag{3.2.7}$$

For a general second-order tensor, it is *not* possible to find two sets of three numbers in the sense explained above for $\boldsymbol{D}$. In other words, a general second-order tensor cannot be written as a single dyad. It can, however, be written as a sum of at least three dyads (in 3-D).

Let us show this for a tensor written with covariant components. Choose any three vectors $\boldsymbol{A}$, $\boldsymbol{B}$ and $\boldsymbol{C}$ with the constraints that the 3×3 matrix formed by their components A_i, B_j, and C_k has a non-zero determinant. We then must find three vectors $\boldsymbol{X}$, $\boldsymbol{Y}$ and $\boldsymbol{Z}$ such that

$$\boldsymbol{F} = \boldsymbol{XA} + \boldsymbol{YB} + \boldsymbol{YC}. \tag{3.2.8}$$

When $\boldsymbol{F}$ is written in terms of its covariant components F_{jl} and all the vectors appearing on the RHS of (3.2.8) are written using their covariant components as well, we can equate the coefficients of the same "basis-dyads" $e^j e^l$ on the LHS and RHS:

$$F_{jl} = X_j A_l + Y_j B_l + Z_j C_l. \tag{3.2.9}$$

In matrix language, the RHS is the matrix product of the matrices formed by the components A_i, B_j, and C_k and the transpose of that formed by the components X_l, Y_m, Z_n:

$$[F] = [\chi]^{\mathrm{T}}[\alpha], \tag{3.2.10}$$

Here, $\chi_{11} = X_1$, $\chi_{12} = X_2$, $\chi_{13} = X_3$, $\chi_{21} = Y_1$, etc., and $\alpha_{11} = A_1$, $\alpha_{12} = A_2$, $\alpha_{13} = A_3$, $\alpha_{21} = B_1$, etc. Since we required that $\det[\alpha] \neq 0$ from the outset, the matrix equation, (3.2.10), can be inverted, and the coefficients of $[\chi]$ can be found.

To write $\boldsymbol{F}$ with more than three dyads is straightforward. For instance, any dyad can be replaced by the sum of a symmetric and an antisymmetric dyad: $\boldsymbol{D} \equiv (\boldsymbol{D} + \boldsymbol{D}^{\mathrm{T}})/2 + (\boldsymbol{D} - \boldsymbol{D}^{\mathrm{T}})/2$. Here, T stands for transpose. Also, if we absorb the coefficients in the second form of (3.2.7) into other vectors, so that e.g., $\boldsymbol{a} = F_{11}\boldsymbol{e}^1, \boldsymbol{b} = \boldsymbol{e}^1, \boldsymbol{c} = \boldsymbol{e}^1, \boldsymbol{d} = \mathrm{F}_{12}\boldsymbol{e}^2, \boldsymbol{e} = \sqrt{F_{13}}\boldsymbol{e}^1, \boldsymbol{f} = \sqrt{F_{13}}\boldsymbol{e}^3, \ldots$, we can write $\boldsymbol{F}$ as a sum of nine dyads

$$\boldsymbol{F} = \boldsymbol{ab} + \boldsymbol{cd} + \boldsymbol{ef} + \cdots. \tag{3.2.11}$$

Because a general second-order tensor can be written as a sum of (at least three) dyads, it is also called a *dyadic*. It is clear that a dyad is a dyadic, but not all dyadics are dyads. The expressions in (3.2.4) and (3.2.7) are called the *nonion* form of the dyad or the dyadic to denote the "nine-component form".

The words "second-order tensor" and "dyadic" refer to one and the same "geometrical object", although they might be regarded as different representations for the same entity. If one deals with the indexed-component form, one could refer to a "tensor"; if, on the other hand, one considers the "geometrical object" as a sum of dyads, one could refer to a "dyadic". This is a matter of taste. Since there is no real difference between the two, we will use the words *dyadic and second order tensor as synonyms.*

We have encountered two other interpretations of dyadics, which we think have no ground. First, Morse and Feshbach (1953) state that the relationship between the components of a dyadic and those of a second-order tensor is analogous to that between the physical components of a vector and its covariant or contravariant components. They essentially state that a dyadic is to a tensor what a real "vector" is to a co- or contravariant "vector". Since they subsequently treat dyadics as if they are common second-order tensors, and since we found no other evidence in the literature to back up this notion, we do not accept this interpretation. Another (restricted) definition is given by Drew (1961), who asserts that dyads and dyadics refer to tensor analysis in Cartesian coordinate systems. Again, we found no other reference stating the same. This interpretation seems too restricted and is actually contradicted by most applications in plasma physics (e.g., the pressure tensor is really a dyad, and the viscous-stress tensor is a dyadic; however, we almost never express them in Cartesian components).

Higher-order tensors can be defined in a "similar" way. A third-order "simple" tensor or a triad $\boldsymbol{ABC}$ is defined as

$$\boldsymbol{Y}\cdot\boldsymbol{ABC} = (\boldsymbol{Y}\cdot\boldsymbol{A})\boldsymbol{BC}\ , \tag{3.2.12}$$

where $\boldsymbol{BC}$ is a dyad, a by now familiar object. Triadics or general third-order tensors can be written as sums of triads. For fourth-order tensorial objects, we talk about tetrads and tetradics. For the n-th order case, we refer to polyads and polyadics.

A vector can be considered as a first-order adic or first-order tensor, and a scalar as a zeroth-order tensor.

3.3 Dot Product, Double Dot Product, Contraction

The dot product of a vector and a tensor is merely an extension of the definition of a dyad.

In this section, we shall stress the index notation, so by straightforward expansions in components:

$$\begin{aligned}
\boldsymbol{X}\cdot\boldsymbol{F} &= X_m e^m\cdot F^{ik}e_i e_k = X_m F^{ik} e^m\cdot e_i e_k = X_i F^{ik} e_k = Y^k e_k = \boldsymbol{Y}\\
&= X^n e_n\cdot F_j^{\cdot k} e^j e_k = X^n F_j^{\cdot k} e_n\cdot e^j e_k = X^j F_j^{\cdot k} e_k = Y^k e_k = \boldsymbol{Y}\\
&= X^n e_n\cdot F_{jl} e^j e^l = X^n F_{jl} e_n\cdot e^j e^l = X^j F_{jl} e^l = Y_l e^l = \boldsymbol{Y}\\
&= X_m e^m\cdot F^i_{\cdot l} e_i e^l = X_m F^i_{\cdot l} e^m\cdot e_i e^l = X_i F^i_{\cdot l} e^l = Y_l e^l = \boldsymbol{Y}
\end{aligned} \tag{3.3.1a}$$

for $\boldsymbol{X}\cdot\boldsymbol{F} = \boldsymbol{Y}$, and

$$\begin{aligned}
\boldsymbol{F}\cdot\boldsymbol{X} &= F^{ik} e_i e_k\cdot X_m e^m = F^{ik} X_m e_i e_k\cdot e^m = F^{ik} X_k e_i = Z^i e_i = \boldsymbol{Z}\\
&= F^i_{\cdot l} e_i e^l\cdot X^n e_n = F^i_{\cdot l} X^n e_i e^l\cdot e_n = F^i_{\cdot l} X^l e_i = Z^i e_i = \boldsymbol{Z}\\
&= F_{jl} e^j e^l\cdot X^m e_m = F_{jl} X^m e^j e^l\cdot e_m = F_{jl} X^l e^j = Z_j e^j = \boldsymbol{Z}\\
&= F_j^{\cdot k} e^j e_k\cdot X_n e^n = F_j^{\cdot k} X_n e^j e_k\cdot e^n = F_j^{\cdot k} X_k e^j = Z_j e^j = \boldsymbol{Z}
\end{aligned} \tag{3.3.1b}$$

for $\boldsymbol{F}\cdot\boldsymbol{X} = \boldsymbol{Z}$.

Here we used the fact that $e^i\cdot e_j = \delta^i_j$. Note that we always performed the dot product between a basis vector with a subscript and one with a superscript. That way we were able to avoid using the $g_{ij} = e_i\cdot e_j$. In component form, the dot product amounts to a *saturation* over one upper index and one lower index:

$$\begin{aligned}
(\boldsymbol{X}\cdot\boldsymbol{F})^k &= X_i F^{ik} = X^j F_j^{\cdot k} = Y^k\\
(\boldsymbol{X}\cdot\boldsymbol{F})_l &= X^j F_{jl} = X_i F^i_{\cdot l} = Y_l
\end{aligned} \tag{3.3.1c}$$

and

$$(\boldsymbol{F}\cdot \boldsymbol{X})^i = F^{ik}X_k = F^i_{\cdot l}X^l = Z^i$$

$$(\boldsymbol{F}\cdot \boldsymbol{X})_j = F_{jl}X^l = F_j^{\cdot k}X_k = Z_j \ . \tag{3.3.1d}$$

It is clear that the index which is the furthest away from where the dot "would have been" is kept unchanged. The index saturation is over the indices closest to the "dot": the first one of $\boldsymbol{F}$ in $\boldsymbol{X}\cdot \boldsymbol{F}$, and the second one in $\boldsymbol{F}\cdot \boldsymbol{X}$. The saturation of the indices is formally called the *contraction* of a tensor: the summation over two identical indices, one upper and one lower. The contraction of a tensor results in a tensor of a rank two lower than that of the original tensor. The dot product of a vector with a tensor can be considered as the contraction of a third-order tensor. If we represent the tensors by their components, we obtain, e.g.,

$$P_i^{\cdot ik} \equiv X_i F^{ik} = Y^k \tag{3.3.2}$$

and

$$P^j_{\cdot jl} \equiv X^j F_{jl} = Y_l \ .$$

The dot product between two second-order tensors is likewise *not* commutative. It can be considered as the contraction of a fourth-order tensor; the saturation is over the middle indices, and the result is a second-order tensor. We have for $\boldsymbol{E} = \boldsymbol{F}\cdot \boldsymbol{V}$

$$F_{ij}V^{jk} = F_i^{\cdot j}V_j^{\cdot k} = E_i^{\cdot k}$$

$$F^{ij}V_{jk} = F^i_{\cdot j}V^j_{\cdot k} = E^i_{\cdot k}$$

$$F_i^{\cdot j}V_{jk} = F_{ij}V^j_{\cdot k} = E_{ik}$$

$$F^i_{\cdot j}V^{jk} = F^{ij}V_j^{\cdot k} = E^{ik} \ . \tag{3.3.3}$$

Considered as the contraction of a fourth-order tensor, we would have, for instance,

$$Q_{ij}^{\cdot\cdot jk} \equiv F_{ij}V^{jk} = E_i^{\cdot k} \ . \tag{3.3.4}$$

Consider now the dot product of a tensor with two vectors, one on its left and one on its right:

$$\begin{aligned} \boldsymbol{A}\cdot \boldsymbol{F}\cdot \boldsymbol{X} &= \boldsymbol{A}\cdot(\boldsymbol{F}\cdot \boldsymbol{X}) = \boldsymbol{A}\cdot \boldsymbol{Z} = A^i(F_{ij}X^j) = A^iZ_i \\ &= (\boldsymbol{A}\cdot \boldsymbol{F})\cdot \boldsymbol{X} = \boldsymbol{B}\cdot \boldsymbol{X} = (A^iF_{ij})X^j = B_jX^j = \text{number} \ . \end{aligned} \tag{3.3.5}$$

Here we have used the covariant components of $\boldsymbol{F}$, and we have set $\boldsymbol{F}\cdot \boldsymbol{X} = \boldsymbol{Z}$ and $\boldsymbol{A}\cdot \boldsymbol{F} = \boldsymbol{B}$. For later reference, we write

$$\boldsymbol{A}\cdot \boldsymbol{F}\cdot \boldsymbol{X} = A^iX^jF_{ij} = X^jA^iF_{ij} = F_{ij}X^jA^i = \text{number} \ . \tag{3.3.6}$$

Because of the non-commutativity of the vector-tensor dot product, the vectors $\boldsymbol{A}$ and $\boldsymbol{X}$ cannot be interchanged, in general. However, since the dot product of *vectors* is commutative, we have that:

$$\boldsymbol{A}\cdot\boldsymbol{F}\cdot\boldsymbol{X} = (\boldsymbol{A}\cdot\boldsymbol{F})\cdot\boldsymbol{X} = \boldsymbol{X}\cdot(\boldsymbol{A}\cdot\boldsymbol{F}) \tag{3.3.7a}$$

$$\boldsymbol{A}\cdot\boldsymbol{F}\cdot\boldsymbol{X} = \boldsymbol{A}\cdot(\boldsymbol{F}\cdot\boldsymbol{X}) = (\boldsymbol{F}\cdot\boldsymbol{X})\cdot\boldsymbol{A}\ . \tag{3.3.7b}$$

It is convenient to write the consecutive dot product in a more compact way, namely, as the *double dot* product of two tensors. Given two second-order tensors $\boldsymbol{P}$ and $\boldsymbol{R}$, we define the double dot product as:

$$\boldsymbol{P}:\boldsymbol{R} \equiv \sum_{ij} P^{ij}R_{ji} \equiv \sum_{ij} P_{ij}R^{ji} \equiv \sum_{ij} P_i^jR_j^i \equiv \boldsymbol{R}:\boldsymbol{P}\ . \tag{3.3.8}$$

Here, $P_i^jR_j^i$ stands for $P_i^{\cdot j}R_j^{\cdot i}$ or $P^j_{\cdot i}R^i_{\cdot j}$. It is important to note that the sequence of indices is in a certain sense symmetric ("reflected" about the position where the double dot would have been).

The literature is not consistent in defining the double product. Most authors use the definition as given in (3.3.8): Krall and Trivelpiece (1973); Bird, Stewart and Lightfoot (1960); Shkarofsky, Johnston and Bachynski (1966); Jancel and Kahan (1966); Chapman and Cowling (1973); Ferziger and Kaper (1972); Milne-Thomson (1968). One important deviation from this convention is used in the standard textbook on classical mechanics by Goldstein (1981), where the tensor indices are "repeated" $(ij\,|\,ij)$ instead of "reflected" $(ij\,|\,ji)$. When the indices are "repeated", the position of the "dot" in the mixed-tensor components is different: $P_i^{\cdot j}R^i_{\cdot j}$ or $P^j_{\cdot i}R_j^{\cdot i}$; one "dot" above, and one underneath. Van Bladel (1964) also uses this latter convention. On page 57 of Morse and Feshbach (1953), the double dot is defined as in our (3.3.8), but in (1.6.29) of M&F, the "Goldstein" convention is used.

From (3.3.6), we see that we can write for (3.3.5):

$$\boldsymbol{A}\cdot\boldsymbol{F}\cdot\boldsymbol{X} \equiv \boldsymbol{X}\boldsymbol{A}:\boldsymbol{F} \equiv \boldsymbol{F}:\boldsymbol{X}\boldsymbol{A}\ . \tag{3.3.9}$$

Observe that the first double-dot expression also resembles the last form in (3.3.7a): the vector $\boldsymbol{A}$ remains closest to the tensor $\boldsymbol{F}$ if it operates on the left. Similarly, the last form of (3.3.7b) looks like the last expression in (3.3.9) and keeps the vector $\boldsymbol{X}$ closest to $\boldsymbol{F}$ if it is on the right.

For a "simple" tensor or dyad of the form $\boldsymbol{ST}$, (3.3.9) leads to:

$$\boldsymbol{XA}:\boldsymbol{ST} = \boldsymbol{A}\cdot\boldsymbol{ST}\cdot\boldsymbol{X} = (\boldsymbol{A}\cdot\boldsymbol{S})(\boldsymbol{T}\cdot\boldsymbol{X}) \tag{3.3.10a}$$

$$\boldsymbol{ST}:\boldsymbol{XA} = \boldsymbol{T}\cdot\boldsymbol{XA}\cdot\boldsymbol{S} = (\boldsymbol{T}\cdot\boldsymbol{X})(\boldsymbol{A}\cdot\boldsymbol{S})\ . \tag{3.3.10b}$$

According to the convention used here, a first dot product is performed between the two vectors next to the double dot symbol; the other dot product is then between the two other vectors. Following the Goldstein (1981) or Van Bladel (1964) convention, we would have instead of (3.3.10) $\boldsymbol{XA}:\boldsymbol{ST} = \boldsymbol{X}\cdot\boldsymbol{ST}\cdot\boldsymbol{A} =$

$(\boldsymbol{X}\cdot\boldsymbol{S})(\boldsymbol{T}\cdot\boldsymbol{A})$, which is clearly a different result because $\boldsymbol{X}$ and $\boldsymbol{A}$ have been interchanged.

The above discussion should make clear how single and double dot products of tensors must be interpreted. Generalizations for three or more dot products are readily made on the basis of the index definition of (3.3.8).

The dot product of a vector or a tensor with a dyad involving the vector operator *del*, or the divergence of a tensor can straightforwardly be computed with the rules given so far. A few important rules should be kept in mind, however:

1) The *order* of the vectors in a dyad is important; $\boldsymbol{A}\boldsymbol{\nabla}$ is different from $\boldsymbol{\nabla}\boldsymbol{A}$.
2) The dot product of vector and dyad or dyadic is *not* commutative.
3) the del operator is most conveniently expressed along a contravariant basis: $\boldsymbol{\nabla} \equiv \boldsymbol{e}^i \partial/\partial u^i$.
4) The basis vectors are position dependent.

In addition, we mention the obvious facts:

5) The del operator operates on everything on its right, except when brackets indicate otherwise.
6) The product rule of differentiation is often a very handy tool.

Regarding cross products of vectors and tensors, we again expand each in components along a basis and perform the operations between the basis vectors. To keep the order of the indices right, the Levi-Civita symbol will be helpful. Similar rules apply to expressions of the form $\boldsymbol{\nabla}\times$ (tensor) or $\boldsymbol{A}\times\boldsymbol{\nabla}\boldsymbol{B}$, etc., as long as we keep in mind the rules involving the $\boldsymbol{\nabla}$ operator, mentioned above.

3.4 The Relationship Between Covariant, Contravariant and Mixed Components

In (3.3.6), we stated that $\boldsymbol{A}\cdot\boldsymbol{F}\cdot\boldsymbol{X} = A^i X^j F_{ij}$. If we now take the numbers A^i equal to zero, except the i-th component which we set equal to 1, and similarly for the X^j, where we take only the j-th component equal to 1 and the others zero, we have $\boldsymbol{A} = A^i\boldsymbol{e}_i \rightarrow \boldsymbol{e}_i$ and $\boldsymbol{X} = X^j\boldsymbol{e}_j \rightarrow \boldsymbol{e}_j$. This leads to

$$F_{ij} \equiv \boldsymbol{e}_i\cdot\boldsymbol{F}\cdot\boldsymbol{e}_j \ . \tag{3.4.1}$$

Analogously, we have for the contravariant and mixed components, respectively,

$$F^{kl} \equiv \boldsymbol{e}^k\cdot\boldsymbol{F}\cdot\boldsymbol{e}^l \tag{3.4.2}$$

$$F_m^{\cdot n} \equiv \boldsymbol{e}_m\cdot\boldsymbol{F}\cdot\boldsymbol{e}^n \tag{3.4.3}$$

$$F^n_{\cdot m} \equiv \boldsymbol{e}^n\cdot\boldsymbol{F}\cdot\boldsymbol{e}_m \ . \tag{3.4.4}$$

Consider now $\boldsymbol{F} = F_{ij}\boldsymbol{e}^i\boldsymbol{e}^j$ and dot multiply on the left by $\boldsymbol{e}^k$ and on the right by

e^l. Because of (3.4.2), we obtain:

$$F^{kl} = e^k \cdot F \cdot e^l = F_{ij} e^k \cdot e^i e^j \cdot e^l$$
$$= F_{ij}(e^k \cdot e^i)(e^j \cdot e^l) \; , \tag{3.4.5}$$

which, after using the definition of the reciprocal metric coefficients g^{ij}, becomes

$$F^{kl} = g^{ki} g^{lj} F_{ij} \; . \tag{3.4.6}$$

By an analogous procedure, we can write:

$$F_i^{\cdot n} = g^{jn} F_{ij} \tag{3.4.7}$$

and

$$F_{ij} = g_{im} g_{jn} F^{mn} \; . \tag{3.4.8}$$

3.5 Special Tensors

3.5.1 The Kronecker Delta and the Metric Tensor

The Kronecker delta δ_j^i is in fact a second-order tensor $\boldsymbol{I}$. Strictly, δ_j^i are the mixed components of the *identity tensor* $\boldsymbol{I}$ (also called the "idemfactor"). $\boldsymbol{I}$ is defined by

$$A \cdot \boldsymbol{I} \equiv A \equiv \boldsymbol{I} \cdot A \; . \tag{3.5.1}$$

An even stronger statement is: δ_j^i are the only legitimate components of the identity tensor (as long as we work in the covariant-contravariant formalism). The symbols δ_{ij} and δ^{ij}, where $\delta_{ij} = \delta^{ij} = 1$ for $i = j$ and 0 for $i \neq j$, which could be thought of as covariant and contravariant components in fact do *not* qualify as such. This can easily be checked by means of the transformation rules of Chap. 13. We can use the *symbols* δ_{ij} and δ^{ij} to indicate that something is 1 when the two indices coincide and zero when they are different. However, these symbols can only be used in one particular coordinate system. The 1, 0 character of the symbols δ_{ij} and δ^{ij} considered as potential components of a tensor is spoiled when we go to another coordinate system.

A second important second-order tensor is the *metric tensor* $\boldsymbol{g}$ with covariant components the metric coefficients $g_{ij} \equiv e_i \cdot e_j$ and with contravariant components the reciprocal metric coefficients $g^{ij} \equiv e^i \cdot e^j$. The transformation properties show that both sorts of components are legitimate.

We now encounter a very important and interesting observation. Since we work in one single Euclidean space, *the metric tensor is identical with the identity tensor*. This particular tensor will be called the *fundamental tensor* (***FT***). It has the following properties:

$$(\boldsymbol{FT})_{ij} \equiv g_{ij} \tag{3.5.2a}$$

$$(\boldsymbol{FT})^{ij} \equiv g^{ij} \tag{3.5.2b}$$

$$(\boldsymbol{FT})^{i}_{j} \equiv \delta^{i}_{j} \ . \tag{3.5.2c}$$

The equivalence can be understood from (3.4.7), where we set F_{ij} on the RHS equal to g_{ij}. We have then

$$g^{n}_{i} = g^{jn} g_{ij} \ . \tag{3.5.3a}$$

On the other hand, we have that $\boldsymbol{e}_i = g_{ij}\boldsymbol{e}^j$. Dot multiplying this expression on both sides by $\boldsymbol{e}^n$ leads to

$$\delta^{n}_{i} = g_{ij} g^{jn} \ . \tag{3.5.3b}$$

Therefore, we have from comparing (3.5.3a) and (3.5.3b):

$$\delta^{n}_{i} \equiv g^{n}_{i} \ ; \tag{3.5.4}$$

which says that the mixed components of the metric (or fundamental) tensor are identical to the Kronecker delta. Because $\boldsymbol{I}$ is equivalent to $\boldsymbol{g}$, there should be a similar relationship to (3.5.1) involving $\boldsymbol{g}$. We, therefore, propose a formal relationship:

$$\boldsymbol{A}\cdot\boldsymbol{g} = \boldsymbol{A} = \boldsymbol{g}\cdot\boldsymbol{A} \ , \tag{3.5.5}$$

given that the $\boldsymbol{A}$ in the dot product is written in terms of the "other" components than the $\boldsymbol{A}$ in the middle. Indeed, we have

$$\begin{aligned} A^m\boldsymbol{e}_m\cdot g_{ij}\boldsymbol{e}^i\boldsymbol{e}^j &= A^m g_{ij}\boldsymbol{e}_m\cdot\boldsymbol{e}^i\boldsymbol{e}^j = A^m g_{ij}\delta^i_m\boldsymbol{e}^j \\ &= A^i g_{ij}\boldsymbol{e}^j = A_j\boldsymbol{e}^j = \boldsymbol{A} \end{aligned} \tag{3.5.6a}$$

and

$$\begin{aligned} A_n\boldsymbol{e}^n\cdot g^{ij}\boldsymbol{e}_i\boldsymbol{e}_j &= A_n g^{ij}\boldsymbol{e}^n\cdot\boldsymbol{e}_i\boldsymbol{e}_j = A_n g^{ij}\delta^n_i\boldsymbol{e}_j \\ &= A_i g^{ij}\boldsymbol{e}_j = A^j\boldsymbol{e}_j = \boldsymbol{A} \ , \end{aligned} \tag{3.5.6b}$$

whereas, for comparison, we can write for (3.5.1):

$$\begin{aligned} A^m\boldsymbol{e}_m\cdot\delta^i_j\boldsymbol{e}_i\boldsymbol{e}^j &= A^m\delta^i_j\boldsymbol{e}_m\cdot\boldsymbol{e}_i\boldsymbol{e}^j = A^m\delta^i_j g_{mi}\boldsymbol{e}^j \\ &= A^m g_{mj}\boldsymbol{e}^j = A_j\boldsymbol{e}^j = \boldsymbol{A} \end{aligned} \tag{3.5.6c}$$

or

$$A_n\boldsymbol{e}^n\cdot\delta^i_j\boldsymbol{e}_i\boldsymbol{e}^j = A_n\delta^i_j\boldsymbol{e}^n\cdot\boldsymbol{e}_i\boldsymbol{e}^j = A_n\delta^i_j\delta^n_i\boldsymbol{e}^j = A_i\boldsymbol{e}^i = \boldsymbol{A} \ . \tag{3.5.6d}$$

The most convenient form of the fundamental tensor is the "sum-of-dyad" form:

$$(\boldsymbol{FT}) \equiv \boldsymbol{g} \equiv \boldsymbol{I} \equiv \boldsymbol{e}_1\boldsymbol{e}^1 + \boldsymbol{e}_2\boldsymbol{e}^2 + \boldsymbol{e}_3\boldsymbol{e}^3 \equiv \boldsymbol{e}_i\boldsymbol{e}^i \tag{3.5.7a}$$

$$\equiv \boldsymbol{e}^1\boldsymbol{e}_1 + \boldsymbol{e}^2\boldsymbol{e}_2 + \boldsymbol{e}^3\boldsymbol{e}_3 \equiv \boldsymbol{e}^i\boldsymbol{e}_i \ . \tag{3.5.7b}$$

From this form, we obtain its coefficients via (3.4.1) through (3.4.4):

$$(\boldsymbol{FT})_{kl} \equiv \boldsymbol{e}_k \cdot (\boldsymbol{FT}) \cdot \boldsymbol{e}_l = \boldsymbol{e}_k \cdot \boldsymbol{e}_i \boldsymbol{e}^i \cdot \boldsymbol{e}_l = g_{ki}\delta_l^i \equiv g_{kl}(\equiv \delta_{kl}) \tag{3.5.8a}$$

$$(\boldsymbol{FT})^{kl} \equiv \boldsymbol{e}^k \cdot (\boldsymbol{FT}) \cdot \boldsymbol{e}^l = \boldsymbol{e}^k \cdot \boldsymbol{e}_i \boldsymbol{e}^i \cdot \boldsymbol{e}^l = \delta_i^k g^{il} \equiv g^{kl}(\equiv \delta^{kl}) \tag{3.5.8b}$$

$$(\boldsymbol{FT})_l^k \equiv \boldsymbol{e}^k \cdot (\boldsymbol{FT}) \cdot \boldsymbol{e}_l = \boldsymbol{e}^k \cdot \boldsymbol{e}_i \boldsymbol{e}^i \cdot \boldsymbol{e}_l = \delta_i^k \delta_l^i \equiv \delta_l^k(\equiv g_l^k) \tag{3.5.8c}$$

The symbols between brackets on the extreme RHS of these expressions may be used if they are interpreted properly, i.e., the co- or contravariant deltas are *not* generally one or zero, and the g_l^k *are* either one or zero. The same results are obtained if we use $(\boldsymbol{FT}) \equiv \boldsymbol{e}^i\boldsymbol{e}_i$ instead of $\boldsymbol{e}_i\boldsymbol{e}^i$.

3.5.2 Levi-Civita and Christoffel Symbols[1]

a) Levi-Civita Symbols

It is important to draw attention to the fact that the *Levi-Civita symbols* used before are *not* the components of a third-order tensor. These symbols were defined as

$$\varepsilon_{ijk} = \frac{\boldsymbol{e}_i \cdot \boldsymbol{e}_j \times \boldsymbol{e}_k}{\sqrt{g}} \tag{3.5.9a}$$

and

$$\varepsilon^{ijk} = \sqrt{g}\, \boldsymbol{e}^i \cdot \boldsymbol{e}^j \times \boldsymbol{e}^k \ . \tag{3.5.9b}$$

They are $+1$ for i, j, k equal to an even permutation of 1, 2, 3; -1 for an odd permutation and zero otherwise. If these symbols *were* the components of a tensor, then we *would* have that the tensor $\vec{\vec{\varepsilon}}$ could be written as $\vec{\vec{\varepsilon}} = \varepsilon_{lmn}\boldsymbol{e}^l\boldsymbol{e}^m\boldsymbol{e}^n = \varepsilon^{lmn}\boldsymbol{e}_l\boldsymbol{e}_m\boldsymbol{e}_n$. We now show that this statement is incorrect.

In analogy with (3.4.2), we can write for the contravariant components of a third-order tensor $\vec{\vec{\boldsymbol{F}}}$:

$$F^{ijk} \equiv \boldsymbol{e}^j \cdot (\boldsymbol{e}^i \cdot \vec{\vec{\boldsymbol{F}}} \cdot \boldsymbol{e}^k) \equiv (\boldsymbol{e}^i \cdot \vec{\vec{\boldsymbol{F}}} \cdot \boldsymbol{e}^k) \cdot \boldsymbol{e}^j \ . \tag{3.5.10}$$

Between the brackets we need the two outer indices of F^{ijk}, here i and k in that order. Since the quantity between the brackets is a vector, commutativity applies, and the position of $\boldsymbol{e}^j$ can be either of those indicated. Similarly, we have that

$$F_{ijk} \equiv \boldsymbol{e}_j \cdot (\boldsymbol{e}_i \cdot \vec{\vec{\boldsymbol{F}}} \cdot \boldsymbol{e}_k) \equiv (\boldsymbol{e}_i \cdot \vec{\vec{\boldsymbol{F}}} \cdot \boldsymbol{e}_k) \cdot \boldsymbol{e}_j \ . \tag{3.5.11}$$

According to (3.5.10), we would have for the contravariant components of $\vec{\vec{\varepsilon}} = \varepsilon^{lmn}\boldsymbol{e}_l\boldsymbol{e}_m\boldsymbol{e}_n$:

[1] Can be skipped without loss of continuity.

$$(\varepsilon)^{ijk} \equiv \varepsilon^{lmn} e^j \cdot (e^i \cdot e_l e_m e_n \cdot e^k) = \varepsilon^{lmn} \delta_l^i \delta_n^k e^j \cdot e_m$$
$$= \varepsilon^{lmn} \delta_l^i \delta_n^k \delta_m^j = \varepsilon^{ijk} \tag{3.5.12a}$$

which is what we expect, being equal to 1, -1, or zero. However, if we try to compute the contravariant components from the form $\overleftrightarrow{\varepsilon} = \varepsilon_{lmn} e^l e^m e^n$, we get instead

$$(\varepsilon)^{ijk} = \varepsilon_{lmn} e^j \cdot (e^i \cdot e^l e^m e^n \cdot e^k) = \varepsilon_{lmn} g^{il} g^{nk} e^j \cdot e^m$$
$$= \varepsilon_{lmn} g^{il} g^{nk} g^{jm} = \varepsilon^{ijk} g^{-1} = \frac{\varepsilon^{ijk}}{g} , \tag{3.5.12b}$$

which is different from (3.5.12a). Here, $g = \det(g_{ij})$ and $\det(g^{ij}) = g^{-1}$. We used the fact that $\sum_{lmn} g^{il} g^{jm} g^{kn} \varepsilon_{lmn}$, with ε_{lmn} the Levi-Civita symbol, is equal to the determinant g for i, j, k an even permutation of 1, 2, 3; $-g$ for i, j, k an odd permutation, and zero if any two indices are equal. Hence, $\overleftrightarrow{\varepsilon}$ *cannot* be a tensor with components ε^{ijk} or ε_{ijk}. Similar results are obtained for covariant components $(\varepsilon)_{ijk}$. Note that one should not convert $\varepsilon_{lmn} g^{il} g^{jm} g^{kn}$ into ε^{ijk} in analogy to (3.4.6). This would lead to the wrong result because the ε_{lmn} is a *symbol* equal to $+1$, -1 or 0 and is not a covariant component of $\overleftrightarrow{\varepsilon}$. In Chap. 13, an alternative proof for the non-tensor character of ε_{ijk} and ε^{ijk} is given via transformation equations.

One could define a tensor called the *permutation tensor*, which is closely related to the Levi-Civita symbol:

$$\overleftrightarrow{\boldsymbol{E}} \equiv \frac{\varepsilon^{lmn}}{\sqrt{g}} e_l e_m e_n \tag{3.5.13a}$$

or

$$\overleftrightarrow{\boldsymbol{E}} \equiv \varepsilon_{lmn} \sqrt{g}\, e^l e^m e^n \ . \tag{3.5.13b}$$

As in (3.5.12a) and (3.5.12b), we have

$$E^{ijk} = \frac{\varepsilon^{lmn}}{\sqrt{g}} \delta_l^i \delta_m^j \delta_n^k = \frac{\varepsilon^{ijk}}{\sqrt{g}} \tag{3.5.14a}$$

or

$$E^{ijk} = \varepsilon_{lmn} \sqrt{g}\, g^{il} g^{jm} g^{kn} = \varepsilon^{ijk} \sqrt{g}\, g^{-1} = \frac{\varepsilon^{ijk}}{\sqrt{g}} \tag{3.5.14b}$$

and

$$E_{ijk} = \frac{\varepsilon^{lmn}}{\sqrt{g}} g_{il} g_{jm} g_{kn} = \frac{\varepsilon_{ijk}}{\sqrt{g}} g = \varepsilon_{ijk} \sqrt{g} \tag{3.5.14c}$$

or

$$E_{ijk} = \varepsilon_{lmn}\sqrt{g}\,\delta_i^l\delta_j^m\delta_k^n = \varepsilon_{ijk}\sqrt{g} \; . \tag{3.5.14d}$$

So we have agreement: $\varepsilon^{lmn}/\sqrt{g}$ and $\varepsilon_{lmn}\sqrt{g}$ are respectively the contravariant and covariant components of $\overleftrightarrow{\boldsymbol{E}}$. In comparison with (3.5.9), we have for the components of $\overleftrightarrow{\boldsymbol{E}}$:

$$E_{ijk} = \boldsymbol{e}_i \cdot \boldsymbol{e}_j \times \boldsymbol{e}_k = \begin{cases} +\sqrt{g} \\ -\sqrt{g} \\ 0 \end{cases} \tag{3.5.15a}$$

and

$$E^{ijk} = \boldsymbol{e}^i \cdot \boldsymbol{e}^j \times \boldsymbol{e}^k = \begin{cases} +1/\sqrt{g} \\ -1/\sqrt{g} \; . \\ 0 \end{cases} \tag{3.5.15b}$$

The transformation properties of E_{ijk} and E^{ijk} are given in Chap. 13.

b) Christoffel Symbols

The Christoffel symbols discussed in Chap. 2 are *not* the components of a third-order tensor as the alternative Γ-notation might suggest. Recall that the Γ-notation identified Γ^i_{jk} with $\{{}_j{}^i{}_k\}$ and $\Gamma_{i,jk}$ with $[i,jk]$. Therefore, the Γ notation should be avoided. The easiest way to prove that $\{{}_j{}^i{}_k\}$ and $[i,jk]$ are not tensor components is via the transformation properties. This is discussed in Chap. 13.

3.6 Tensor and Dyadic Identities

In this section we give a summary of the most important identities involving vectors and second-order tensors. Most of these formulae can be found in the more pragmatic texts mentioned in the next section. However, this collection is mainly based on Shkarofsky, Johnston and Bachynski (1966); Bird, Stewart and Lightfoot (1960) and Wimmel (1982). The last is especially recommended for its tensor formulae.

$$\boldsymbol{I}\cdot\boldsymbol{A} = \boldsymbol{A}\cdot\boldsymbol{I} = \boldsymbol{A} \tag{T.1}$$

$$\boldsymbol{I}\cdot\boldsymbol{AB} = \boldsymbol{AB} \tag{T.2}$$

$$\boldsymbol{I}\cdot\nabla\boldsymbol{B} = \nabla\boldsymbol{B} \tag{T.3}$$

$$\boldsymbol{AB}:\boldsymbol{CD} = (\boldsymbol{B}\cdot\boldsymbol{C})(\boldsymbol{A}\cdot\boldsymbol{D}) = \boldsymbol{D}\cdot\boldsymbol{AB}\cdot\boldsymbol{C} \tag{T.4}$$

$$\boldsymbol{F}:\boldsymbol{AB} = (\boldsymbol{F}\cdot\boldsymbol{A})\cdot\boldsymbol{B} = \boldsymbol{B}\cdot\boldsymbol{F}\cdot\boldsymbol{A} \tag{T.5}$$

$$\boldsymbol{I}: \boldsymbol{AB} = \boldsymbol{A}\cdot\boldsymbol{B} \tag{T.6}$$

$$\boldsymbol{I}: \nabla\boldsymbol{B} = \nabla\cdot\boldsymbol{B} \tag{T.7}$$

$$\boldsymbol{I}: \boldsymbol{F} = F_i^i = \operatorname{tr}\boldsymbol{F} \tag{T.8}$$

$$\nabla\cdot(a\boldsymbol{I}) = \nabla a \tag{T.9}$$

$$\nabla\cdot\boldsymbol{AB} = \boldsymbol{A}\cdot\nabla\boldsymbol{B} + (\nabla\cdot\boldsymbol{A})\boldsymbol{B} \tag{T.10}$$

$$\boldsymbol{I}\times\boldsymbol{A} = \boldsymbol{A}\times\boldsymbol{I} \tag{T.11}$$

$$(\boldsymbol{A}\times\boldsymbol{B})\times\boldsymbol{I} = \boldsymbol{I}\times(\boldsymbol{A}\times\boldsymbol{B}) = \boldsymbol{BA} - \boldsymbol{AB} \tag{T.12}$$

$$(\boldsymbol{A}\times\boldsymbol{B})\times\boldsymbol{C} = \boldsymbol{C}\times(\boldsymbol{B}\times\boldsymbol{A}) = \boldsymbol{B}(\boldsymbol{C}\cdot\boldsymbol{A}) - \boldsymbol{A}(\boldsymbol{C}\cdot\boldsymbol{B}) = \boldsymbol{C}\cdot(\boldsymbol{AB} - \boldsymbol{BA}) \tag{T.13}$$

$$\boldsymbol{A}\times(\boldsymbol{BC}) = (\boldsymbol{A}\times\boldsymbol{B})\boldsymbol{C} \tag{T.14}$$

$$\nabla\cdot(\boldsymbol{I}\times\boldsymbol{C}) = \nabla\times\boldsymbol{C} \tag{T.15}$$

$$\nabla\times(\boldsymbol{A}\times\boldsymbol{B}) = \nabla\cdot(\boldsymbol{BA} - \boldsymbol{AB}) \tag{T.16}$$

$$\boldsymbol{A}\times(\nabla\times\boldsymbol{B}) = (\nabla\boldsymbol{B})\cdot\boldsymbol{A} - \boldsymbol{A}\cdot\nabla\boldsymbol{B} \tag{T.17}$$

$$\begin{aligned}\nabla(\boldsymbol{A}\cdot\boldsymbol{B}) &= \boldsymbol{A}\times(\nabla\times\boldsymbol{B}) + \boldsymbol{B}\times(\nabla\times\boldsymbol{A}) + \boldsymbol{A}\cdot\nabla\boldsymbol{B} + \boldsymbol{B}\cdot\nabla\boldsymbol{A}\\ &= (\nabla\boldsymbol{A})\cdot\boldsymbol{B} + (\nabla\boldsymbol{B})\cdot\boldsymbol{A}\end{aligned} \tag{T.18}$$

$$\nabla(\boldsymbol{A}\times\boldsymbol{B}) = \nabla\boldsymbol{A}\times\boldsymbol{B} - \nabla\boldsymbol{B}\times\boldsymbol{A} \tag{T.19}$$

$$\nabla\times(\boldsymbol{AB}) = (\nabla\times\boldsymbol{A})\boldsymbol{B} - \boldsymbol{A}\times\nabla\boldsymbol{B} \tag{T.20}$$

$$\boldsymbol{A}\times\nabla\boldsymbol{B} - (\boldsymbol{A}\times\nabla\boldsymbol{B})^{\mathrm{T}} = [(\nabla\cdot\boldsymbol{B})\boldsymbol{A} - \nabla\boldsymbol{B}\cdot\boldsymbol{A}]\times\boldsymbol{I} \tag{T.21}$$

$$\nabla\boldsymbol{B}\times\boldsymbol{A} - (\nabla\boldsymbol{B}\times\boldsymbol{A})^{\mathrm{T}} = [(\nabla\cdot\boldsymbol{B})\boldsymbol{A} - \boldsymbol{A}\cdot\nabla\boldsymbol{B}]\times\boldsymbol{I} \tag{T.22}$$

$$(\nabla\cdot\boldsymbol{B})\boldsymbol{A} - \nabla\boldsymbol{B}\cdot\boldsymbol{A} = -(\boldsymbol{A}\times\nabla)\times\boldsymbol{B} \tag{T.23}$$

$$(\nabla\cdot\boldsymbol{B})\boldsymbol{A} - \boldsymbol{A}\cdot\nabla\boldsymbol{B} = \boldsymbol{A}\times(\nabla\times\boldsymbol{B}) - (\boldsymbol{A}\times\nabla)\times\boldsymbol{B} \tag{T.24}$$

$$(\boldsymbol{A}\times\nabla)\cdot\boldsymbol{B} = \boldsymbol{A}\cdot(\nabla\times\boldsymbol{B}) \tag{T.25}$$

$$\boldsymbol{A}\times\nabla\boldsymbol{B} + (\nabla\boldsymbol{B}\times\boldsymbol{A})^{\mathrm{T}} = [\boldsymbol{A}\cdot(\nabla\times\boldsymbol{B})]\boldsymbol{I} - (\nabla\times\boldsymbol{B})\boldsymbol{A} \tag{T.26}$$

$$\nabla\boldsymbol{B}\times\boldsymbol{A} + (\boldsymbol{A}\times\nabla\boldsymbol{B})^{\mathrm{T}} = [\boldsymbol{A}\cdot(\nabla\times\boldsymbol{B})]\boldsymbol{I} - \boldsymbol{A}(\nabla\times\boldsymbol{B}) \tag{T.27}$$

$$\begin{aligned}\boldsymbol{A}\times\nabla\boldsymbol{B} + \nabla\boldsymbol{B}\times\boldsymbol{A} &= \boldsymbol{I}\times[(\nabla\cdot\boldsymbol{B})\boldsymbol{A} - \nabla\boldsymbol{B}\cdot\boldsymbol{A}]\\ &\qquad + [\boldsymbol{A}\cdot(\nabla\times\boldsymbol{B})]\boldsymbol{I} - \boldsymbol{A}(\nabla\times\boldsymbol{B})\\ &= \boldsymbol{I}\times[(\nabla\cdot\boldsymbol{B})\boldsymbol{A} - \boldsymbol{A}\cdot\nabla\boldsymbol{B}]\\ &\qquad + [\boldsymbol{A}\cdot(\nabla\times\boldsymbol{B})]\boldsymbol{I} - (\nabla\times\boldsymbol{B})\boldsymbol{A}\end{aligned} \tag{T.28}$$

$$(\boldsymbol{A} \times \boldsymbol{C}) \cdot \boldsymbol{F} = \boldsymbol{A} \cdot (\boldsymbol{C} \times \boldsymbol{F}) = -\boldsymbol{C} \cdot (\boldsymbol{A} \times \boldsymbol{F}) \tag{T.29}$$

$$\boldsymbol{F} \cdot (\boldsymbol{A} \times \boldsymbol{C}) = (\boldsymbol{F} \times \boldsymbol{A}) \cdot \boldsymbol{C} = -(\boldsymbol{F} \times \boldsymbol{C}) \cdot \boldsymbol{A} \tag{T.30}$$

$$\boldsymbol{A} \cdot \boldsymbol{F} \times \boldsymbol{C} = (\boldsymbol{A} \cdot \boldsymbol{F}) \times \boldsymbol{C} = \boldsymbol{A} \cdot (\boldsymbol{F} \times \boldsymbol{C}) = -\boldsymbol{C} \times (\boldsymbol{A} \cdot \boldsymbol{F}) \tag{T.31}$$

$$\boldsymbol{A} \times \boldsymbol{F} \cdot \boldsymbol{C} = \boldsymbol{A} \times (\boldsymbol{F} \cdot \boldsymbol{C}) = (\boldsymbol{A} \times \boldsymbol{F}) \cdot \boldsymbol{C} = -(\boldsymbol{F} \cdot \boldsymbol{C}) \times \boldsymbol{A} \tag{T.32}$$

$$\boldsymbol{A} \cdot \nabla \boldsymbol{B} \times \boldsymbol{C} + \boldsymbol{C} \times \nabla \boldsymbol{B} \cdot \boldsymbol{A} = \boldsymbol{C} \times [\boldsymbol{A} \times (\nabla \times \boldsymbol{B})] \tag{T.33}$$

$$\boldsymbol{A} \cdot \nabla \boldsymbol{B} \times \boldsymbol{C} - \boldsymbol{C} \cdot \nabla \boldsymbol{B} \times \boldsymbol{A} = [(\nabla \cdot \boldsymbol{B}) \boldsymbol{I} - \nabla \boldsymbol{B}] \cdot (\boldsymbol{A} \times \boldsymbol{C}) \tag{T.34}$$

$$\boldsymbol{A} \times \nabla \boldsymbol{B} \cdot \boldsymbol{C} - \boldsymbol{C} \times \nabla \boldsymbol{B} \cdot \boldsymbol{A} = (\boldsymbol{A} \times \boldsymbol{C}) \cdot [(\nabla \cdot \boldsymbol{B}) \boldsymbol{I} - \nabla \boldsymbol{B}] \tag{T.35}$$

$$\boldsymbol{A} \cdot \nabla \boldsymbol{B} \cdot \boldsymbol{C} - \boldsymbol{C} \cdot \nabla \boldsymbol{B} \cdot \boldsymbol{A} = (\boldsymbol{A} \times \boldsymbol{C}) \cdot (\nabla \times \boldsymbol{B}) \tag{T.36}$$

$$\nabla \boldsymbol{R} = \boldsymbol{I}\,, \qquad \boldsymbol{R} = \text{position vector}\ . \tag{T.37}$$

3.7 Suggestions for Further Reading

This chapter was a minimal discussion of tensors. For more on this powerful "technique", we must refer to the vast literature in mathematics and mathematical physics. It is difficult to suggest one single reference because none of those known to us treats tensors correctly *and* generally enough *and* appropriately for plasma physics.

A substantial number of physics textbooks devote a chapter or an appendix to tensors or polyadics. These books cover fields like plasma physics, mechanics, electromagnetics, transport, elasticity theory and fluid dynamics. In particular, we would like to mention Symon (1971); Goldstein (1981); Bird, Stewart and Lightfoot (1960); Milne-Thomson (1968); Stratton (1941); Van Bladel (1964); Krall and Trivelpiece (1973); Seshadri (1973); Shkarofsky, Johnston and Bachynski (1966); Chapman and Cowling (1973); Ferziger and Kaper (1972); Jancel and Kahan (1966); Woodson and Melcher (1968).

In the more practical mathematics and mathematical-physics books, tensors are often defined by means of their transformation laws; in some of them, dyads and dyadics are discussed (or parenthetically referred to). Dual spaces are not mentioned. In this category, we recommend Wrede (1963); Borisenko and Tarapov (1968); Morse and Feshbach (1953); Mathews and Walker (1970); Margenau and Murphy (1976); Butkov (1968); Spiegel (1959); Menzel (1961); Hoffmann (1977).

On the subject of tensor calculus in the context of differential geometry and its relationship with differential forms and Grassmann algebra, see Stoker (1969); Guggenheimer (1977); Spivak (1970); Choquet-Bruhat (1982); Behnke (1974); Flanders (1963); Arnold (1978); Abraham and Marsden (1978); Deschamps (1970); Schutz (1980), Misner, Thorne and Wheeler (1973); Burke (1985).

4. Magnetic-Field-Structure-Related Concepts

4.1 The Equation of a Magnetic-Field Line

A magnetic *field line* (or "line of force") is by definition a curve whose tangent at any point is parallel to the magnetic field-vector $\boldsymbol{B}$. This property can be translated into mathematical terms as follows:

$$\boldsymbol{B} \propto d\boldsymbol{R} \qquad \text{or} \qquad \boldsymbol{B} = c\, d\boldsymbol{R} \;, \tag{4.1.1}$$

where c is a proportionality constant and $d\boldsymbol{R}$ is a differential vector tangent to the field line ($\boldsymbol{R}$ is the position vector from a chosen origin to the point in question on the curve). Equivalently,

$$\boldsymbol{B} \times d\boldsymbol{R} = 0 \;. \tag{4.1.2}$$

Using the contravariant components of both $\boldsymbol{B}(B^1, B^2, B^3)$ and $d\boldsymbol{R}(du^1, du^2, du^3)$, either of the above equations leads to:

$$\frac{B^1}{du^1} = \frac{B^2}{du^2} = \frac{B^3}{du^3} = c \;. \tag{4.1.3}$$

If we recall that $B^i = \boldsymbol{B} \cdot \nabla u^i$, we can also write:

$$\frac{\boldsymbol{B} \cdot \nabla u^1}{du^1} = \frac{\boldsymbol{B} \cdot \nabla u^2}{du^2} = \frac{\boldsymbol{B} \cdot \nabla u^3}{du^3} \;. \tag{4.1.4}$$

If we parameterize the field-line curve by the arc length l, the tangent vector $d\boldsymbol{R}/dl$ is a unit vector, commonly denoted by $\hat{\boldsymbol{t}}$ or $\hat{\boldsymbol{B}} \equiv \boldsymbol{B}/B$:

$$\frac{\boldsymbol{B}}{B} = \left.\frac{d\boldsymbol{R}}{dl}\right|_{\text{along } \boldsymbol{B}} \;. \tag{4.1.5}$$

From (4.1.1), then, the constant c equals B/dl, so that the *equation of a magnetic-field line* is, in component form:

$$\frac{B}{dl} = \frac{B^1}{du^1} = \frac{B^2}{du^2} = \frac{B^3}{du^3} \tag{4.1.6}$$

or

$$\frac{B}{dl} = \frac{\boldsymbol{B}\cdot\nabla u^1}{du^1} = \frac{\boldsymbol{B}\cdot\nabla u^2}{du^2} = \frac{\boldsymbol{B}\cdot\nabla u^3}{du^3} \ . \tag{4.1.7}$$

The form of (4.1.1) implies that a choice for contravariant components of $d\boldsymbol{R}$, $du^i \equiv d\boldsymbol{R}\cdot\nabla u^i$, requires the use of the contravariant components of $\boldsymbol{B}$, the B^i.

4.2 The Frozen-Flux Theorem

The frozen-flux theorem states that the magnetic flux threading an open surface in a perfectly conducting (MHD) plasma is constant. Even when the surface moves with the plasma, the flux remains the same. This is equivalent to saying that the flux is frozen within the plasma or that the field lines move with the plasma. This theorem is proven in most plasma-physics textbooks. We will follow closely the concise but clear discussion given by Schmidt (1979).

The starting point to prove the theorem is a combination of Ampère's law, Faraday's law and Ohm's law:

$$\nabla\times\boldsymbol{B} = \mu_0\boldsymbol{J} \tag{4.2.1}$$

$$\frac{\partial\boldsymbol{B}}{\partial t} = -\nabla\times\boldsymbol{E} \tag{4.2.2}$$

$$\boldsymbol{J} = \sigma(\boldsymbol{E} + \boldsymbol{V}\times\boldsymbol{B}) \ . \tag{4.2.3}$$

Here, $\boldsymbol{V}$ represents the plasma-fluid velocity, σ is the electrical conductivity, and the other symbols have the usual meanings. For consistency with MHD, we neglected the displacement current in Ampère's law, (4.2.1). This statement is equivalent to $\nabla\cdot\boldsymbol{J} = 0$. This is a particular version of charge conservation. The equations are written in *MKS units.*

After substituting (4.2.1) for $\boldsymbol{J}$ into Ohm's law, (4.2.3), and solving for the electric field $\boldsymbol{E}$, which in turn is substituted into (4.2.2), we obtain an equation for $\partial\boldsymbol{B}/\partial t$:

$$\frac{\partial\boldsymbol{B}}{\partial t} = -\nabla\times\left[\frac{\nabla\times\boldsymbol{B}}{\mu_0\sigma} - \boldsymbol{V}\times\boldsymbol{B}\right] . \tag{4.2.4}$$

With the help of the vector identity $\nabla\times(\nabla\times\boldsymbol{A}) = \nabla(\nabla\cdot\boldsymbol{A}) - \nabla^2\boldsymbol{A}$, this becomes

$$\frac{\partial\boldsymbol{B}}{\partial t} = \frac{1}{\mu_0\sigma}\nabla^2\boldsymbol{B} - \nabla\left(\frac{1}{\mu_0\sigma}\right)\times(\nabla\times\boldsymbol{B}) + \nabla\times(\boldsymbol{V}\times\boldsymbol{B}) \ , \tag{4.2.5}$$

since $\nabla\cdot\boldsymbol{B} = 0$. Equation (4.2.5) is a diffusion equation for the magnetic field in a moving plasma (see, e.g., Schmidt, 1979, or Golant, Zhilinsky and Sakharov, 1980).

We are interested in the perfectly conducting limit, where $\sigma \to \infty$:

$$\frac{\partial \boldsymbol{B}}{\partial t} = \nabla \times (\boldsymbol{V} \times \boldsymbol{B}) \ . \tag{4.2.6}$$

We now integrate this equation over an arbitrary open but fixed surface S inside the plasma. With Stokes' law, we obtain

$$\iint\limits_S \frac{\partial \boldsymbol{B}}{\partial t} \cdot d\boldsymbol{S} = \iint\limits_S \nabla \times (\boldsymbol{V} \times \boldsymbol{B}) \cdot d\boldsymbol{S} = \oint\limits_\Gamma \boldsymbol{V} \times \boldsymbol{B} \cdot d\boldsymbol{l} \ . \tag{4.2.7}$$

$d\boldsymbol{l}$ represents a differential length element along the boundary curve Γ of S (see Fig. 4.1). The last form of (4.2.7) can be rewritten using the triple-product identity $\boldsymbol{a} \cdot \boldsymbol{b} \times \boldsymbol{c} = -\boldsymbol{c} \cdot (\boldsymbol{b} \times \boldsymbol{a})$, while the $\partial/\partial t$ operator can be moved outside the first integral. The surface S is stationary, and $\partial/\partial t$ is taken at constant $\boldsymbol{R}$. With the magnetic flux Ψ through S defined as usual, $\Psi = \iint \boldsymbol{B} \cdot d\boldsymbol{S}$, we can write

$$\frac{\partial \Psi}{\partial t} + \oint\limits_\Gamma \boldsymbol{B} \cdot (\boldsymbol{V} \times d\boldsymbol{l}) \equiv 0 \ . \tag{4.2.8}$$

To interpret this integral relation, we consider Fig. 4.1. The open surface S is fixed in space, regardless of the plasma velocity. Since the velocity $\boldsymbol{V}$ represents the plasma velocity, it is clear that the open surface S' is the equivalent of surface S, but now "carried" with the plasma. Between the open surfaces S (at time $t = 0$), and S' (at time $t = \Delta t$), there is a "side"-surface created by the moving boundary Γ'. A differential area element of this side-surface is $|\boldsymbol{V} \times d\boldsymbol{l}|\Delta t$ (cross-hatched in the figure). The first term of (4.2.8) is the time rate of change of the magnetic flux through the fixed open surface S. The second term represents the increment of flux through the side-surface swept out by the moving boundary Γ'. Because $\boldsymbol{B}$ is a solenoidal field, we know that the total flux through the closed surface $S + S_{\text{side}} + S'$ must be zero. Since (4.2.8) states that the change in flux through $S + S_{\text{side}}$ is zero, we must have that the change of flux through the moving surface S' is zero as well. This implies that the total time derivative of the flux through a (moving) surface (such as S') must be zero (see, e.g., Jackson, 1975, Chap. 6):

$$\frac{d\Psi}{dt} \equiv 0 \ . \tag{4.2.9}$$

The surface S (or S') in the above proof was arbitrary. It follows that magnetic-field lines are tied to the plasma, which makes the abstract concept of a field line more concrete in an ideal MHD plasma. This allows us to speak about the motion of field lines (see Schmidt 1979). The motion of magnetic-field lines has been studied by Newcomb (1958).

The notion of moving field lines is used throughout fusion-plasma physics, even when the fluid is not perfectly conducting, and the plasma and the field lines are no longer tied together. However, we must be careful with the motion of field

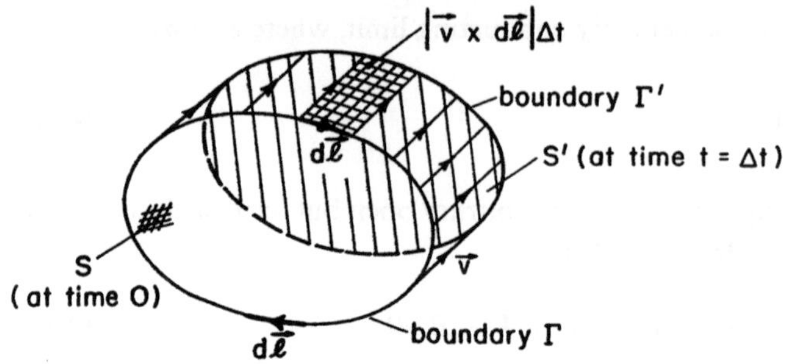

Fig. 4.1. The motion of a boundary curve Γ, which is tied to the plasma. See text

lines when a parallel electric field is present (Newcomb, 1958; Kerst, private communication). Perpendicular $\boldsymbol{E}$ fields lead to perpendicular $\boldsymbol{E} \times \boldsymbol{B}$ fluid motion and do not give rise to problems.

4.3 The Magnetic Field-Line Curvature

Consider a magnetic-field line parameterized by its arc length l. The vector $d\boldsymbol{R}/dl$, where $\boldsymbol{R}$ is the position vector of a point on the curve, is a unit tangent vector. It is denoted by $\hat{\boldsymbol{l}}$ or $\hat{\boldsymbol{t}}$ or $\hat{\boldsymbol{B}} \equiv \boldsymbol{B}/B$. (We assume for convenience that the arc length which parameterizes the field line increases in the direction in which $\boldsymbol{B}$ points. Otherwise, we would have that $\hat{\boldsymbol{B}} = -\hat{\boldsymbol{l}}$.) The dot product of $\hat{\boldsymbol{t}}$ with itself is unity, $\hat{\boldsymbol{t}} \cdot \hat{\boldsymbol{t}} = 1$, so that $\hat{\boldsymbol{t}} \cdot d\hat{\boldsymbol{t}}/dl = 0$. Thus the vector $d\hat{\boldsymbol{t}}/dl$ is perpendicular to the tangent vector $\hat{\boldsymbol{t}}$. We now define:

$$\left.\frac{d\hat{\boldsymbol{t}}}{dl}\right|_{\text{along }\hat{\boldsymbol{t}}} \equiv \kappa\hat{\boldsymbol{n}} \equiv \frac{1}{R_c}\hat{\boldsymbol{n}} \equiv \boldsymbol{\kappa} \equiv \left.\frac{d\hat{\boldsymbol{B}}}{dl}\right|_{\text{along }\hat{\boldsymbol{B}}} . \tag{4.3.1}$$

Here, $\hat{\boldsymbol{n}}$ is the principal normal to the field line. The parameter κ is called the *curvature*, and the vector $\boldsymbol{\kappa}$ the *curvature vector*. The quantity $R_c = 1/\kappa$ is the *local radius of curvature*. That (4.3.1) is in agreement with intuition can be shown geometrically from the definition:

$$\boldsymbol{\kappa} = \lim_{\Delta l \to 0} [\hat{\boldsymbol{t}}(l + \Delta l) - \hat{\boldsymbol{t}}(l)]/\Delta l = \lim_{\Delta l \to 0} \frac{\Delta \hat{\boldsymbol{t}}}{\Delta l} .$$

Sometimes one also uses the *radius-of-curvature vector* $\boldsymbol{R}_c$, pointing in the negative $\hat{\boldsymbol{n}}$ direction. It is defined by

$$\boldsymbol{\kappa} = -\frac{\boldsymbol{R}_c}{R_c^2} . \tag{4.3.2}$$

The derivative $d/dl|_{\text{along } \hat{\boldsymbol{B}}}$ is the directional derivative and is equal to the dot product of $\hat{\boldsymbol{B}}$ with $\boldsymbol{\nabla}$:

$$\hat{\boldsymbol{B}} \cdot \boldsymbol{\nabla} \equiv \frac{\partial}{\partial l} = \frac{d}{dl}\bigg|_{\text{along } \hat{\boldsymbol{B}}} . \tag{4.3.3}$$

From (4.3.1) and (4.3.3), we can write the curvature in the form most commonly used in plasma physics:

$$\boldsymbol{\kappa} \equiv (\hat{\boldsymbol{B}} \cdot \boldsymbol{\nabla})\hat{\boldsymbol{B}} = \hat{\boldsymbol{B}} \cdot \boldsymbol{\nabla}\hat{\boldsymbol{B}} . \tag{4.3.4}$$

The second form reminds us that it is basically the dot product of a vector $\hat{\boldsymbol{B}}$ with a dyad $\boldsymbol{\nabla}\hat{\boldsymbol{B}}$. Note that the dot-product rule from Chap. 3, on dyads, is apparently "violated". The result of $\boldsymbol{A} \cdot (\boldsymbol{C}\boldsymbol{D}) = (\boldsymbol{A} \cdot \boldsymbol{C})\boldsymbol{D}$ is normally a vector proportional to the vector not affected by the dot, here $\boldsymbol{D}$. Because of the special character of the *operator* $\boldsymbol{\nabla}$, $(\hat{\boldsymbol{B}} \cdot \boldsymbol{\nabla})\hat{\boldsymbol{B}}$ is not along the vector $\hat{\boldsymbol{B}}$, but orthogonal to $\hat{\boldsymbol{B}}$ according to (4.3.1).

An alternative expression for $\boldsymbol{\kappa}$ can be found by applying the *bac-cab* rule to $\hat{\boldsymbol{B}} \times (\boldsymbol{\nabla} \times \hat{\boldsymbol{B}})$, which yields $\hat{\boldsymbol{B}} \times (\boldsymbol{\nabla} \times \hat{\boldsymbol{B}}) = \boldsymbol{\nabla}(\hat{\boldsymbol{B}} \cdot \hat{\boldsymbol{B}})/2 - (\hat{\boldsymbol{B}} \cdot \boldsymbol{\nabla})\hat{\boldsymbol{B}} = -(\hat{\boldsymbol{B}} \cdot \boldsymbol{\nabla})\hat{\boldsymbol{B}}$. Hence, we obtain

$$\boldsymbol{\kappa} = \hat{\boldsymbol{B}} \cdot \boldsymbol{\nabla}\hat{\boldsymbol{B}} = -\hat{\boldsymbol{B}} \times (\boldsymbol{\nabla} \times \hat{\boldsymbol{B}}) . \tag{4.3.5}$$

There is yet another commonly used expression for the curvature vector, where it is related to the gradient of $|\boldsymbol{B}|$. However, this expression is only valid *for low β plasmas*. With $\boldsymbol{J}$ set equal to $\boldsymbol{\nabla} \times \boldsymbol{B}/\mu_0$, and with the help of the vector identity $\boldsymbol{A} \times (\boldsymbol{\nabla} \times \boldsymbol{B}) = (\boldsymbol{\nabla}\boldsymbol{B}) \cdot \boldsymbol{A} - \boldsymbol{A} \cdot \boldsymbol{\nabla}\boldsymbol{B}$, we obtain from the equilibrium equation $\boldsymbol{\nabla}p = \boldsymbol{J} \times \boldsymbol{B}$:

$$\boldsymbol{\nabla}p = \frac{1}{\mu_0}(\boldsymbol{\nabla} \times \boldsymbol{B}) \times \boldsymbol{B} = \frac{1}{\mu_0}(\boldsymbol{B} \cdot \boldsymbol{\nabla})\boldsymbol{B} - \frac{1}{2\mu_0}\boldsymbol{\nabla}(B)^2 . \tag{4.3.6}$$

The "2" in $(B)^2$ represents a square, not a superscript. When the first term on the right-hand side is written as $\boldsymbol{B} = B\hat{\boldsymbol{B}}$, it becomes equal to

$$\frac{B^2}{\mu_0}(\hat{\boldsymbol{B}} \cdot \boldsymbol{\nabla})\hat{\boldsymbol{B}} + \frac{B\hat{\boldsymbol{B}}}{\mu_0} \cdot (\boldsymbol{\nabla}B)\hat{\boldsymbol{B}} . \tag{4.3.7}$$

Recall that the order of the dyad in this second term must be kept, first $\boldsymbol{\nabla}$ and then $\hat{\boldsymbol{B}}$. Then

$$\boldsymbol{\nabla}p = \frac{B^2}{\mu_0}\boldsymbol{\kappa} + \hat{\boldsymbol{B}}\left(\hat{\boldsymbol{B}} \cdot \boldsymbol{\nabla}\frac{B^2}{2\mu_0}\right) - \boldsymbol{\nabla}\frac{B^2}{2\mu_0} . \tag{4.3.8}$$

We define the parallel and perpendicular (to $\hat{\boldsymbol{B}}$) $\boldsymbol{\nabla}$ operators as

$$\boldsymbol{\nabla}_{\parallel} = \hat{\boldsymbol{B}}(\hat{\boldsymbol{B}} \cdot \boldsymbol{\nabla}) \text{ and } \boldsymbol{\nabla}_{\perp} = \boldsymbol{\nabla} - \boldsymbol{\nabla}_{\parallel} = \boldsymbol{\nabla} - \hat{\boldsymbol{B}}(\hat{\boldsymbol{B}} \cdot \boldsymbol{\nabla}) = -\hat{\boldsymbol{B}} \times (\hat{\boldsymbol{B}} \times \boldsymbol{\nabla}) ,$$

respectively. Since $\hat{\boldsymbol{B}} \cdot \boldsymbol{\nabla}p = 0$, we see that $\boldsymbol{\nabla}p = \boldsymbol{\nabla}_{\perp}p$, so that (4.3.8) can be written as

$$\nabla_\perp\left(p+\frac{B^2}{2\mu_0}\right)=\frac{B^2}{\mu_0}\boldsymbol{\kappa}\ . \qquad (4.3.9)$$

In low-pressure plasmas, when $\beta \equiv p/(B^2/2\mu_0) \ll 1$, we can neglect the plasma pressure p in comparison with the magnetic pressure $B^2/2\mu_0$, and obtain

$$\nabla_\perp \frac{B^2}{2} \simeq B^2\boldsymbol{\kappa}\ . \qquad (4.3.10)$$

Alternatively,

$$\boldsymbol{\kappa} \simeq \frac{\nabla_\perp B}{B} = \nabla_\perp \ln B, \quad \text{for} \quad \beta \ll 1\ . \qquad (4.3.11)$$

This expression shows that the curvature (in low β plasmas) is effectively the reciprocal of the scale length for (perpendicular) variation of $|\boldsymbol{B}|$.

So far, we have been dealing with the curvature of a curve (e.g., a magnetic-field line) in isolation. When the curve under consideration lies within a surface, it is customary to define curvature components with respect to the surface. The component of $\boldsymbol{\kappa}$ normal to the surface is called the *normal curvature*, whereas the component of $\boldsymbol{\kappa}$ tangent to the surface is referred to as the *geodesic curvature* (Goetz 1970; Guggenheimer 1977).

As we will discuss below, in magnetic fusion the magnetic-field lines are often chosen to be coordinate curves. As indicated on Fig. 4.2, we identify the

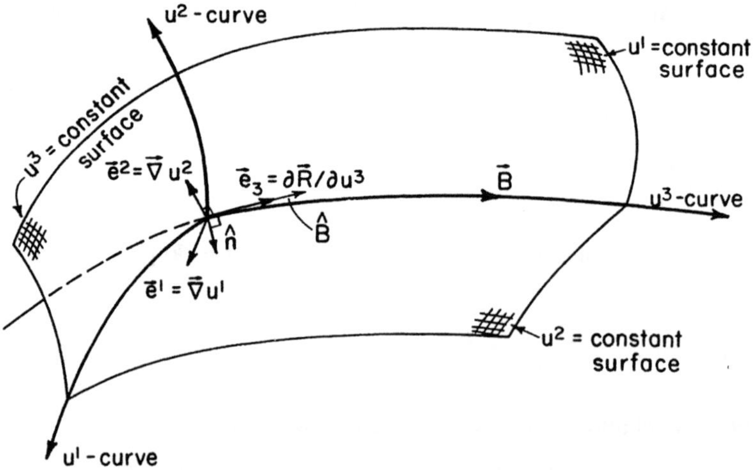

Fig. 4.2. The basis vectors $\boldsymbol{e}^1$, $\boldsymbol{e}^2$, and $\boldsymbol{e}_3$ in relation to the principal normal to the u^3-curve. The vectors $\hat{\boldsymbol{n}}$, $\nabla u^1 = \boldsymbol{e}^1$ and $\nabla u^2 = \boldsymbol{e}^2$ are perpendicular to $\boldsymbol{B}$ and lie in the same plane

third coordinate curve, the u^3-curve with a magnetic-field line. This lies at the intersection of the coordinate surfaces u^1 = constant and u^2 = constant. By construction, the tangent-basis vector $\boldsymbol{e}_3$ (which is proportional to $\hat{\boldsymbol{B}}$) is perpendicular to the reciprocal-basis vectors $\boldsymbol{e}^1 = \nabla u^1$ and $\boldsymbol{e}^2 = \nabla u^2$. To make the discussion somewhat less abstract, one could take u^3 as the length l along the field line.

"Natural" surfaces to which magnetic-field lines belong are magnetic-flux surfaces (which will be defined rigorously below). We could think of the surfaces u^1 = constant as flux surfaces. For some applications in fusion-plasma physics, it is desirable to know the components of the curvature $\boldsymbol{\kappa}$ perpendicular and tangent to the flux surface u^1 = constant. The curvature lies along the principal normal, $\hat{\boldsymbol{n}}$, to the field line. The vectors, $\hat{\boldsymbol{n}}$, ∇u^1 and ∇u^2 are perpendicular to $\boldsymbol{B}$ and lie in the same plane as each other.

We want to decompose $\hat{\boldsymbol{n}}$ into a component along ∇u^1 (i.e., a component perpendicular to the (flux) surface u^1 = constant), and a component tangent to the u^1 = constant surface. ∇u^2 is (in general) not tangential to the u^1 = constant surface (see Fig. 4.2). Also, the tangent-basis vector, $\boldsymbol{e}_2 = \partial \boldsymbol{R}/\partial u^2$, which is tangential to the u^1 = constant surface, is not acceptable for our purposes, since it has a component along $\boldsymbol{B}$ (in general, tangent-basis vectors are not orthogonal to each other, $\boldsymbol{e}_i \cdot \boldsymbol{e}_j \neq 0$). This is equivalent to saying that $\boldsymbol{e}_2$ does not lie in the plane to which $\boldsymbol{\kappa}$ and ∇u^1 belong. As mentioned above, ∇u^2 does lie in the plane perpendicular to $\boldsymbol{B}$, and all we must do is find the component of ∇u^2 parallel to the ∇u^1 direction and subtract it from ∇u^2. Then we have a vector both tangent to the flux surface and perpendicular to $\boldsymbol{B}$. The parallel and perpendicular components of the vector ∇u^2 with respect to a vector normal to the u^1 = constant surface, $\hat{\boldsymbol{N}} \equiv \nabla u^1/|\nabla u^1|$ are given by

$$(\nabla u^2)_{\parallel} = \hat{\boldsymbol{N}}(\hat{\boldsymbol{N}} \cdot \nabla u^2) \tag{4.3.12a}$$

$$(\nabla u^2)_{\perp} = -\hat{\boldsymbol{N}} \times (\hat{\boldsymbol{N}} \times \nabla u^2) \ . \tag{4.3.12b}$$

When the curvature vector $\boldsymbol{\kappa}$ is expanded in its covariant components as usual, we have

$$\boldsymbol{\kappa} = \kappa_1 \nabla u^1 + \kappa_2 \nabla u^2 \ ; \qquad \kappa_i \equiv \boldsymbol{\kappa} \cdot \boldsymbol{e}_i \ , \tag{4.3.13}$$

since $\kappa_3 \equiv \boldsymbol{\kappa} \cdot \boldsymbol{e}_3 = 0$, recalling that $\boldsymbol{e}_3 \propto \boldsymbol{B}$. After setting ∇u^2 in (4.3.13) equal to $(\nabla u^2)_{\parallel} + (\nabla u^2)_{\perp}$ of (4.3.12), we obtain:

$$\boldsymbol{\kappa} = [\kappa_1 |\nabla u^1| + \kappa_2 (\hat{\boldsymbol{N}} \cdot \nabla u^2)] \hat{\boldsymbol{N}} + \kappa_2 (\hat{\boldsymbol{N}} \times \nabla u^2) \times \hat{\boldsymbol{N}} \ . \tag{4.3.14}$$

The first term is the *normal-curvature vector* (along the normal $\hat{\boldsymbol{N}}$ to the surface u^1 = constant). The second term is the *geodesic-curvature vector*. In summary, we have

$$\kappa_{\mathrm{N}} = \kappa_1 |\nabla u^1| + \kappa_2 |\nabla u^2| \cos \measuredangle (\nabla u^1, \nabla u^2) \ , \quad \text{normal curvature} \tag{4.3.15a}$$

$$\kappa_{\mathrm{G}} = \kappa_2 |\nabla u^2| \sin \measuredangle (\nabla u^1, \nabla u^2) \ , \quad \text{geodesic curvature} \ . \tag{4.3.15b}$$

These are the correct definitions of normal and geodesic curvature. Sometimes, however, this terminology is used loosely, and κ_1 in (4.3.13) is called the normal curvature, while κ_2 is designated as the geodesic curvature. In magnetic geometries with a high degree of symmetry, these distinctions become irrelevant.

Below we will identify u^1 with a radial variable and represent it by symbols such as ρ, Ψ, or α. u^2 will be considered as an angle-like variable and will be labeled β or ν. Choosing $u^1 = \alpha$, and $u^2 = \beta$, (4.3.15a, b) are then in a more familiar form:

$$\kappa_N = \kappa_\alpha |\nabla\alpha| + \kappa_\beta |\nabla\beta| \cos \sphericalangle (\nabla\alpha, \nabla\beta) \tag{4.3.16a}$$

$$\kappa_G = \kappa_\beta |\nabla\beta| \sin \sphericalangle (\nabla\alpha, \nabla\beta) \ . \tag{4.3.16b}$$

In some cases it may be convenient to use the tangent vector $\hat{\boldsymbol{t}}$ and the principal normal $\hat{\boldsymbol{n}}$ to a curve as basis vectors of a coordinate system. The third basis vector is the binormal $\hat{\boldsymbol{b}}$ and is defined by

$$\hat{\boldsymbol{b}} \equiv \hat{\boldsymbol{t}} \times \hat{\boldsymbol{n}} \ . \tag{4.3.17}$$

Often the curve on which this frame is attached is the magnetic axis (to be defined rigorously below). See, for instance, Shafranov (1968), Solov'ev and Shafranov (1970) and Solov'ev (1975). This type of system has gained renewed interest in fusion-plasma physics, since the development of the helical-axis stellarators and heliacs.

The coordinate frame $\hat{\boldsymbol{t}}$, $\hat{\boldsymbol{n}}$, $\hat{\boldsymbol{b}}$ changes from point to point along its "base"-curve and is known as the moving *Frenet-Serret* frame. (Other names are F-S trihedron, F-S trihedral or F-S triad. Note, however, that we used the word "triad" with a different meaning as a third-order "simple" tensor.) It is necessary to introduce a second scalar function τ, known as the *torsion*, in order to relate the change in $\hat{\boldsymbol{t}}$, $\hat{\boldsymbol{n}}$ and $\hat{\boldsymbol{b}}$ via the F-S formulae:

$$\frac{d\hat{\boldsymbol{t}}}{dl} = \kappa \hat{\boldsymbol{n}} \tag{4.3.18a}$$

$$\frac{d\hat{\boldsymbol{n}}}{dl} = -\kappa \hat{\boldsymbol{t}} + \tau \hat{\boldsymbol{b}} \tag{4.3.18b}$$

$$\frac{d\hat{\boldsymbol{b}}}{dl} = -\tau \hat{\boldsymbol{n}} \ . \tag{4.3.18c}$$

The first expression, (4.3.18a) is our definition of the curvature κ encountered earlier in (4.3.1). The last equation defines the torsion τ. To show (4.3.18c), we first take the derivative of (4.3.17) and apply (4.3.18a). Hence, $d\hat{\boldsymbol{b}}/dl = \hat{\boldsymbol{t}} \times d\hat{\boldsymbol{n}}/dl$, implying that $d\hat{\boldsymbol{b}}/dl \perp \hat{\boldsymbol{t}}$. Furthermore, from $\hat{\boldsymbol{b}} \cdot \hat{\boldsymbol{b}} = 1$, we have that $\hat{\boldsymbol{b}} \cdot d\hat{\boldsymbol{b}}/dl = 0$ or $d\hat{\boldsymbol{b}}/dl \perp \hat{\boldsymbol{b}}$. Therefore, $d\hat{\boldsymbol{b}}/dl$ is perpendicular to both $\hat{\boldsymbol{t}}$ and $\hat{\boldsymbol{b}}$, meaning that it is directed along $-\hat{\boldsymbol{n}}$, since $\hat{\boldsymbol{t}}$, $\hat{\boldsymbol{n}}$, $\hat{\boldsymbol{b}}$ are mutually orthogonal and form a right-handed system. τ is the proportionality constant. The middle F-S equation is easily proved by taking the derivative of $\hat{\boldsymbol{n}} = \hat{\boldsymbol{b}} \times \hat{\boldsymbol{t}}$ and subsequently substituting (4.3.18a) and (4.3.18c). Then, from $\hat{\boldsymbol{b}} = \hat{\boldsymbol{t}} \times \hat{\boldsymbol{n}}$ and $\hat{\boldsymbol{t}} = \hat{\boldsymbol{n}} \times \hat{\boldsymbol{b}}$, (4.3.18b) follows.

4.4 Magnetic Pressure and Magnetic Tension

It is instructive to investigate how one can assign certain forces to magnetic-field lines embedded in a current-carrying medium. This allows us to elucidate commonly used concepts such as magnetic pressure and magnetic tension (Schmidt 1979; Cross 1989).

The Lorentz force density $\boldsymbol{f}$ (force per unit volume) in a current-carrying magnetized medium is given by

$$\boldsymbol{f} = \frac{d\boldsymbol{F}}{dV} = \boldsymbol{J} \times \boldsymbol{B} \ . \tag{4.4.1}$$

However, since the current density is curl $\boldsymbol{B}$, we can rewrite this in terms of $\boldsymbol{B}$ only:

$$\boldsymbol{f} = \frac{1}{\mu_0}(\boldsymbol{\nabla} \times \boldsymbol{B}) \times \boldsymbol{B} = \frac{1}{\mu_0}(\boldsymbol{B} \cdot \boldsymbol{\nabla})\boldsymbol{B} - \frac{1}{2\mu_0}\boldsymbol{\nabla} B^2 \ . \tag{4.4.2}$$

The RHS is identical to (4.3.6). Following Schmidt (1979), we decompose $\boldsymbol{f}$ into its components along the Frenet-Serret trihedron tied to a field-line curve. The manipulations needed have already been performed in (4.3.7) and (4.3.8). With the RHS of (4.3.8) and (4.3.3), we have

$$\boldsymbol{f} = \frac{B^2}{\mu_0}\kappa\hat{\boldsymbol{n}} + \hat{\boldsymbol{t}}\frac{\partial}{\partial l}\left(\frac{B^2}{2\mu_0}\right) - \boldsymbol{\nabla}\left(\frac{B^2}{2\mu_0}\right) , \tag{4.4.3}$$

where we have replaced $\hat{\boldsymbol{B}}$ by $\hat{\boldsymbol{t}}$ (we still assume that $\hat{\boldsymbol{B}} = +\hat{\boldsymbol{t}}$) for notational aesthetics. Recall that κ represents the curvature $\kappa = 1/R_c$, where R_c is the local radius of curvature, $\hat{\boldsymbol{n}}$ is the principal normal and $\boldsymbol{\kappa} \equiv (\hat{\boldsymbol{B}} \cdot \boldsymbol{\nabla})\hat{\boldsymbol{B}} = \kappa\hat{\boldsymbol{n}}$.

Expansion of $\boldsymbol{\nabla}$ along the Frenet-Serret trihedron by means of the partial derivatives $\partial/\partial l$, $\partial/\partial n$, and $\partial/\partial b$ gives

$$\boldsymbol{\nabla}\left(\frac{B^2}{2\mu_0}\right) = \hat{\boldsymbol{t}}\frac{\partial}{\partial l}\left(\frac{B^2}{2\mu_0}\right) + \hat{\boldsymbol{n}}\frac{\partial}{\partial n}\left(\frac{B^2}{2\mu_0}\right) + \hat{\boldsymbol{b}}\frac{\partial}{\partial b}\left(\frac{B^2}{2\mu_0}\right) . \tag{4.4.4}$$

Here, $\hat{\boldsymbol{b}}$ is the binormal and *not* the unit vector along $\boldsymbol{B}$, as is sometimes used in the literature. This implies that the component along $\boldsymbol{B}$ (or $\hat{\boldsymbol{t}}$) in the expression for $\boldsymbol{f}$ vanishes:

$$\boldsymbol{f} = 0\hat{\boldsymbol{t}} + \left[\frac{B^2}{\mu_0}\kappa - \frac{\partial}{\partial n}\left(\frac{B^2}{2\mu_0}\right)\right]\hat{\boldsymbol{n}} - \frac{\partial}{\partial b}\left(\frac{B^2}{2\mu_0}\right)\hat{\boldsymbol{b}} \ . \tag{4.4.5}$$

We can interpret this result in the following way. In the plane perpendicular to the field line (i.e., the normal plane which contains the vectors $\hat{\boldsymbol{n}}$ and $\hat{\boldsymbol{b}}$), the force density can be thought of as a superposition of a *pressure*, as if the magnetic field were a fluid with the "magnetic pressure" $B^2/(2\mu_0)$, and a *tension* as if the field lines were elastic cords each under a tension B/μ_0, pressing against

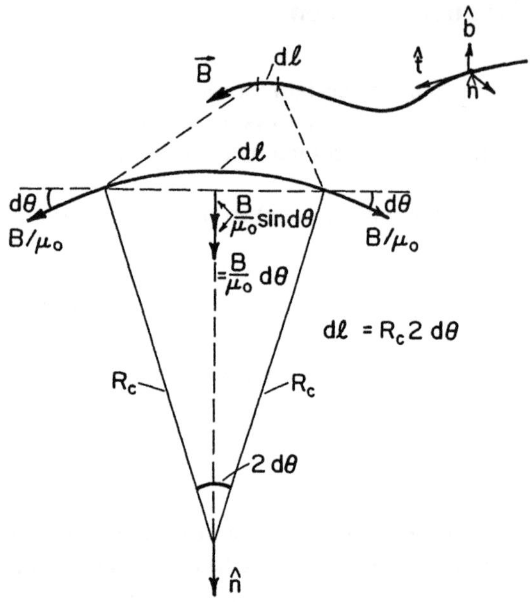

Fig. 4.3. Computation of the tension on a field line. See text

the fluid (Schmidt 1979, page 89). The part of the force density representing pressure is the negative gradient of the magnetic pressure:

$$-(\hat{\boldsymbol{n}}\partial/\partial n + \hat{\boldsymbol{b}}\partial/\partial b)\,(B^2/(2\mu_0)).$$

The tension needs more explanation. In this context, "tension" must be interpreted as a *force* pulling tangentially on a single field line. (In elasticity, by contrast, "tension" stands for pulling-*force-per-unit area* (Symon 1971).) The force density representing this "tension" is $B^2\kappa/\mu_0 = B^2/(\mu_0 R_c)$. Conventionally, the number of field lines per unit area is taken to be equal to the local value of $|\boldsymbol{B}| = B$ so that the area associated with a single field line is $1/B$. Therefore, the force per unit length (of one field line) in the $\hat{\boldsymbol{n}}$ direction equals $(B^2/(\mu_0 R_c))(1/B) = B/(\mu_0 R_c)$. How this force per unit length is related to a tension of B/μ_0 is made plausible in Fig. 4.3. The force in the $\hat{\boldsymbol{n}}$ direction due to an elemental length dl equals $2Bd\theta/\mu_0$ (for $\sin d\theta \simeq d\theta$). With $dl = R_c 2d\theta$, this is then $B/(\mu_0 R_c)dl$, leading to the above force per unit length (due to one field line) $B/(\mu_0 R_c)$.

4.5 Magnetic Surfaces

Any surface that is traced out by a series of magnetic-field lines is called a *magnetic surface*. In practice, however, the name is reserved for "closed" surfaces of a cylindrical or toroidal type. In that context, a magnetic surface is also called a *flux surface*.

a) Toroidal Systems

In a toroidal system, a magnetic surface (or flux surface) is a surface which is ergodically covered by non-closing magnetic-field lines. The structure of magnetic fields and surfaces in various devices has been discussed instructively by Morozov and Solov'ev (1966). The *existence* of magnetic surfaces has been investigated in the above review paper and by many others. We shall say a few words on the existence question in Chap. 9, where a literature survey is included.

The equation of a field line can be cast in a Hamiltonian form. Hamiltonian theory guarantees the existence of magnetic surfaces in devices with a symmetry direction: an ideal, axisymmetric tokamak; a helically symmetric straight stellarator; etc. In non-symmetric devices, on the other hand, magnetic surfaces do not exist rigorously everywhere because ergodic regions of finite volume exist (see Chap. 9). In practice, we can get away with considering approximate magnetic surfaces provided the perturbations away from symmetry are not too large. In principle, one should either measure the surfaces experimentally or follow field lines on a computer to demonstrate the existence of suitable surfaces. Note that in a tokamak, a toroidal (or parallel) plasma current is necessary for the existence of magnetic surfaces; in a stellarator, no such current is required, and it is meaningful to consider vacuum magnetic surfaces.

Between the magnetic surfaces traced out by non-closing field lines, we can define surfaces that contain field lines that close upon themselves after one or several transits around the machine (we have to "construct" or define these surfaces, since they are not traced out by a single field line). Such surfaces are called *rational surfaces*, because the ratio of the number of poloidal transits (the short way around the torus) to the number of toroidal transits (the long way around) of a field line is a rational number. The other surfaces are called *irrational surfaces*. Since the set of real numbers is dense, every irrational number can be approximated infinitesimally closely by a rational number and vice versa. The same holds for magnetic surfaces. However, the rational numbers are denumerable whereas the irrationals are not; there are infinitely more irrationals than rationals. Nevertheless, the important point is that every rational surface can be approximated very accurately by a nearby irrational surface, and the other way around.

Suitable magnetic surfaces are usually taken to be a requirement for adequate plasma equilibrium as described by the MHD equilibrium equation $\boldsymbol{J} \times \boldsymbol{B} = \nabla p$. These magnetic surfaces are constant pressure surfaces, since $\boldsymbol{B} \cdot \nabla p = 0$, and the current-density lines lie on these surfaces because $\boldsymbol{J} \cdot \nabla p = 0$. A consequence of the simple force balance, $\boldsymbol{J} \times \boldsymbol{B} = \nabla p$, and the current-continuity equation, $\nabla \cdot \boldsymbol{J} = 0$, is the "current-closure condition" on rational surfaces: the closed loop integral $\oint dl/B$ must be constant on rational surfaces, no matter on which field line we start (we shall discuss this condition below). Violation of this condition leads to the formation of an island-chain configuration and possibly to ergodic behavior of field lines in a volume in the vicinity of the rational surfaces.

In axisymmetric devices, the loop integral is constant and no problems arise. In non-axisymmetric devices, however, $\oint dl/B$ is not constant on rational surfaces, and according to ideal MHD the rational surfaces are unstable and break up, leading to the formation of fiber-like structures with many magnetic axes (i.e., the creation of magnetic islands) (Solov'ev and Shafranov 1970).

It is questionable, however, whether the simple MHD equilibrium model based on $\boldsymbol{J} \times \boldsymbol{B} = \nabla p$ is not too crude an approximation. The model neglects potentially important effects such as resistivity, non-scalar pressure (including viscosity), finite flow velocities (inertial forces) and finite Larmor radius effects. In reality, one must solve a time evolution problem; the system is likely to evolve to a self-consistent stationary state that probably has some magnetic surfaces, but not those predicted by ideal MHD. It can be expected that some of the requirements of the idealized model will be washed out by dissipative and other effects.

From a pragmatic point of view, *we assume that (even in non-axisymmetric systems) a suitable set of nested toroidal magnetic surfaces is present, whose cross-section in a poloidal plane forms a set of smooth closed curves.* This "imaginary" structure is actually the zero-th order approximation to the real structure. The magnetic surfaces in ideal tokamaks are symmetric about the major axis; the toroidal direction is ignorable. In stellarators, the surfaces have no symmetry direction and can take many forms. As an example, consider a surface of an $l = 2$ stellarator (Fig. 4.4). The toroidal magnetic surface is produced approximately by making an elliptic curve rotate in the poloidal direction, while moving toroidally along the minor axis.

The degenerate magnetic surface with the limiting value of zero volume is called the *magnetic axis*. It is formed by a field line that we assume closes upon itself after one toroidal transit. It is a line around which the magnetic surfaces are nested. A magnetic axis does not necessarily have to be circular, but can have, for example, a helical form ($l = 1$ stellarator, helical-axis stellarator). (The scope of our treatment is limited to situations with only one magnetic axis; thus, we disregard magnetic islands.)

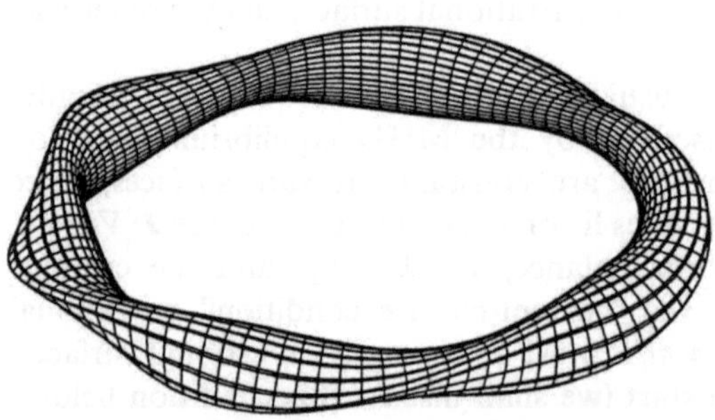

Fig. 4.4. Three-dimensional picture of a flux surface of an $l = 2$ stellarator. The elliptical cross section rotates poloidally when traversing toroidally. (Provided by H. Wobig and J. Kisslinger, Max-Planck-Institut für Plasmaphysik, Garching bei München, Fed. Rep. Germany.)

b) Open-Ended Systems

Magnetic-mirror devices do not possess magnetic surfaces that are traced out by one single field line. This leads, in principle, to a considerable degree of arbitrariness. One must *define* the shape of the surfaces, and a natural choice seems to be that of a circular cross-section in the center of the device, i.e., at $z = 0$ if the device is bounded axially by $z = \pm L$. For a simple baseball mirror, this circle fans out into an ellipse at one end, at $z = +L$, and into another ellipse rotated by 90-degrees at $z = -L$. In tandem mirrors with thermal barriers, the magnetic-surface structure is essentially a connection of a regular cylinder and baseball-coil generated surfaces.

4.6 Curvilinear Coordinate Systems in Confinement Systems with "Simple" Magnetic Surfaces

In this section we undertake a qualitative exploration of the kind of curvilinear coordinate systems suitable for the description of magnetically confined plasmas. We start the section with a brief reminder of the terminology in general curvilinear systems. Then we attack successively coordinates suitable for toroidal confinement systems on the one hand and for open-ended devices on the other. In this first category, we begin with a discussion of the cylindrical-toroidal hybrid coordinate system commonly used in elementary considerations, which we shall call the "elementary" toroidal system. It is based on "ideal" or "perfect" tori, by which are meant tori characterized by *circles* upon intersection with vertical and horizontal planes. Then we describe a generalization of that hybrid coordinate system. The main coordinate surfaces will now be the toroidal magnetic surfaces, "toroids" (which are topologically equivalent to regular tori). The usual concept of angle is generalized as well in order to find a coordinate system in which the magnetic-field lines appear as straight lines. To clarify the issue, we present a simple example next. To end the exposition on toroidal systems, we treat the (ρ, β, l) system, which has the arc length along a magnetic-field line as its third coordinate. This system is later identified as a particular type of Clebsch system. The disadvantages of its use in toroidal devices are stressed. The section is concluded with a note that Clebsch-type systems are most appropriate for open-ended systems.

A coordinate system is a tool to locate a point uniquely. In three-dimensional space, it consists of three suitable families of surfaces such that a point is defined by the intersection of three surfaces, one for each family. The *coordinate curves* are the intersection lines of two *coordinate surfaces.* We label the three coordinates by u^1, u^2 and u^3; the coordinate surfaces are labeled by u^i = constant while we designate a coordinate curve by its running variable u^j (where it is understood that this curve lies on the intersection of u^i = constant and u^k = constant, where $i \neq k, i \neq j$). A point $P(c^1, c^2, c^3)$ is completely determined by the

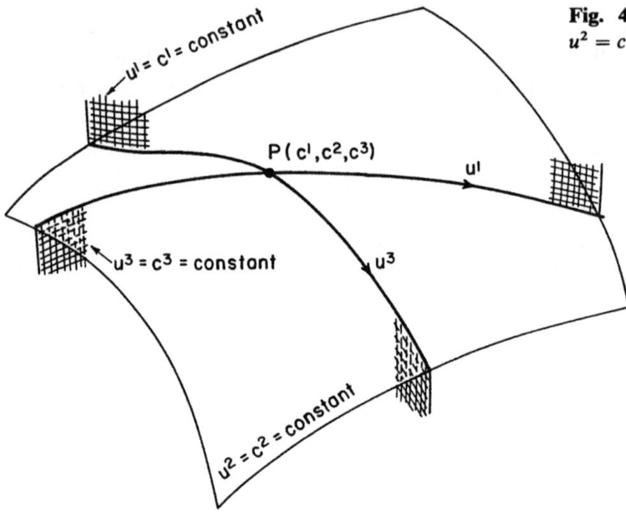

Fig. 4.5. Position location on a $u^2 = c^2 =$ constant surface

intersection of the three coordinate surfaces, $u^1 = c^1$, $u^2 = c^2$, $u^3 = c^3$ or equivalently by the intersection of two coordinate curves lying in the appropriate coordinate surface. For example, Fig. 4.5 shows that P is at the crossing point of the u^1 curve ($u^2 = c^2$ and $u^3 = c^3$) and the u^3 curve ($u^1 = c^1$ and $u^2 = c^2$), where evidently both curves lie in the $u^2 = c^2$ surface.

We shall use only *right-handed* coordinate systems. This implies that the triple product of the basis vectors $\boldsymbol{e}_i \cdot \boldsymbol{e}_j \times \boldsymbol{e}_k$ or $\boldsymbol{e}^i \cdot \boldsymbol{e}^j \times \boldsymbol{e}^k$ is positive as long as i, j and k are cyclic permutations of 1, 2, 3. It is clear that the *order* in which we take coordinates as well as their orientation is important: there is a first, a second and a third coordinate. (Of course, the order can be changed at will, but often then the orientation of a coordinate must be changed.)

4.6.1 Toroidal Systems

In toroidal plasma confinement the coordinate systems are based on toroidal topology. "Proper" toroidal coordinates, being the three-dimensional version of bipolar coordinates, have been used only occasionally in fusion-plasma equilibrium calculations; see, e.g., Shafranov (1960) and Fielding and Hitchon (1980). These coordinates are discussed briefly in Chap. 10.

a) The "Cylindrical-Toroidal" or "Elementary" Toroidal System

The best known coordinate system is the "pseudo-toroidal" or "hybrid cylindrical-toroidal" or the "elementary" toroidal coordinate system. This coordinate system is essentially a "toroidalized" cylindrical system with coordinates

$$(R, \zeta_c, z),$$

as shown in Fig. 4.6. The cylindrical angle is measured from the x-axis towards

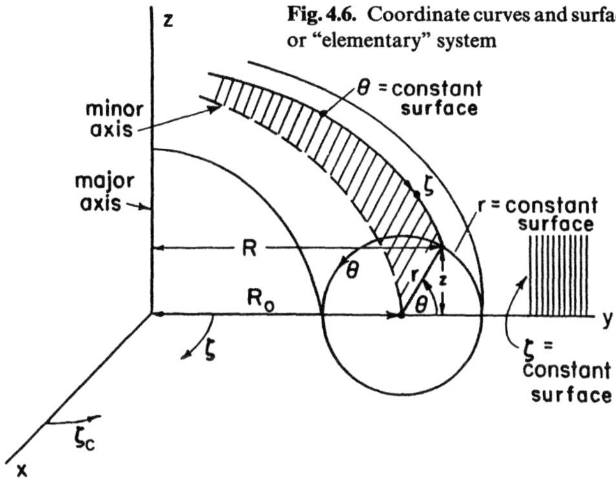

Fig. 4.6. Coordinate curves and surfaces of the "cylindrical-toroidal" or "elementary" system

the y-axis. The "toroidalization" is done by "replacing" R and z by r and θ. Because conventionally, θ and ζ are taken as the second and third coordinates, respectively, whereas ζ_c is the second in the cylindrical system, the angle ζ is measured in the positive sense from the y-axis towards the x-axis (i.e., $\zeta = \pi/2 - \zeta_c$). Only then has the trihedron of the unit vectors $\hat{r}$, $\hat{\theta}$, $\hat{\zeta}$ a right-handed orientation. Thus the coordinates are $u^1 = r$, $u^2 = \theta$, $u^3 = \zeta$.

The coordinate surfaces r = constant are nested perfect *tori* with *major radius* R_o. The circle with radius R_o is called the *minor axis*, while the axis in the center of the doughnut hole is the *major axis* (in Fig. 4.6, this is the z-axis). r is called the *minor radius* and measures the distance from the minor axis. The coordinate surfaces θ = constant touch the minor axis and are partial *cones*. The third family of coordinate surfaces, ζ = constant, are planes touching the major axis. An r-coordinate curve is formed by the intersection of a θ = constant "cone" and a ζ = constant plane. It is a *straight line*, with origin on the minor axis. The θ-coordinate curve is a *circle* obtained from the intersection of a torus r = constant and a plane ζ = constant. The third coordinate curve ζ, is a *circle* traced out by the intersection of the torus r = constant and the "cone" θ = constant. When moving around the minor axis (the short way around), we move in the *poloidal* direction. The θ-coordinate curve encircles the minor axis, hence the designation that θ is the poloidal angle. The plane in which the poloidal-angle coordinate curve lies is called the poloidal plane (ζ = constant is a poloidal plane). Analogously, going around the major axis corresponds to moving in the *toroidal* direction. ζ is the toroidal angle. The relationship between x, y, z and r, θ, ζ is as follows:

$$x = (R_o + r\cos\theta)\sin\zeta$$

$$y = (R_o + r\cos\theta)\cos\zeta$$

$$z = r\sin\theta \ . \qquad (4.6.1)$$

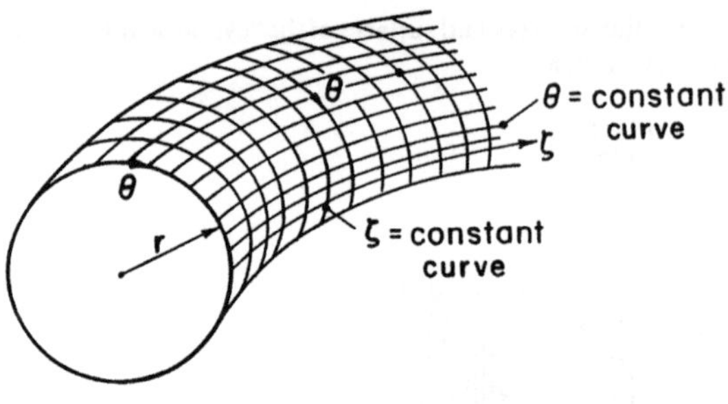

Fig. 4.7. Position location on a toroidal surface via a θ and ζ grid

The usual procedure for specifying the position of a point in space is to first choose a coordinate surface, r = constant, and then to lay out a grid of θ and ζ curves on this torus. The intersection of two curves, one from each grid, fixes the position of the point in question (Fig. 4.7). Note that the curves of the grid satisfy the properties that any two members of the same family never intersect each other, and that every curve of one family intersects every curve of the other family at precisely one point. This observation here, of course, is plausible, since the two sorts of coordinate curves are circles intersecting each other at right angles. The angles θ and ζ are multiple-valued functions with regard to the position on the torus. θ and ζ increase by 2π after encircling the minor axis and the major axis, respectively. Every point on the toroidal surface is thus labeled by $\theta + 2\pi m$ and $\zeta + 2\pi n$ for all integers m and n. Below, we shall denote the poloidal and toroidal angles in the "elementary" toroidal system by θ_e and ζ_e.

b) Generalized Cylindrical-Toroidal Coordinates; Flux Coordinates

In theoretical fusion-plasma physics, it is customary to choose as the first family of coordinate surfaces the *magnetic surfaces*, regardless of their shape. We label these magnetic surfaces by ϱ = constant, thus $u^1 \equiv \varrho$. Below, we shall identify this general label ϱ with more specific measures such as the magnetic flux enclosed by the surface, its volume or the pressure.

The coordinate surfaces for the angles θ and ζ are generalized as well. The u^2 = constant surfaces are now based on the *magnetic axis* and, in general, are partial-cone like surfaces. (Note that the magnetic axis does not normally coincide with the geometrical minor axis and can be non-circular or even non-planar.) These surfaces can be thought of as being produced by bending, stretching or squeezing the partial cones of the "elementary" toroidal coordinate system (of Fig. 4.6), provided that each surface crosses the toroidal magnetic surface only once. Figure 4.8 shows an example of the cross-section in the yz Cartesian plane of a simple tokamak with non-concentric circular toroidal magnetic surfaces, where the u^2 = constant surfaces are chosen such that they intersect the tori at right angles. Because the coordinate surfaces u^2 = constant have topologically the same structure as the θ_e = constant cones in the "elementary" toroidal system, it is convenient to denote u^2 by the symbol θ. It is clear that θ is a poloidal

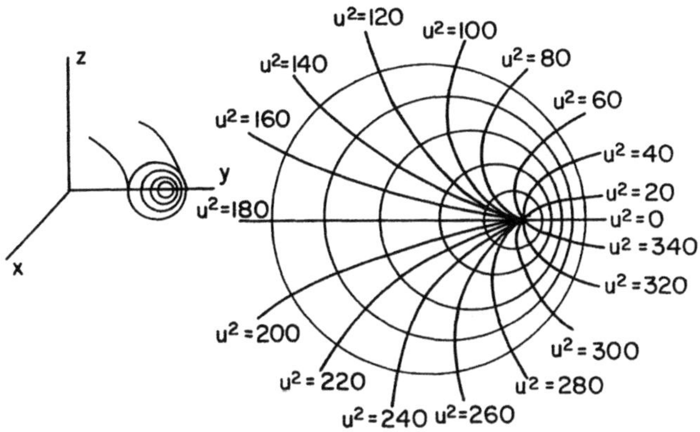

Fig. 4.8. u^2 = constant surfaces which are perpendicular to the Shafranov-shifted flux-surfaces in a simple tokamak

angle-like variable or a generalized poloidal angle. In short, we shall call θ the "poloidal" angle, although it is understood that it is a generalization of the usual concept of "angle". Likewise, the u^3 = constant coordinate surfaces are, in general, a deformed version of the ζ_e = constant planes. The u^3 = constant surfaces touch the major axis and fan out to cut through the magnetic surface. Because of the analogy with the variable ζ_e, we denote u^3 by ζ and call this angle-like variable the "toroidal" angle. We also say that a surface ζ = constant makes a "poloidal" cross-section of the torus. From now on, the words, "toroidal" and "poloidal" are related to the curvilinear coordinates θ and ζ, which do not necessarily coincide with the "elementary" angles θ_e and ζ_e. The context should make it clear what is meant.

Coordinate curves for θ and ζ are constructed as usual, via intersections of coordinate surfaces. With the coordinate surfaces constructed as explained above, we obtain *closed curves* for the coordinate curves as shown in Fig. 4.9. We require that the functions $\theta(x, y, z)$ and $\zeta(x, y, z)$, which are transformations from x, y, z to θ and ζ, have continuous first derivatives, implying that the tangents to the curves and the gradients to the surfaces are continuous vector fields. $\theta(x, y, z)$ and $\zeta(x, y, z)$ are multiple-valued functions since they change by 2π after each complete circuit. Each point $P(x, y, z)$ can have θ coordinates that can differ by multiples of 2π and similarly for ζ. θ and ζ must be single valued, however, on any other path on the surface which does not encircle the major or minor axes. On every magnetic surface, the θ and ζ coordinate curves lay out a grid as shown in Fig. 4.10. It should be noted that it is sufficient for *place* positioning to isolate a ϱ = constant surface and lay out a grid of θ and ζ curves. However, this is not enough for a proper description of *vectors*. With the grid of contours, the covariant-(tangent-) basis vectors are determined at every point, but the contravariant-(reciprocal-) basis vectors ∇u^i can only be found if the u^2 = constant and u^3 = constant *surfaces* are known.

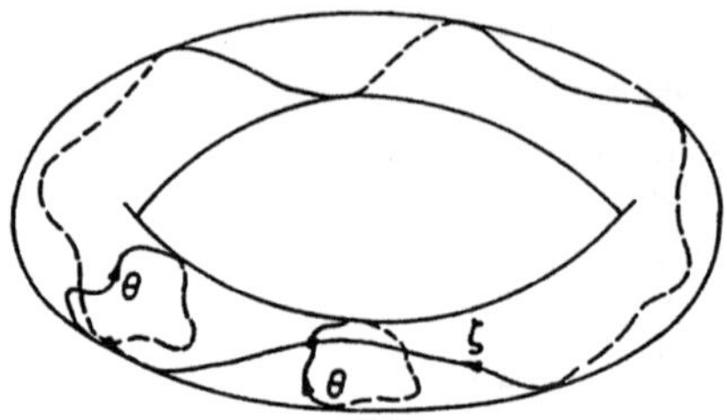

Fig. 4.9. Coordinate curves for the generalized poloidal and toroidal "angles"

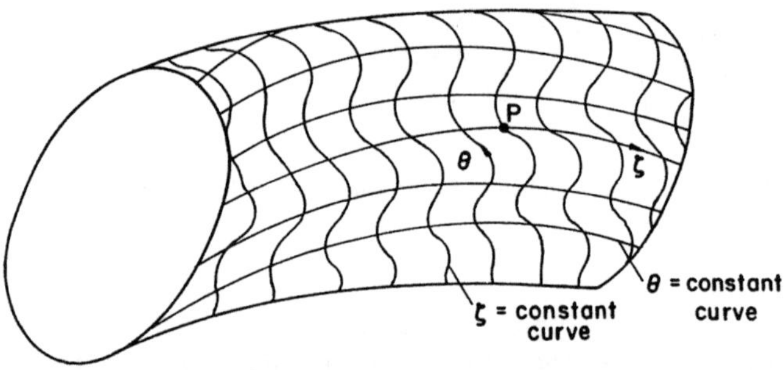

Fig. 4.10. Position location on a toroidal surface based on a grid constructed for generalized "angles" θ and ζ

We can always construct a suitable set of "new" angle functions (or variables) $\tilde{\theta} \equiv \tilde{\theta}(\boldsymbol{R})$ and $\tilde{\zeta} \equiv \tilde{\zeta}(\boldsymbol{R})$ by adding functions which are periodic with a period of 2π to the original θ and ζ:

$$\tilde{\theta} = \theta + f(\varrho, \theta, \zeta) \tag{4.6.2a}$$

$$\tilde{\zeta} = \zeta + g(\varrho, \theta, \zeta) \ . \tag{4.6.2b}$$

The periodicity of f and g in both θ and ζ is equivalent to stating that f and g are single-valued functions of θ and ζ. Because of this single valuedness, the (θ, ζ) coordinate curves are deformed in a such a way that the new coordinates $(\tilde{\theta}, \tilde{\zeta})$ have proper curves as well, making them acceptable coordinates.

The reason for working in these coordinate systems is straightforward. To perform analytic calculations, it is convenient to choose a "natural" coordinate system, such that in that coordinate system the geometry of the problem under consideration becomes simple. Then the complex geometrical features that exist in a "neutral" (e.g., Cartesian) coordinate frame are taken care of in the transformation equations between the coordinate systems. In theoretical plasma-physics computations, it is very helpful to choose the $\theta(x, y, z)$ and $\zeta(x, y, z)$ functions so that the magnetic-field lines (which oscillate in the "elementary" toroidal coordinate system) appear as *straight lines*. This means that, along a field line, $d\theta/d\zeta$ is a constant (more precisely, is a flux function), for that choice of θ and ζ functions. A particular choice for the θ, ζ coordinates makes both the $\boldsymbol{B}$-field lines and the $\boldsymbol{J}$ current-density lines straight. It is customary to call coordinates in which the magnetic field is straight *magnetic coordinates* or *flux coordinates*. When we deal with angles, these coordinates will be labeled with an index f, θ_f and ζ_f. (The flux-surface label ϱ is exempted from this index, since we want to make field lines that lie in $u^1 = \varrho =$ constant surfaces straight in the remaining u^2 and u^3 coordinates.)

To summarize, given that the magnetic field $\boldsymbol{B}$ is known everywhere, and consequently, so is $\boldsymbol{J}$, we try to find suitable poloidal and toroidal coordinates that make $\boldsymbol{B}$ and/or $\boldsymbol{J}$ straight in (θ_f, ζ_f). One can start with any acceptable grid of θ and ζ curves as a function of which the magnetic field is known (for example, the "elementary" angles, θ_e and ζ_e), and deform the coordinate system until it possesses the desired properties.

c) An Illustrative Example

In Fig. 4.11a, we show the constant flux-surface contours of a circular tokamak, where the surfaces are non-concentrically nested. Also shown are the cross-sections of constant θ_e surfaces. The coordinate system used here is very similar to the "elementary" toroidal system, except that it is based on the magnetic axis, rather than the minor axis of the torus. Because of this close relationship we designate the angle by θ_e.

Consider one of the larger surfaces (shown dashed in Fig. 4.11a) and its (θ_e, ζ_e) *grid* (shown in Fig. 4.11b). The magnetic field is assumed to traverse this surface with constant pitch everywhere. Let us assume that the rotational transform $\iota\!\!\!-$ equals 3 on this particular surface ($\iota\!\!\!-$ will be defined rigorously below). This means that the magnetic field makes 3 poloidal transits for every toroidal transit. If we mark the points where the $\boldsymbol{B}$ line crosses the $\theta_e =$ constant curves in Fig. 4.11b and transfer them to a (θ_e, ζ_e) *coordinate system*, we observe an oscillatory behavior of the field line as seen in Fig. 4.11c. To make the $\boldsymbol{B}$ lines straight in a new (θ_f, ζ_f) coordinate system, it is obvious that the arc lengths on the constant flux-surface contours cut off by the $\theta_f =$ constant contours must be equal. This can be accomplished by constructing poloidal angle surfaces as shown in Fig. 4.11d. (This procedure is equivalent to isolating the surface under consideration and constructing angles as "usual" based on the center of the circle, not the magnetic axis. Still another method is to isolate the surface and disregard the structure of the coordinate surfaces, but simply construct an equidistant (θ_f, ζ_f) grid on the surface. This secondary procedure does not allow the determination of the gradient $\nabla\theta_f$, however.) The net result of all this is that the $\boldsymbol{B}$ lines are straight in (θ_f, ζ_f), but that our new θ_f has an oscillatory behavior as a function of the old θ_e (Fig. 4.11e).

d) (ϱ, β, l) Coordinates in Toroidal Systems

An alternative (but complicated) set of flux coordinates in toroidal devices is the so-called (ϱ, β, l) coordinate system. Again, ϱ is a magnetic-surface label, β labels

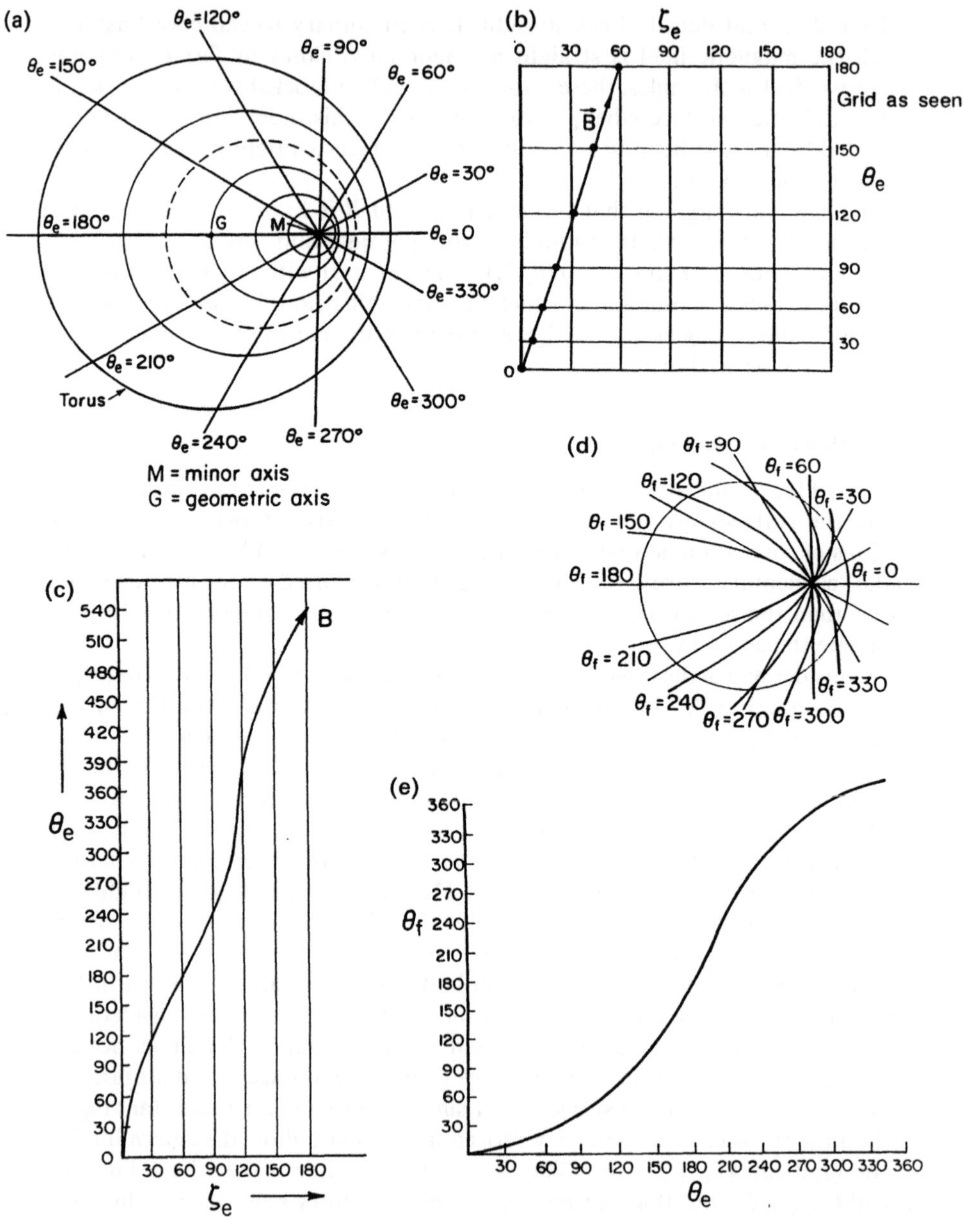

Fig. 4.11. (**a**) The usual poloidal angle surfaces (planes) θ_e = constant, fanning out from the magnetic axis of a Shafranov-shifted tokamak plasma. (**b**) The (θ_e, ζ_e) grid *as seen* on a folded-open toroidal surface. The magnetic-field line is seen as a straight line. (**c**) A magnetic-field line "wiggles" in a (ζ_e, θ_e) coordinate system. (**d**) The constant coordinate surfaces of the flux-coordinate angle θ_f cut each toroidal surface equidistantly. (**e**) The flux-coordinate angle θ_f, in which the ***B***-lines appear as straight lines, wiggles with respect to the old (usual) coordinate θ_e

the field lines in a surface, and l is the length along a field line. A β = constant surface is a "ribbon" starting at the magnetic axis and passing through the trajectory of a field line on a given surface, that twists around with the same pitch as the field line on that surface (see Fig. 4.12). The issue of l = constant surfaces is not straightforward. First, one could choose an arbitrary origin, $l = 0$ for all field lines (this could be a common poloidal-plane cut), and then constant l surfaces must be constructed. The length l and thus these l surfaces are influenced by the pitch of the field lines. If the field lines twist around more on the outer surfaces, the l surfaces will look like paraboloids. These will not be regular paraboloids, in general, however, since the rate of twisting of a field line is, in general, not constant on a surface. An important consequence of such l = constant surface structure is that in most cases, ∇l *is not tangential* (*i.e., not parallel*) to $\boldsymbol{B}$.

On the other hand, an irrational surface can be thought of as being constructed from a single field line. Thus, on that flux surface, there would be only one point which we could assign a value $l = 0$. So what then are the l = constant surfaces?

Similarly, what are the β = constant surfaces given this last point of view? Since the magnetic lines lie on β = constant ribbons, and one single field line covers the whole magnetic surface, it might appear that β is zero everywhere on the surface. Although in principle the entire surface can be parameterized by l-values, with $\beta = 0$, this is not very useful in practice. Rational surfaces also present problems. If a field line closes after, say, n toroidal transits, there are n (probably overlapping) β ranges ($\beta = 0 \rightarrow \beta = \beta_{\text{final}}$) with jumps between them.

A possible way to make sense out of this coordinate system would be to introduce cuts in the torus and limit l to one toroidal transit (which is effectively what we did when we constructed the l = constant "paraboloids"). A complication arises, however, because of the multiple valuedness of both β and l in that

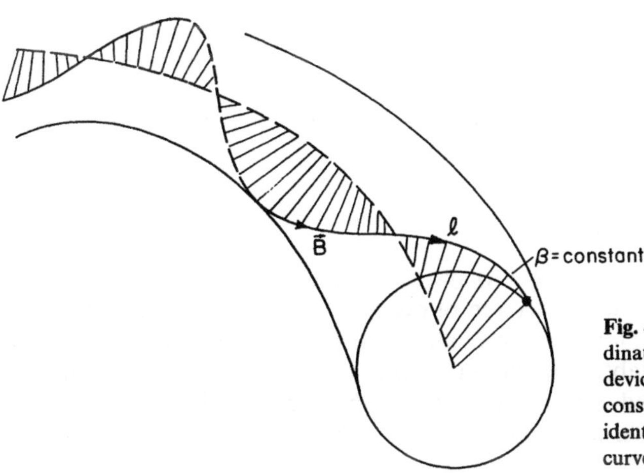

Fig. 4.12. A β = constant coordinate surface in a toroidal device. The intersection of β = constant and ϱ = constant is identified with the l-coordinate curve

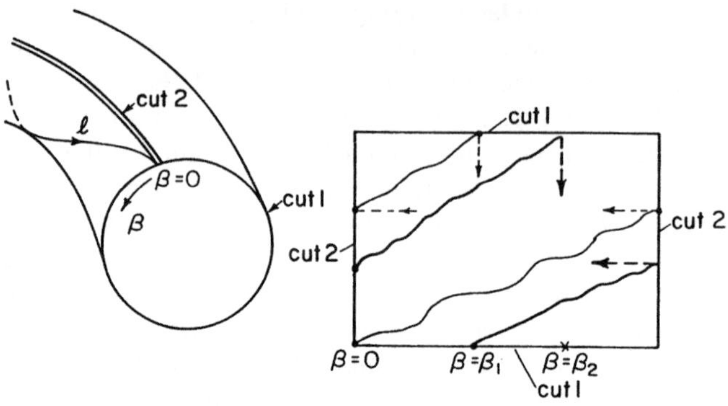

Fig. 4.13. Field-line labeling by means of the β Clebsch coordinate. The figure on the right is the topologically equivalent folded-open surface of the toroidal surface on the left. See text

picture. One considers every field line in this cut as an independent entity (independent of the fact that all irrational lines are generated by one single field line); β still labels the field lines. In Fig. 4.13, we show the cuts and the topologically equivalent folded-open surface. The opposite sides of the rectangle are topologically identical. If we pick a field line at the crossing of the cuts and label it by $\beta = 0$, it will arrive at a point where $\beta = \beta_1$. However, β is still 0 because β is constant along the field line. For single-valuedness, we must thus require that $\beta(l = L, \varrho) = \beta(l = 0, \varrho)$. But if we start with a field line labeled by $\beta = \beta_1$, we will arrive at $\beta = \beta_2$, which must be identified with β_1, which, in turn, was equal to 0. This means that every point on the flux surface can be labeled by *every* value of β, ranging from 0 to 2π, if we take $\beta = 2\pi$ the last label at the other end of the cut. These problems arise because the l coordinate curve does not close upon itself after one toroidal or poloidal transit, in contrast to the poloidal and toroidal angles θ and ζ, whose coordinate curves do close after their respective transits.

The (ϱ, β, l) coordinates, therefore, may be useful for formal reasoning, but lead to severe problems in practice. The $\boldsymbol{B} \cdot \boldsymbol{\nabla}$ operator reduces to $B(\partial/\partial l)$ in this coordinate system, which may be useful in calculations involving local properties of the surface.

4.6.2 Open-Ended Systems

Unlike toroidal topology, the (ϱ, β, l) system is very well suited to open-ended systems. It is precisely the connection of one end of the device onto the other, which caused problems for the single valuedness of β and l in toroidal systems, that is avoided here. ϱ labels the chosen "cylindrical" flux surfaces, β is a suitable angle-like variable, and l measures the length of a field line as before. A (ϱ, β, l) system is known as a Clebsch system. It will be discussed in detail below.

4.7 Magnetic-Surface Labeling

Before we can develop flux-coordinate systems in detail, we must find adequate measures that represent magnetic surfaces. This section provides these quantities. We shall again differentiate between toroidal and open-ended systems. For toroidal systems, we first consider the magnetic fluxes enclosed by and outside of the magnetic surfaces. We distinguish between toroidal magnetic flux, poloidal ribbon flux and poloidal disk flux. Then other magnetic-surface labels such as volume and pressure are mentioned. Finally, the magnetic axis is defined. For open-ended systems, axial and azimuthal magnetic fluxes and scalar pressure are considered.

4.7.1 Toroidal Systems

A convenient way to label the magnetic surfaces is by means of the magnetic flux. Because the magnetic-field lines are by definition everywhere tangent to a magnetic surface, the flux inside or outside a magnetic surface is conserved. For this reason, we can also call magnetic surfaces constant flux surfaces or, in short, *flux surfaces*.

There are several magnetic fluxes which label a flux surface in a toroidal device uniquely. Assume that an arbitrary $(\theta(x, y, z), \zeta(x, y, z))$ coordinate system has been established. First, there is the *toroidal flux* (Fig. 4.14a):

$$\Psi_{\mathrm{tor}} \equiv \iint_{S_{\mathrm{tor}}} \boldsymbol{B} \cdot d\boldsymbol{S} \ . \tag{4.7.1}$$

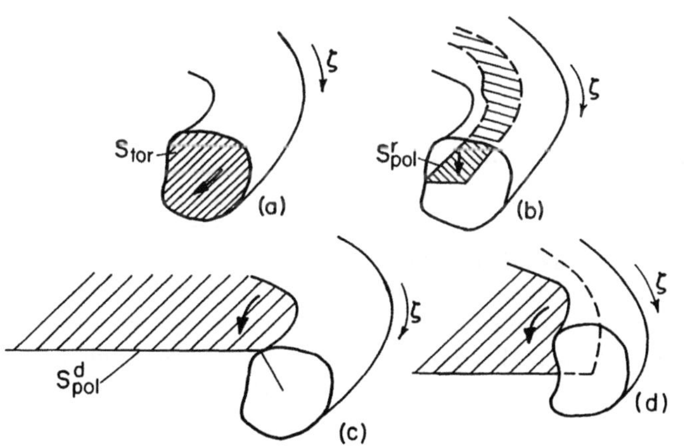

Fig. 4.14a-d. Cross sections through which the magnetic fluxes characterizing a magnetic-flux surface are defined. **(a)** The field lines intersecting S_{tor} determine the toroidal flux within a flux surface. **(b)** $S^{\mathrm{r}}_{\mathrm{pol}}$ is a ribbon bounded by the magnetic axis and the flux surface. It "contains" the poloidal (ribbon) flux which resides inside the flux surface, $\Psi^{\mathrm{r}}_{\mathrm{pol}}$. **(c)** $S^{\mathrm{d}}_{\mathrm{pol}}$ is a disk touching the magnetic surface. It picks up the poloidal (disk) flux outside the flux surface, $\Psi^{\mathrm{d}}_{\mathrm{pol}}$. **(d)** A special case of a disk is shown. It touches the magnetic surface where $\theta = \pi$, i.e., it lies in the equatorial plane

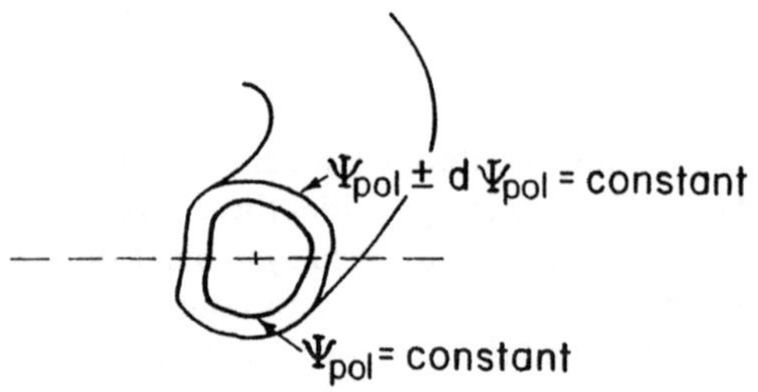

Fig. 4.15. The flux-surface volume between two infinitesimally close flux surfaces. + sign for Ψ^{r}_{pol}, − sign for Ψ^{d}_{pol}

S_{tor} is the surface produced by cutting the flux surface with a poloidal surface ζ = constant. Next, the *poloidal fluxes* inside and outside a flux surface may be used. The poloidal flux *inside* the flux surface is taken, e.g., through a ribbon at constant θ between the magnetic axis and the surface, and is given by (Fig. 4.14b):

$$\Psi^{r}_{pol} \equiv \iint_{S^{r}_{pol}} \boldsymbol{B} \cdot d\boldsymbol{S} \,. \tag{4.7.2}$$

The superscript r refers to the *r*ibbon. S^{r}_{pol} is the part of a θ = constant surface that lies within the flux surface under consideration. The flux *outside* the flux surface is obtained by integrating $\boldsymbol{B}$ over a disk-like surface that is tangent to the flux surface everywhere (Fig. 4.14c), and is bounded by a ζ coordinate curve on the flux surface:

$$\Psi^{d}_{pol} \equiv \iint_{S^{d}_{pol}} \boldsymbol{B} \cdot d\boldsymbol{S} \tag{4.7.3}$$

The poloidal-disk flux includes the tokamak transformer flux, and no reference to the magnetic axis is required. The flux obtained from (4.7.3) is fixed as long as the disk is tangent to the flux surface.

To compare the two poloidal fluxes, it is convenient to choose a disk that is part of the approximately circular surface bounded by the magnetic axis (see Fig. 4.14d). Comparing Figs. 4.14b, d, it is clear that the fluxes are complementary with respect to the total flux in the system. The flux from the magnetic axis out to infinity has the same value as the flux through $S^{r}_{pol} + S^{d}_{pol}$. This remains true even if there is an additional vertical field component to keep the magnetic axis (the point where the net vertical field component is zero) from expanding. In every steady-state device which has fixed magnetic surfaces, such as a stellarator, the spatial changes in poloidal flux satisfy the relationship (see Figs. 4.14 and 4.15):

$$d\Psi^{d}_{pol} = -d\Psi^{r}_{pol} \,. \tag{4.7.4}$$

$d\Psi_{pol}$ is the differential flux between the two flux surfaces Ψ_{pol} = constant and $\Psi_{pol} \pm d\Psi_{pol}$ = constant. Because

$$d\Psi_{pol} \equiv \boldsymbol{\nabla}\Psi_{pol} \cdot d\boldsymbol{R} \tag{4.7.5}$$

for every arbitrary differential vector $\boldsymbol{dR}$, (4.7.4) is equivalent to:

$$\nabla \Psi^{\mathrm{d}}_{\mathrm{pol}} = -\nabla \Psi^{\mathrm{r}}_{\mathrm{pol}} \ . \tag{4.7.6}$$

Of course, both sides of (4.7.5) and (4.7.6) must be evaluated at the same position.

In tokamaks, the situation is more complicated because the equilibrium is dynamic. Suppose we want to label the flux surfaces by means of the poloidal-disk flux, $\Psi^{\mathrm{d}}_{\mathrm{pol}} =$ constant. In Chap. 11, we show that the increasing transformer flux forces the flux surfaces, labeled by $\Psi^{\mathrm{d}}_{\mathrm{pol}} =$ constant, to shrink with a velocity $\boldsymbol{v}$ such that $\boldsymbol{v} \cdot \nabla \Psi^{\mathrm{d}}_{\mathrm{pol}} = -\partial \Psi^{\mathrm{d}}_{\mathrm{pol}}/\partial t$. This velocity turns out to be equal to the inward pinch $\boldsymbol{E} \times \boldsymbol{B}$ velocity observed in a *fully-ionized ideal*-MHD plasma ($\sigma \to \infty$); since the infinite conductivity requires $E_{\parallel}$ to be zero, the (induced) electric field has only a perpendicular (to $\boldsymbol{B}$) component $\boldsymbol{E}_{\perp}$. Because both the plasma and the flux surfaces $\Psi^{\mathrm{d}}_{\mathrm{pol}} =$ constant move with the same speed, this is in effect a statement of the *frozen-flux theorem* in "ideal" plasmas. From the standpoint of the poloidal ribbon flux, the same conclusion can be drawn: because of the frozen flux theorem, the surfaces labeled by $\Psi^{\mathrm{r}}_{\mathrm{pol}} =$ constant move at the same speed as the plasma. The conclusion is that *it does not matter which label we use for the poloidal flux.*

Since we are primarily concerned with spatial variations of the flux functions, we can ignore the time evolution because it does not change the magnetic topology relative to the plasma. This is similar to taking a photograph and assuming that it reflects the time independent equilibrium state. Once this time evolution is disregarded, we can claim that (4.7.4) and (4.7.6) are still valid. (We stress that it is the spatial part of $d\Psi_{\mathrm{pol}}$ in (4.7.4) that is of concern. We should say $\delta \Psi^{\mathrm{d}}_{\mathrm{pol}} = -\delta \Psi^{\mathrm{r}}_{\mathrm{pol}}$, where $\delta \Psi_{\mathrm{pol}} = \nabla \Psi_{\mathrm{pol}} \cdot \boldsymbol{dR}$, since, when the time t is retained as a variable, $d\Psi_{\mathrm{pol}} = \partial \Psi_{\mathrm{pol}}/\partial t \, dt + \nabla \Psi_{\mathrm{pol}} \cdot \boldsymbol{dR}$. However, with the time dependence suppressed, we have that $d\Psi_{\mathrm{pol}} = \delta \Psi_{\mathrm{pol}}$.)

In a non-ideal plasma, the story is somewhat different. The finite resistivity allows magnetic-field diffusion to take place, while it also gives rise to transport processes such as classical and neoclassical particle diffusion fluxes across the flux surfaces. This leads to a self-consistent evolution of the magnetic structure and the plasma. We shall not deal with the time evolution of flux surfaces in this monograph, but refer the reader to the literature. (Plasma-equilibrium evolution on the resistive time scale in the context of the closure of moment equations in transport theory is discussed by Hinton and Hazeltine (1976), Chap. VII, and by Hirshman and Sigmar (1981), Chap. IX. For more detailed information, the (tokamak) theory has been developed in Grad and Hogan (1970), Grad (1974), Grad, et al. (1975), Pao (1978), Pereversev, et al. (1978), Hirshman and Jardin (1979), Blum and Le Foll (1984) and Blum (1989) among others.)

To develop the flux-coordinate formalism, we need more specific expressions for the toroidal and poloidal fluxes. In particular, we wish to derive a form-invariant expression with respect to the choice of θ and ζ functions (see, e.g., Kruskal and Kulsrud 1958). Consider the integral

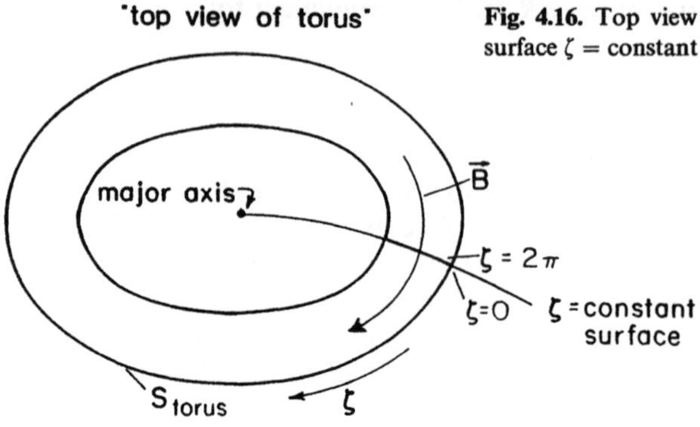

Fig. 4.16. Top view of a toroidal surface, cut by a surface ζ = constant

$$\iiint_V d^3R\, \boldsymbol{B}\cdot \nabla\zeta = \iiint_V d^3R\, \nabla\cdot(\boldsymbol{B}\zeta) \ . \tag{4.7.7}$$

Here, d^3R is an elementary volume element and V is the volume enclosed by the flux surface. The second form follows from $\nabla\cdot\boldsymbol{B} = 0$. In Fig. 4.16 we show a top view of the flux surface together with a ζ = constant (poloidal) surface. In particular, we choose the $\zeta = 0 + 2\pi k$ surface. This surface cuts the torus open, allowing us to consider it as a volume enclosed by the surfaces S_{torus}, $S_{\zeta=0}$ and $S_{\zeta=2\pi}$. If we now apply Gauss' theorem to the integral of (4.7.7) over the region of the cut-open torus, we obtain

$$\oint\!\!\oint_S \zeta\boldsymbol{B}\cdot d\boldsymbol{S} = \iint_{S_{\text{torus}}} \zeta\boldsymbol{B}\cdot d\boldsymbol{S} + \iint_{S_{\zeta=2\pi}} \zeta\boldsymbol{B}\cdot d\boldsymbol{S} + \iint_{S_{\zeta=0}} \zeta\boldsymbol{B}\cdot d\boldsymbol{S} \ . \tag{4.7.8}$$

The first integral on the RHS vanishes by definition: $\boldsymbol{B}$ is tangent everywhere to the magnetic surface, $d\boldsymbol{S}\perp\boldsymbol{B}$. ζ is constant on the surfaces $S_{\zeta=2\pi}$ and $S_{\zeta=0}$, meaning that ζ can be replaced by 2π and 0 in the second and third terms, respectively. Then (4.7.7) and (4.7.8) give

$$\iiint_V d^3R\, \boldsymbol{B}\cdot \nabla\zeta = 2\pi \iint_{S_{\zeta=2\pi}} \boldsymbol{B}\cdot d\boldsymbol{S} \ . \tag{4.7.9}$$

$S_{\zeta=2\pi}$ (or $S_{\zeta=0}$) is what we previously called S_{tor}. A combination of (4.7.1) and (4.7.9) leads to the form-invariant expression for Ψ_{tor}:

$$\Psi_{\text{tor}} = \frac{1}{2\pi}\iiint_V \boldsymbol{B}\cdot \nabla\zeta\, d^3R \ . \tag{4.7.10}$$

Analogously, the poloidal ribbon flux $\Psi^{\text{r}}_{\text{pol}}$ can be written as

$$\Psi^{\text{r}}_{\text{pol}} = \frac{1}{2\pi}\iiint_V \boldsymbol{B}\cdot \nabla\theta\, d^3R \ . \tag{4.7.11}$$

There exists a variety of flux-surface labels, other than the fluxes. Any function or variable that is uniform over a magnetic surface is referred to as a *flux function* or a *flux-surface quantity*. Such functions are independent of θ and ζ and depend only on the flux-surface label; hence, we can use these functions as surface labels. *Any* flux function can be used to label the flux surfaces. An obvious surface quantity is the *volume* enclosed by the magnetic surface:

$$V = \iiint d^3R = \iiint_{\text{Volume}} dx\,dy\,dz = \iiint \sqrt{g}\,d\varrho\,d\theta\,d\zeta \ . \tag{4.7.12}$$

Here, g is the determinant of the metric tensor, and $\sqrt{g}$ is the Jacobian, $\sqrt{g} = [\nabla\varrho\cdot(\nabla\theta \times \nabla\zeta)]^{-1}$.

Another surface quantity (used in the plasma literature) is the scalar pressure p. From $\nabla p = \boldsymbol{J} \times \boldsymbol{B}$ which implies that $\boldsymbol{B}\cdot\nabla p = 0$, it follows that $\boldsymbol{B}$ lines lie in constant pressure surfaces. Since $\boldsymbol{B}$ also lies in flux surfaces, it follows that $p \equiv p(\Psi)$ only, although this may break down at points where $dp/d\Psi = 0$.

Two of the flux labels that we have referred to have the disadvantage of decreasing as we go radially outward; the pressure and the poloidal disk flux. To keep a right-handed system (and a positive Jacobian), we would have to change the orientation of, say, the toroidal angle ζ. Furthermore, this implies that positive integration along coordinate curves would give rise to a flux-surface volume and inner fluxes that are zero at the plasma boundary and maximum (in absolute value) at the magnetic axis, in contrast to common conventions. Although it is possible to keep track of these adaptations, it is quite confusing and may overload expressions with absolute value bars and/or $(\pm)$ choices, to obtain the proper results for a general flux label ϱ. Therefore, we use only *radially increasing flux labels*. Then no angles need be changed and the formulae remain simple. It requires, however, that we use the *negative* of the flux functions that decrease radially. Thus, we can have $\varrho \equiv \Psi_{\text{tor}}$, $\varrho \equiv \Psi^{\text{r}}_{\text{pol}}$, $\varrho \equiv V$, $\varrho \equiv -p$, $\varrho \equiv -\Psi^{\text{d}}_{\text{pol}}$.

The label Ψ_{pol} is used most frequently in the tokamak literature, while Ψ_{tor} is more popular for stellarators. This is in part because the tokamak literature uses the safety factor q, whereas for stellarators the rotational transform, $\iota\!\!\!- = 1/q$, is used as a measure of the rate of twisting of a field line. The volume was chosen by Hamada (1962) when he developed his coordinate system in which both $\boldsymbol{B}$ and $\boldsymbol{J}$ lines are straight.

The defining equation of a flux surface with flux-surface label ϱ, is:

$$\boldsymbol{B}\cdot\nabla\varrho \equiv 0 \ . \tag{4.7.13}$$

It simply states that $\boldsymbol{B}$ lies in the surface $\varrho =$ constant everywhere. Its form is more familiar when $\varrho \equiv \Psi$.

$$\boldsymbol{B}\cdot\nabla\Psi \equiv 0 \ . \tag{4.7.14}$$

The degenerate magnetic surface, namely the magnetic axis, is properly defined by:

$$\nabla \varrho \equiv 0 \quad \text{or} \quad \nabla \Psi \equiv 0 \ . \tag{4.7.15}$$

This equation states that there is no singularity in the gradient of the flux (label) for V approaching 0. Its physical meaning is easily understood. In the limit of $V \to 0$, we have that $\Psi_{tor} \propto r^2$, where r is an average minor radius. Then $\nabla \Psi_{tor} = (d\Psi_{tor}/dr)\hat{r}$ allows us to conclude that $\nabla \Psi_{tor}$, evaluated at $r = 0$ is zero. (When taking Ψ_{tor}, the dependence of the area on r^2 must be invoked to prove (4.7.15); when one wishes to prove (4.7.15) for Ψ_{pol}, one must make use of the fact that $B_\theta = 0$ at the magnetic axis.) For every other surface label $\varrho(\Psi)$, we must make use of the property that:

$$\nabla \varrho(\Psi) = \frac{d\varrho}{d\Psi} \nabla \Psi \tag{4.7.16}$$

with the condition that $d\varrho/d\Psi$ is non-singular at the axis.

4.7.2 Open-Ended Systems

In mirrors, similar flux labels as in toroidal systems are employed. The flux volume does not require further comment. Concerning magnetic fluxes, we are now limited to fluxes contained within the flux surface. By analogy to Ψ_{tor} and Ψ_{pol} in a torus, we have the axial flux Ψ_{ax} and the azimuthal flux Ψ_{az}, respectively. A scalar pressure p seems to be less convenient because the pressure is often anisotropic and not constant along a magnetic-field line. Depending on the choice of flux label, $\hat{l}$ of the (ϱ, β, l) coordinate system points in the $\boldsymbol{B}$ direction or its reverse: $\hat{l} = \pm \hat{\boldsymbol{B}}$.

4.8 The Rotational Transform in Toroidal Systems

The total poloidal angle $\Delta\theta$ traversed by a field line after one toroidal transit ($\Delta\zeta = 2\pi$), is called the rotational transformation angle ι (Greek small letter iota). The average rotational transformation angle is defined as (Morozov and Solov'ev 1966):

$$\iota = \lim_{n \to \infty} \frac{\sum_{k=1}^{n} \iota_k}{n} \ . \tag{4.8.1}$$

Here ι_k represents the rotation angle resulting from the k-th toroidal transit. n is the product of the number of toroidal transits traced out by a single field line times the number of independent field lines on that flux surface (for an irrational surface, the latter is obviously one). The angle ι of (4.8.1) is an average over the flux-surface and is by definition a flux-surface quantity.

It is more customary to work with the quantity called the *rotational transform* $\iota\!\!-$:

$$\iota\!\!- = \frac{\iota}{2\pi} \; . \tag{4.8.2}$$

$\iota\!\!-$ (iota-bar) equals the average number of poloidal transits for one toroidal transit. (In the paper by Solov'ev and Shafranov (1970), the symbol μ is used instead of $\iota\!\!-$. Their Eqs. (1.7) and (1.8) should read m/n instead of n/m). A rational surface has $\iota\!\!-$ equal to a rational number; on an irrational surface, $\iota\!\!-$ is irrational.

As mentioned above, in the tokamak literature, one prefers to work with the "toroidal winding number," which is the reciprocal of $\iota\!\!-$:

$$q = \frac{1}{\iota\!\!-} \; . \tag{4.8.3}$$

Because of its importance in stability calculations, q is called the *safety factor*.

There is an equivalent and more useful expression for the (average) rotational transform $\iota\!\!-$:

$$\iota\!\!- = \frac{d\Psi^{\mathrm{r}}_{\mathrm{pol}}}{d\Psi_{\mathrm{tor}}} = \frac{\dot{\Psi}^{\mathrm{r}}_{\mathrm{pol}}}{\dot{\Psi}_{\mathrm{tor}}} \; , \tag{4.8.4a}$$

or

$$\iota\!\!- = -\frac{d\Psi^{\mathrm{d}}_{\mathrm{pol}}}{d\Psi_{\mathrm{tor}}} = -\frac{\dot{\Psi}^{\mathrm{d}}_{\mathrm{pol}}}{\dot{\Psi}_{\mathrm{tor}}} \; . \tag{4.8.4b}$$

The sign of $\iota\!\!-$ is chosen such that it is positive if the toroidal field and the poloidal field are in the positive ζ and θ directions, respectively (then $\boldsymbol{B}\cdot\nabla\zeta > 0$, and $\boldsymbol{B}\cdot\nabla\theta > 0$). The dot represents the derivative with respect to the chosen flux-surface label ϱ:

$$\dot{f}(\varrho) \equiv \frac{df}{d\varrho} \; . \tag{4.8.5}$$

We should note that for time dependent problems, $\iota\!\!-$ is defined by

$$\iota\!\!- = \frac{\delta\Psi^{\mathrm{r}}_{\mathrm{pol}}}{\delta\Psi_{\mathrm{tor}}} = \frac{\nabla\Psi^{\mathrm{r}}_{\mathrm{pol}}\cdot d\boldsymbol{R}}{\nabla\Psi_{\mathrm{tor}}\cdot d\boldsymbol{R}} \; . \tag{4.8.6}$$

If we consider a functional relationship, $\Psi^{\mathrm{r}}_{\mathrm{pol}} \equiv \Psi^{\mathrm{r}}_{\mathrm{pol}}(\Psi_{\mathrm{tor}}(\boldsymbol{R}, t), t)$, where $\boldsymbol{R}$ is an arbitrary position vector, the chain rule of calculus leads to

$$\nabla\Psi^{\mathrm{r}}_{\mathrm{pol}} = \left.\frac{\partial\Psi^{\mathrm{r}}_{\mathrm{pol}}}{\partial\Psi_{\mathrm{tor}}}\right|_t \nabla\Psi_{\mathrm{tor}} \; . \tag{4.8.7}$$

Then $\iota\!\!\!-$, as defined in (4.8.6) becomes $\iota\!\!\!- = \partial \Psi^{\mathrm{T}}_{\mathrm{pol}}/\partial \Psi_{\mathrm{tor}}|_t$ (Hinton and Hazeltine (1976), Chap. 7).

The validity of (4.8.4) cannot easily be proved yet, so we shall give an outline which will be justified below. (For simple examples which show the equivalence of our definitions for $\iota\!\!\!-$, see Bateman (1978), p. 6 and Morozov and Solov'ev (1966), p. 18.) The crucial point is that the expressions for Ψ_{tor} and $\Psi^{\mathrm{T}}_{\mathrm{pol}}$, given in (4.7.10) and (4.7.11), respectively, are valid for any toroidal and poloidal angle ζ and θ, and in particular for a (θ, ζ) coordinate system in which the field lines are straight. When $\boldsymbol{B}$ is straight in (θ, ζ), it follows from the definition of $\iota\!\!\!-$ ($\Delta\theta = \iota$ for $\Delta\zeta = 2\pi$) that

$$\frac{d\theta}{d\zeta} = \iota\!\!\!- (\varrho) \qquad \text{or} \qquad \frac{d\zeta}{d\theta} = q(\varrho) \; . \tag{4.8.8}$$

As indicated, $\iota\!\!\!-$ and q are flux functions (because of its straightness, $\boldsymbol{B}$ always makes the same angles with the coordinate curves for θ and ζ). The equation of a field line, (4.1.7), requires that we always have

$$\frac{d\theta}{d\zeta} = \frac{B^\theta}{B^\zeta} = \frac{\boldsymbol{B}\cdot\nabla\theta}{\boldsymbol{B}\cdot\nabla\zeta} \; . \tag{4.8.9}$$

In an upcoming section we shall show that, in a straight-$\boldsymbol{B}$ (θ, ζ) coordinate system:

$$\frac{B^\theta}{B^\zeta} = \frac{\dot{\Psi}^{\mathrm{T}}_{\mathrm{pol}}}{\dot{\Psi}_{\mathrm{tor}}} \; . \tag{4.8.10}$$

A combination of (4.8.8), (4.8.9) and (4.8.10) gives the required result (4.8.4).

A final comment on the rotational transform. The "well-known" expression for the safety factor q

$$q = \frac{rB_{\mathrm{T}}}{RB_{\mathrm{P}}} \tag{4.8.11}$$

is actually a lowest order approximation. Equation (4.8.11) is only rigorously correct in an axisymmetric tokamak with circular cross-section flux surfaces where r is measured from the center of the circular flux surface. R is the distance from the major axis to a point on the surface, $B_{\mathrm{T}} \equiv (\boldsymbol{B}\cdot\hat{\zeta}_{\mathrm{e}})$, and $B_{\mathrm{P}} \equiv (\boldsymbol{B}\cdot\hat{\theta}_{\mathrm{e}})$, where θ_{e} and ζ_{e} are the "elementary" angles based on the center of the surface.

In almost all toroidal devices, the rotational transform changes from flux surface to flux surface. To measure the amount of change, one uses the concept of *shear* defined simply as

$$\sigma(\varrho) \equiv \frac{d\iota\!\!\!-(\varrho)}{d\varrho} = \dot{\iota\!\!\!-}(\varrho) \; . \tag{4.8.12}$$

4.9 The Flux-Surface Average

The flux surface average of a function $\Phi(\boldsymbol{R})$ is defined by the *volume average* over an infinitesimally small shell with volume ΔV, where ΔV lies between two neighboring flux surfaces with volumes V and $V+\Delta V$. It is denoted by $\langle\Phi(\boldsymbol{R})\rangle$, and is equal to:

$$\langle\Phi(\boldsymbol{R})\rangle \equiv \lim_{\Delta V\to 0}\frac{\iiint_{\Delta V}\Phi(\boldsymbol{R})\,d^3R}{\iiint_{\Delta V} d^3R} = \lim_{\Delta V\to 0}\frac{1}{\Delta V}\iiint_{\Delta V}\Phi\, d^3R\;. \tag{4.9.1}$$

There are several ways to write $\langle\Phi\rangle$ in a more practical form. The elemental volume element d^3R appearing under the integral sign in (4.9.1) is by definition equal to

$$d^3R = |d\boldsymbol{R}(\varrho)\cdot d\boldsymbol{S}(\varrho)|\;. \tag{4.9.2}$$

The flux surface is assumed to be labeled by $\varrho = \text{constant}$; $d\boldsymbol{R}(\varrho)$ is the infinitesimal change in the position vector along the ϱ-coordinate curve, while $d\boldsymbol{S}(\varrho)$ is the elemental area vector of a $\varrho = \text{constant}$ surface. If we orient $d\boldsymbol{S}$ so that with $d\boldsymbol{R}(\varrho)$ in the positive ϱ-direction, the dot product of (4.9.2) leads to a positive volume element, we can omit the absolute value bars. $d\boldsymbol{S}(\varrho)$ is perpendicular to the surface, and then we have

$$d\boldsymbol{S}(\varrho) = \frac{\nabla\varrho}{|\nabla\varrho|}dS\;. \tag{4.9.3}$$

Thus (4.9.2) reads:

$$d^3R = d\boldsymbol{R}(\varrho)\cdot\frac{\nabla\varrho}{|\nabla\varrho|}dS \tag{4.9.4}$$

or with $d\varrho \equiv \nabla\varrho\cdot d\boldsymbol{R}(\varrho)$:

$$d^3R = d\varrho\frac{dS}{|\nabla\varrho|}\;. \tag{4.9.5}$$

Substitution of this expression for d^3R in the integral of (4.9.1) leads to a *surface integral* as follows:

$$\langle\Phi\rangle = \lim_{\substack{\Delta V\to 0\\ \Delta\varrho\to 0}}\frac{\Delta\varrho}{\Delta V}\iint_{S(\varrho)}\Phi\frac{dS}{|\nabla\varrho|}\;, \tag{4.9.6}$$

or

$$\langle\Phi\rangle = \frac{d\varrho}{dV}\iint_{S(\varrho)}\Phi\frac{dS}{|\nabla\varrho|}\;. \tag{4.9.7}$$

dV stands for an elemental volume that is infinitesimal in only one direction, here the ϱ-direction. It is the differential volume between two shells after integration over θ and ζ. In contrast, d^3R is infinitesimal in all three coordinates as the notation suggests. (In fact, we should have used the symbol d^2R instead of dS to be consistent. However, since we deal with only one kind of area element, we kept the symbol as simple as possible, hence dS.)

This form of the flux-surface average, particularly with the pressure as flux-surface label has been used often in the earlier fusion literature. When the flux-surface label is chosen to be V, $d\varrho/dV = 1$, and

$$\langle \Phi \rangle = \iint\limits_{S(V)} \Phi \frac{dS}{|\nabla V|} , \qquad \varrho \equiv V . \tag{4.9.8}$$

4.9.1 Toroidal Systems

In arbitrary toroidal coordinates (ϱ, θ, ζ) with Jacobian $\sqrt{g} = (\nabla\varrho \cdot \nabla\theta \times \nabla\zeta)^{-1}$, the surface integral leads to an even more transparent expression. The elemental surface element dS equals $\sqrt{g}\,|\nabla\varrho|\,d\theta\,d\zeta$ according to Chap. 2. Equation (4.9.7) then yields

$$\langle \Phi \rangle = \frac{d\varrho}{dV} \int\limits_0^{2\pi} \int\limits_0^{2\pi} d\theta\, d\zeta \sqrt{g}\, \Phi . \tag{4.9.9}$$

This equation provides an expression for $dV/d\varrho$ in terms of an integral over $d\theta$ and $d\zeta$ if we set $\Phi \equiv 1$

$$\frac{dV}{d\varrho} = \int\limits_0^{2\pi} \int\limits_0^{2\pi} d\theta\, d\zeta \sqrt{g} . \tag{4.9.10}$$

Of course, this also follows from

$$\frac{dV}{d\varrho} = \lim_{\Delta\varrho \to 0} \frac{\Delta V}{\Delta\varrho} \quad \text{and} \quad \Delta V = \Delta\varrho \iint d\theta\, d\zeta \sqrt{g} .$$

The flux-surface average of a quantity Φ, expressed as an integral over the poloidal and toroidal angles, becomes:

$$\langle \Phi \rangle = \frac{\int\limits_0^{2\pi} \int\limits_0^{2\pi} d\theta\, d\zeta \sqrt{g}\, \Phi}{\int\limits_0^{2\pi} \int\limits_0^{2\pi} d\theta\, d\zeta \sqrt{g}} . \tag{4.9.11}$$

This reduces to a very simple expression if the flux-surface label is the flux-surface volume, $\varrho = V$:

$$\langle \Phi \rangle = \int_0^{2\pi} \int_0^{2\pi} d\theta \, d\zeta \sqrt{g} \, \Phi \,, \qquad \varrho \equiv V \,. \tag{4.9.12}$$

This clearly also follows from (4.9.9).

In an *axisymmetric* device, Φ and $\sqrt{g}$ are independent of the toroidal angle ζ; the integration over ζ provides a factor of 2π in the numerator and the denominator of (4.9.11). For such devices, (4.9.11) simplifies to:

$$\langle \Phi \rangle = \frac{\int_0^{2\pi} d\theta \sqrt{g} \, \Phi}{\int_0^{2\pi} d\theta \sqrt{g}} \,. \tag{4.9.13}$$

Thus in an axisymmetric device, θ integration is sufficient to sample conditions on the whole surface.

In axisymmetric machines, the flux-surface average is sometimes referred to as a $\oint dl/B$ average, where the integral is supposed to be performed along closed field lines. In particular, we have

$$\langle \Phi \rangle = \frac{\oint \left(\frac{dl}{B} \right) \Phi}{\oint \frac{dl}{B}} \,. \tag{4.9.14}$$

To show that (4.9.14) is equivalent to (4.9.13), we must anticipate an important "property" of the magnetic field which will be justified below. The above equations are all valid for arbitrary angles θ and ζ, including those in which the magnetic field lines are straight, whose existence will be shown below. In such a system, we recall from (4.8.8) the simple relationship $d\zeta/d\theta = q(\varrho)$. After integration, this leads to:

$$q\theta - \zeta = c(\varrho) = q\theta_0 \tag{4.9.15}$$

where $c(\varrho)$, is only a function of ϱ. θ_0 is the value of the poloidal angle when $\zeta = 0$ (Fig. 4.17).

On a particular flux surface, $\varrho =$ constant, (4.9.15) represents the equation of a straight line in (θ, ζ). However, if we let ϱ vary, in general $q(\varrho)$ varies as well, and (4.9.15) becomes the equation of a "ruled" surface[1] traced out by the "straight" $\boldsymbol{B}$ lines (such as the hatched surface in Fig. 4.17). Thus, we have the situation where the field lines lie in the $q\theta - \zeta = q\theta_0$ surfaces. As a consequence, $\boldsymbol{B}$ is perpendicular to $\nabla(q\theta - \zeta)$. From (4.7.13), we recall that $\boldsymbol{B}$ is also perpendicular to $\nabla\varrho$. A vector perpendicular to these two gradients is their vector product. Hence, we can conclude that $\boldsymbol{B}$ must lie along the cross product of $\nabla\varrho$ and

[1] A ruled surface is one that is swept out by non-parallel straight lines. The quotation marks on "ruled" are there to indicate that in our case, the surface is not made up of straight lines in the literal sense, but by (field) lines, whose *equation* is that of a straight line.

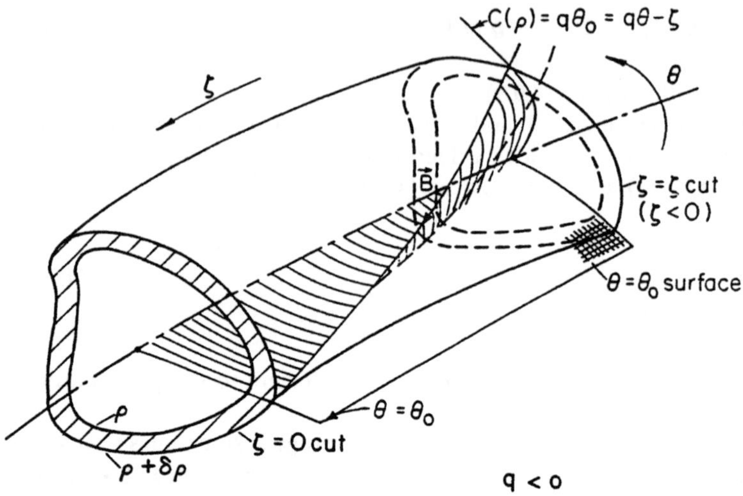

Fig. 4.17. Three-dimensional picture of an angular section of a toroidal surface between $\zeta = 0$ and $\zeta = \zeta$. The surfaces ζ = constant, θ = constant and $\zeta - q\theta$ = constant are shown

$\nabla(q\theta - \zeta)$:

$$\boldsymbol{B} \propto \nabla\varrho \times \nabla(q\theta - \zeta) \ . \tag{4.9.16}$$

The proportionality constant will be shown below to be a function of ϱ only. The dot product of $\nabla\theta$ with the $\boldsymbol{B}$ expression of (4.9.16) leads to

$$\begin{aligned} \boldsymbol{B}\cdot\nabla\theta &\propto [(\nabla\varrho \times \nabla(q\theta)) + (\nabla\zeta \times \nabla\varrho)]\cdot\nabla\theta \\ &\propto \{[\nabla\varrho \times (\dot{q}\theta\,\nabla\varrho + q\,\nabla\theta)]\cdot\nabla\theta + \nabla\varrho\cdot\nabla\theta \times \nabla\zeta\} \ . \end{aligned} \tag{4.9.17}$$

The first term vanishes because $\nabla\varrho \times (\dot{q}\theta)\,\nabla\varrho \equiv 0$, and $q(\nabla\varrho \times \nabla\theta)\cdot\nabla\theta \equiv 0$. The surviving term can be identified with the reciprocal of the Jacobian:

$$\boldsymbol{B}\cdot\nabla\theta \propto \frac{1}{\sqrt{g}} \ . \tag{4.9.18}$$

We now go back to the $\oint dl/B$ expression in (4.9.14). The equation of a field line given in (4.1.7) provides us with the equality $dl/B = d\theta/\boldsymbol{B}\cdot\nabla\theta$. If the loop in (4.9.14) closes after N poloidal transits, the expression for $\langle\Phi\rangle$ can be written as

$$\langle\Phi\rangle = \frac{N\int_0^{2\pi}(d\theta/\boldsymbol{B}\cdot\nabla\theta)\Phi}{N\int_0^{2\pi}d\theta/\boldsymbol{B}\cdot\nabla\theta} = \frac{\int_0^{2\pi}d\theta\sqrt{g}\,\Phi}{\int_0^{2\pi}d\theta\sqrt{g}} \ . \tag{4.9.19}$$

The proportionality constant of (4.9.18) cancels in this last expression, since it appears in both numerator and denominator and does not depend on θ.

So, in an *axisymmetric* device, a closed-loop integral $\oint dl/B$ is a true flux-surface average. Since it is equivalent to an integration over θ, according to (4.9.13), it is not even necessary to close the loop; covering one period in θ suffices. Therefore, a $\int \Phi\, dl/B \sim \int_0^{2\pi} d\theta \sqrt{g}\Phi$ flux-surface average can be evaluated on *any* flux surface. If one insists on performing a *closed*-loop integral, however, it is clear that it can only be done on rational surfaces. In that case, one can invoke the fact that every irrational surface can be approximated arbitrarily closely by a rational one, so that *closed* loop flux-surface averaging can still be used for "essentially" any flux surface.

The situation is different in *non-axisymmetric* systems, however. A *closed-loop* integral performed on a particular rational surface leads to different results depending on the field line we start from. Hence it cannot be a flux-surface average. If we insist on using a dl/B integration to find the flux-surface average, we have two options. Either we take an average of several closed-loop integrals, each "belonging to" different closed field lines; or, we perform a $\int (dl/B)\Phi$ integration on an irrational surface and follow the non-closed field line long enough until, for all practical purposes, we have adequately covered the whole surface. Because rational and irrational surfaces can approximate each other infinitesimally closely, the methods are equivalent.

4.9.2 Open-Ended Systems

In mirror devices, a $\int dl/B$ average is insufficient because a field line does not cover the whole flux surface; it merely runs from L_l to L_r (l for left, and r for right). In addition to the l integration, it is necessary to integrate over the angular coordinate as well. The latter is commonly denoted by β and is usually a Clebsch coordinate (see below). In a (ϱ, β, l) system, we have $dS = \sqrt{g_C}\,|\nabla\varrho|\,dl\,d\beta$, where $\sqrt{g_C} = (\nabla\varrho \cdot \nabla\beta \times \nabla l)^{-1} = (\nabla l \cdot \nabla\varrho \times \nabla\beta)^{-1} = 1/B$. The C subscript on g stands for Clebsch, and the $1/B$ result will be shown below in the section on Clebsch coordinates. Starting from (4.9.7), we have that for mirror devices:

$$\langle\Phi\rangle = \frac{d\varrho}{dV}\iint_{S(\varrho)} \Phi\sqrt{g_C}\,dl\,d\beta = \frac{d\varrho}{dV}\int_0^{2\pi} d\beta \int_{L_l}^{L_r} \frac{dl}{B}\Phi \tag{4.9.20}$$

with

$$dV = d\varrho \iint_{S(\varrho)} \sqrt{g_C}\,dl\,d\beta = d\varrho \int_0^{2\pi} d\beta \int_{L_l}^{L_r} \frac{dl}{B}\,. \tag{4.9.21}$$

We then find that in mirror devices:

$$\langle\Phi\rangle = \frac{\int_0^{2\pi} d\beta \int_{L_l}^{L_r} \Phi\, dl/B}{\int_0^{2\pi} d\beta \int_{L_l}^{L_r} dl/B} = \frac{1}{dV/d\varrho}\int_0^{2\pi} d\beta \int_{L_l}^{L_r} \Phi\frac{dl}{B}. \tag{4.9.22}$$

In general, L_l and L_r are dependent on both ϱ and β.

4.9.3 Properties of the Flux-Surface Average

The flux surface average of the divergence of a vector A can be computed by applying Gauss' law to an infinitesimally thin shell of flux volume. In *toroidal devices*, this is done after making a cut at a certain ζ-constant surface (similarly to Fig. 4.16)

$$\langle \nabla\cdot A\rangle = \lim_{\Delta V\to 0}\frac{1}{\Delta V}\iiint_{\Delta V}(\nabla\cdot A)d^3R = \lim_{\Delta V\to 0}\frac{1}{\Delta V}\oint\!\!\oint A\cdot dS\ . \tag{4.9.23}$$

The closed surface integral consists of three pieces. Firstly, there is the surface $S_{\Delta V}$ made up of the two toroidal flux surfaces enclosing the volume ΔV, $S_{\Delta V} = S(V) + S(V+\Delta V)$. Secondly, there are two surfaces in the poloidal "plane" ζ = constant. The contribution of the last two vanishes because their normals are opposite in direction such that the flux of A that "flows out" of one surface "flows in" through the other.

When we computed expressions for the flux in (4.7.8), the contributions were reversed. There was no contribution through the toroidal surface in (4.7.8) because the vector in question was B and there the contribution of the cuts survived because the integrand was multiple valued, containing a ζ. Here, in (4.9.23), on the other hand, the vector A must be single valued, since it represents a physical quantity. Thus, the integrand in (4.9.23) has the same value at $\zeta = \zeta_0$ and $\zeta = \zeta_0 + 2\pi$, as a consequence of which the "cut"-contributions vanish.

If we also replace dS in (4.9.23), using (4.9.3), but with $\varrho = V$, we obtain

$$\langle \nabla\cdot A\rangle = \lim_{\Delta V\to 0}(\pm)\frac{1}{\Delta V}\iint_{S_{\Delta V}}\frac{A\cdot\nabla V}{|\nabla V|}dS\ . \tag{4.9.24}$$

The integral over $S_{\Delta V}$ in this expression stands for the sum of two integrals, one over $S(V)$ and one over $S(V+\Delta V)$. For the outer surface, we must choose the $+$ sign, since ∇V points outward; for the inner surface, we must take the $-$ sign (dS in Gauss' theorem is outward with respect to the closed surface to which it applies, i.e., dS is outward for $S(V+\Delta V)$ but inward for $S(V)$). It actually gives a difference rather than a sum in this case. The limit of this small difference represents the derivative of an integral over $S(V)$ with respect to V

$$\langle \nabla\cdot A\rangle = \frac{d}{dV}\iint_{S(V)}\frac{A\cdot\nabla V}{|\nabla V|}dS\ . \tag{4.9.25}$$

Comparing this expression with (4.9.8), we find

$$\langle \nabla\cdot A\rangle = \frac{d}{dV}\langle A\cdot\nabla V\rangle\ . \tag{4.9.26}$$

In terms of an arbitrary flux label ϱ, this becomes

$$\langle \nabla\cdot \boldsymbol{A}\rangle = \frac{d\varrho}{dV}\frac{d}{d\varrho}\left\{\frac{dV}{d\varrho}\langle \boldsymbol{A}\cdot\nabla\varrho\rangle\right\}, \tag{4.9.27}$$

where we made use of the fact that

$$\frac{d}{dV} = \left(\frac{d\varrho}{dV}\right)\frac{d}{d\varrho} \quad \text{and} \quad \nabla V = \frac{dV}{d\varrho}\nabla\varrho \ .$$

We also made use of the obvious fact that a flux-surface quantity can be taken out of the $\langle\ \rangle$ bracket.

The proof for *open-ended systems* proceeds analogously, with an important distinction. In contrast to toroidal systems, the "end contributions" to the surface integral in (4.9.23) do not vanish, as least in the general case.

The surface integrals at the end of the infinitesimal shell volume are most easily computed in the (ϱ, β, l) coordinate system. If we bound the shell volume by $l =$ constant surfaces, then (2.5.49) is the suitable expression for dS, in which $(g_c)^{1/2} = 1/B$. This gives for (4.9.23):

$$\begin{aligned}\langle \nabla\cdot \boldsymbol{A}\rangle &= \frac{d\varrho}{dV}\frac{d}{d\varrho}\left\{\frac{dV}{d\varrho}\langle \boldsymbol{A}\cdot\nabla\varrho\rangle\right\}\\ &+ \lim_{\Delta V\to 0}\frac{1}{\Delta V}\int_{\varrho}^{\varrho+d\varrho} d\varrho\int_0^{2\pi}\frac{d\beta}{B}\boldsymbol{A}\cdot\frac{\nabla l}{|\nabla l|}\bigg|_{\text{at } l=L_r}\\ &- \lim_{\Delta V\to 0}\frac{1}{\Delta V}\int_{\varrho}^{\varrho+d\varrho} d\varrho\int_0^{2\pi}\frac{d\beta}{B}\boldsymbol{A}\cdot\frac{\nabla l}{|\nabla l|}\bigg|_{\text{at } l=L_l}\end{aligned} \tag{4.9.28}$$

Written somewhat more compactly, we have

$$\langle \nabla\cdot \boldsymbol{A}\rangle = \frac{1}{\dot V}\frac{d}{d\varrho}\{\dot V\langle \boldsymbol{A}\cdot\nabla\varrho\rangle\} + \frac{1}{\dot V}\int_0^{2\pi}\frac{d\beta}{B}\boldsymbol{A}\cdot\frac{\nabla l}{|\nabla l|}\bigg|_{l=L_l}^{l=L_r} \tag{4.9.29}$$

Here $\dot V \equiv dV/d\varrho$. Upon using the definition of the flux-surface average in open-ended devices, (4.9.22), we obtain

$$\langle \nabla\cdot \boldsymbol{A}\rangle = \frac{1}{\dot V}\frac{d}{d\varrho}\{\dot V\langle \boldsymbol{A}\cdot\nabla\varrho\rangle\} + \left\langle B\frac{\partial}{\partial l}\left(\frac{\boldsymbol{A}}{B}\cdot\frac{\nabla l}{|\nabla l|}\right)\right\rangle, \tag{4.9.30}$$

a somewhat more general form than that given by Catto and Myra (1986) (P.J. Catto, private communication).

In general, ∇l is different from $\hat{\boldsymbol{B}}$. However, if the geometry is such that the $l =$ constant surfaces (at the ends of the device) can be constructed to be perpendicular to the field lines, then $\boldsymbol{\nabla} l = \hat{\boldsymbol{B}}$ and one gets:

$$\langle \boldsymbol{\nabla}\cdot\boldsymbol{A}\rangle = \frac{1}{\dot{V}}\frac{d}{d\varrho}\{\dot{V}\langle\boldsymbol{A}\cdot\boldsymbol{\nabla}\varrho\rangle\} + \left\langle B\frac{\partial}{\partial l}\left(\frac{\boldsymbol{A}\cdot\hat{\boldsymbol{B}}}{B}\right)\right\rangle . \tag{4.9.31}$$

The equality of $\boldsymbol{\nabla} l$ and $\hat{\boldsymbol{B}}$, in order for (4.9.31) to hold, is perhaps not necessary. Since the second term in (4.9.30) and (4.9.31) concerns averaging over the "angle" β, it might be that the averages are identical in certain specific cases, even though $\hat{\boldsymbol{B}}$ differs from $\boldsymbol{\nabla} l$ on the surfaces on which the integrals are performed.

Furthermore, for a particular vector $\boldsymbol{A}$ and/or geometry, this second term might be shown to vanish. If the flux of the vector field $\boldsymbol{A}$ over the infinitesimal annular surface at $l = L_l$ exactly cancels that at $l = L_r$, the end contributions disappear and (4.9.27) applies. (For more on the relationship between strict quadrupole symmetry and end losses in tandem mirrors, the reader should consult the papers by Myra and Catto (1982a) and by Myra, Catto and Hazeltine (1982b).

By far the most important property of the flux-surface average is that it annihilates the $(\boldsymbol{B}\cdot\boldsymbol{\nabla})$ operator:

$$\langle \boldsymbol{B}\cdot\boldsymbol{\nabla}\Phi\rangle \equiv 0 . \tag{4.9.32}$$

This property may be used to define the $\langle\ \rangle$ operation. Here, we prove it with the help of the expression derived above for $\langle\text{div}\rangle$, (4.9.27). It follows from $\boldsymbol{\nabla}\cdot\boldsymbol{B} = 0$ and (4.9.27) that

$$\langle \boldsymbol{B}\cdot\boldsymbol{\nabla}\Phi\rangle = \langle\boldsymbol{\nabla}\cdot(\boldsymbol{B}\Phi)\rangle \propto \langle\Phi\boldsymbol{B}\cdot\boldsymbol{\nabla}\varrho\rangle , \tag{4.9.33}$$

which vanishes identically because $\boldsymbol{B}\cdot\boldsymbol{\nabla}\varrho = 0$ is the equation of a flux surface.

More intuitively, in a toroidal device,

$$\langle \boldsymbol{B}\cdot\boldsymbol{\nabla}\Phi\rangle = \left\langle B\frac{\partial\Phi}{\partial l}\right\rangle \sim \int\frac{dl}{B}B\frac{\partial\Phi}{\partial l} = \int dl\frac{\partial\Phi}{\partial l} = 0 . \tag{4.9.34}$$

For an axisymmetric device, $\int dl$ stands for $\oint dl$ on the nearest rational surface so the null result is obvious. For a non-axisymmetric device, we assume that Φ is continuous over the surface, we pick an irrational surface and go around as many times as required to sample the surface well enough for practical purposes and we reach a point that is sufficiently close to the starting point. The result can be made as small as desired, with zero being the limit.

For mirrors, $\langle\boldsymbol{B}\cdot\boldsymbol{\nabla}\Phi\rangle$ vanishes only if the end contributions cancel. We have:

$$\langle \boldsymbol{B}\cdot\boldsymbol{\nabla}\Phi\rangle = \left\langle B\frac{\partial\Phi}{\partial l}\right\rangle \sim \int_0^{2\pi} d\beta\int_{L_l}^{L_r} dl\frac{\partial\Phi}{\partial l}$$

$$\sim \int_0^{2\pi}\Phi(\beta,L_r)d\beta - \int_0^{2\pi}\Phi(\beta,L_l)d\beta . \tag{4.9.35}$$

Equality implies that the average of Φ over β must be the same at $l = L_l$ and $l = L_r$.

Note that although $\langle \boldsymbol{B} \cdot \nabla\Phi \rangle \equiv 0$ (in toroidal devices), we have, in general that $\langle \hat{\boldsymbol{B}} \cdot \nabla\Phi \rangle \neq 0$, where $\hat{\boldsymbol{B}} = \boldsymbol{B}/B$ is the unit vector along $\boldsymbol{B}$. Thus $\langle (\boldsymbol{B} \cdot \nabla\Phi)/B \rangle \neq 0$. Also, in general, the following expression is non zero:

$$\langle \gamma(\boldsymbol{R}) \boldsymbol{B} \cdot \nabla\Phi(\boldsymbol{R}) \rangle = \langle \gamma \boldsymbol{B} \cdot \nabla\Phi \rangle \neq 0 \ . \tag{4.9.36}$$

$\gamma(\boldsymbol{R})$ is a scalar field analogous to $\Phi(\boldsymbol{R})$. However, (4.9.36) reduces identically to zero if $\gamma = \Phi^{-1}$

$$\left\langle \frac{\boldsymbol{B} \cdot \nabla\Phi}{\Phi} \right\rangle = \langle \boldsymbol{B} \cdot \nabla \ln \Phi \rangle = \langle \boldsymbol{B} \cdot \nabla\Omega \rangle \equiv 0 \ , \tag{4.9.37}$$

where $\Omega = \ln \Phi$.

A final expression concerning the flux-surface average is the identity (for toroidal devices)

$$\langle \nabla\varrho \cdot \nabla \times \boldsymbol{W} \rangle \equiv 0 \ , \tag{4.9.38}$$

in which ϱ is a flux-surface label. This equation is a consequence of the identity $\nabla \cdot (\boldsymbol{W} \times \nabla\varrho) = \nabla\varrho \cdot \nabla \times \boldsymbol{W} - \boldsymbol{W} \cdot \nabla \times \nabla\varrho = \nabla\varrho \cdot \nabla \times \boldsymbol{W}$, and the $\langle$div$\rangle$ result of (4.9.27)

4.10 The Magnetic Differential Equation

An equation of the form

$$\boldsymbol{B} \cdot \nabla f = S \tag{4.10.1}$$

is the prototype magnetic differential equation (Kruskal and Kulsrud 1958; Newcomb 1959; see also, Hazeltine and Meiss 1985, and Solov'ev and Shafranov 1970). S is a source term, and f is the unknown single-valued function that we must solve for. Both f and S are scalar fields: $f \equiv f(\boldsymbol{R})$ and $S \equiv S(\boldsymbol{R})$.

We limit our discussion to toroidal devices because this type of equation poses fewer problems in linear devices.

A first condition on S in order that f be single valued follows from an integration over l:

$$\boldsymbol{B} \cdot \nabla f = B \frac{\partial f}{\partial l} = S \ , \tag{4.10.2}$$

which implies

$$f = f_0 + \int_0^l \frac{dl}{B} S \ . \tag{4.10.3}$$

Here, f_0 is the value of f at $l = 0$, an arbitrary origin. If we happen to perform the integral in (4.10.3) on a rational surface, we get a closed integral, and the single valuedness of f requires that

$$\oint \frac{dl}{B} S = 0 \tag{4.10.4}$$

on rational surfaces. Equation (4.10.4) must be satisfied for every closed field line on the rational surface.

A second solvability condition follows from taking the flux-surface average of both sides of the magnetic differential equation and applying the identity (4.9.32)

$$\langle \boldsymbol{B} \cdot \nabla f \rangle = \langle S \rangle \equiv 0 \ . \tag{4.10.5}$$

A careful examination shows that these two conditions are not independent (if one assumes, which we do throughout this monograph, that the rational surfaces are also a dense set in non-axisymmetric devices, as they are for the "imaginary" lowest-order field structure with "good" surfaces, that we assume throughout this monograph). Equation (4.10.5) follows from (4.10.4). If the requirement given in (4.10.4) is satisfied for *every* field line on the surface, then it must be concluded that the flux-surface average of S is automatically zero on that surface. This is obvious if we write the flux-surface average on the rational surface as

$$\langle S \rangle = \frac{\oint\limits_1 \frac{dl}{B} S + \oint\limits_2 \frac{dl}{B} S + \oint\limits_3 \frac{dl}{B} S + \cdots + \oint\limits_k \frac{dl}{B} S + \cdots}{\oint\limits_1 \frac{dl}{B} + \oint\limits_2 \frac{dl}{B} + \oint\limits_3 \frac{dl}{B} + \cdots + \oint\limits_k \frac{dl}{B} + \cdots} = 0 \ , \tag{4.10.6}$$

where the summation runs ad infinitum over all possible field lines in the surface. Since we then have that $\langle S \rangle$ is zero on every rational surface, continuity and the fact that the real numbers are everywhere dense mean that $\langle S \rangle$ vanishes everywhere. The reverse statement is not valid. The fact that the average of S vanishes does not, in general, guarantee that all the pieces that make up that average vanish. In other words, the fact that $\sum \oint S\, dl/B$ vanishes does not imply that every $\oint S\, dl/B$ vanishes. This would be true only for an axisymmetric device such as a tokamak which has $\langle S \rangle = \oint S\, dl/B / \oint dl/B$. Thus, (4.10.5) cannot be a sufficient condition to guarantee a solution of the differential equation.

The definition of the flux-surface average allows us to write an alternate expression for (4.10.5),

$$\lim_{\Delta V \to 0} \frac{1}{\Delta V} \iiint\limits_{\Delta V} S\, d^3R \equiv 0 \ . \tag{4.10.7}$$

It was in this form that the condition was written by Kruskal and Kulsrud (1958), and that it was discussed by Newcomb (1959). Although (4.10.5) and (4.10.7) follow from (4.10.4), it is a convenient condition to impose in order to eliminate certain features of S, after which the application of (4.10.4) is made easier.

It was Newcomb (1959) who showed that (4.10.4) is a sufficient condition for solving the magnetic differential equation. We shall demonstrate this by finding a solution for f after imposing the condition.

As usual in the theory of differential equations, the general solution of (4.10.1) consists of two parts, a homogeneous solution and a particular solution

$$f = f_{\mathrm{h}} + f_{\mathrm{p}} \ , \tag{4.10.8}$$

where f_{h} satisfies the homogeneous equation without a source term

$$\boldsymbol{B}\cdot\nabla f_{\mathrm{h}} = 0 \ . \tag{4.10.9}$$

This is equivalent to

$$\frac{\partial f_{\mathrm{h}}}{\partial l} = 0 \ , \tag{4.10.10}$$

which means that f_{h} must be a flux function. (This conclusion can only be drawn in toroidal machines; in mirrors, all we can say is that f_{h} is independent of l.) Hence, the total solution of (4.10.1) equals

$$f = f_{\mathrm{h}}(\varrho) + f_{\mathrm{p}} \ . \tag{4.10.11}$$

Any flux-surface quantity satisfies (4.10.10) and is thus eligible as a homogeneous solution for the magnetic differential equation.

To find a particular solution, the differential equation may be rewritten in terms of partial derivatives $\partial/\partial\theta$ and $\partial/\partial\zeta$. It is most convenient to do so in flux coordinates $(\theta_{\mathrm{f}}, \zeta_{\mathrm{f}})$ in which the magnetic field is straight. At this moment, we do not yet have the tools to actually solve the equation. However, the process is straightforward, and we can give an outline. Details can be found in the review by Solov'ev and Shafranov (1970).

The $\boldsymbol{B}\cdot\nabla f$ term becomes $B^i\partial f/\partial u^i$, where u^i stands for θ_{f} or ζ_{f}, since $B^\varrho \equiv \boldsymbol{B}\cdot\nabla\varrho = 0$. The B^i in flux coordinates have a $\sqrt{g_{\mathrm{f}}}$ in the denominator (see below) which upon bringing it to the right-hand side, leads to a source term $\sqrt{g_{\mathrm{f}}}S$. Next, the functions f and $\sqrt{g_{\mathrm{f}}}S$, which are both single valued, are expanded in a double Fourier series in θ_{f} and ζ_{f}. The two integrands in the two solvability conditions (4.10.4) and (4.10.5) contain the factor $\sqrt{g_{\mathrm{f}}}S$ as well. (In the integral $\oint(dl/B)S$, we use the equation of a field line $dl/B = d\zeta_{\mathrm{f}}/B^{\zeta_f} \propto \sqrt{g_{\mathrm{f}}}$, and in the volume integral, $d^3R = \sqrt{g_{\mathrm{f}}}\,d\varrho\,d\zeta_f\,d\theta_f$.) Then we use the equation for a field line on a flux surface in flux coordinates $\theta_{\mathrm{f}0} = \theta_{\mathrm{f}} - \iota\zeta_f$ (see (4.9.15)) to eliminate the θ_{f} variable and impose the two integral constraints.

These conditions have consequences for the expansion coefficients of $\sqrt{g_{\mathrm{f}}}S$, which are denoted by $a_{mn}(\varrho)$. The volume-integral condition requires that

$a_{00}(\varrho)$ be zero on any flux surface. The $\oint S\,dl/B = 0$ condition states that $\sum_{mn} a_{mn}(\varrho)/(m\iota\!\!\!- - n) = 0$, so that the $a_{mn}(\varrho_{\text{rational}})$ must vanish on their respective rational surfaces where $m\iota\!\!\!- - n = 0$. When these conditions are satisfied, the $a_{mn}(\varrho_{\text{rational}})$ are factored as $(m\iota\!\!\!- - n)\tilde{a}_{mn}(\varrho_{\text{rational}})$, and the solution is found by equating the respective harmonics of the Fourier expansion. Thus, $f_{mn} = (\ )\tilde{a}_{mn}$, where ( ) represents a constant factor independent of m or n. This completes the proof because f can be reconstructed from its harmonics.

From this Fourier-expansion procedure, it is also evident that the $\langle S \rangle = 0$ condition follows from the $\oint S\,dl/B = 0$ requirement. The latter restriction, $\sum_{mn} a_{mn}(\varrho)/(m\iota\!\!\!- - n) = 0$, not only forces $a_{mn}(\varrho_{\text{rational}}) = 0$, but also sets $a_{00}(\varrho) = 0$ if $m = n = 0$.

In a tokamak, f and S are independent of ζ, so that their Fourier expansion is only with respect to θ. Therefore, $\sum_m a_{m0}(\varrho)/m\iota\!\!\!- = 0$, requires only that $a_{00}(\varrho) = 0$ (independent of the value of $\iota\!\!\!-$), leading to the observation that both conditions reduce to one and the same thing.

For the numerical solution of the magnetic differential equation, the reader is referred to the paper by Reiman and Greenside (1988).

Part II

Flux Coordinates

From a purely mathematical standpoint, Part II, on flux coordinates, can be considered merely as an extensive example of non-orthogonal curvilinear coordinate systems with certain particular properties. To recall, *magnetic-flux coordinates* are curvilinear coordinates, so chosen that *the equation of a magnetic-field line is that of a straight line.* Somebody familiar with the topics covered in Part I should have no difficulties in this part, at least from a computational standpoint. Conversely, it is likely to be very difficult to obtain an adequate understanding of this part if the previous material has not been mastered.

Providing a discussion of the computational *tools* for magnetic flux coordinate descriptions is the primary reason for the writing of this monograph. However, for plasma and fusion scientists, the physical interpretation of the subject matter is equally important i.e., "In what physical situation do we use which coordinate system?" Therefore, besides explaining most mathematical steps, we have tried to emphasize the relationships between the various coordinate systems. Thus, it is even instructive to discuss links with a few non-flux coordinate systems that are frequently used or have their own intrinsic value.

The emphasis in this flux-coordinate treatment is on *toroidal devices.* From a conceptual standpoint, understanding the relationships between the different types of flux coordinates in toroidal devices is quite intricate; in contrast, in open-ended systems there is only one fairly simple system that is used, namely the Clebsch system.

We nevertheless discuss *mirrors* whenever we feel this is helpful. The very general Clebsch system is one of the topics we will discuss in considerable detail. Since it is based on the properties of the magnetic field $\boldsymbol{B}$, independent of a particular device, it is useful not only in mirrors and toroids in a fusion context, but also for the description of extraterrestrial magnetic fields in *space physics, astronomy,* and *astrophysics.*

As far as toroidal devices are concerned, the discussions are kept completely general with regard to the type of device. To make the treatment less abstract, however, we interpret the general results by means of two prototypes. For machines with a symmetry direction, we consider the axisymmetric tokamak. For general three-dimensional devices, we relate the results to the stellarator family of devices.

In Chap. 5, we discuss Clebsch-related coordinates. There we define stream functions, treat the Clebsch coordinate system itself, point out its relation to the contra- and covariant component formalism, and discuss a closely-related system that we give the name "Boozer-Grad". Toroidal flux coordinates are considered in Chap. 6. Here the stage is set for straight-$\boldsymbol{B}$-line coordinates, non-flux coordinates in tokamaks, straight-$\boldsymbol{J}$-line coordinates, covariant-$\boldsymbol{B}$ components and the relationship to the Boozer-Grad form, Boozer's toroidal flux coordinates, the ideal-MHD equilibrium conditions for toroidally confined plasmas, and Hamada coordinates. In Chap. 7, the focus is on the conversion of Clebsch-type coordinates to toroidal flux coordinates. We do this for the Clebsch and Boozer-Grad systems. The following Chap. 8 is on the establishment of the flux-coordinate transformation. It summarizes how flux coordinates are found and allows us to say a few words about equilibrium calculations, including useful references. Finally, in Chap. 9, we treat a more general but related topic, namely, canonical coordinates. These coordinates superficially resemble the flux-coordinate representation and are intimately related to the existence of flux surfaces. Flux coordinates are reconsidered in a slightly different form to make their evaluation readily comparable with that of the canonical coordinates. The canonical coordinates and the magnetic field-line Hamiltonian are presented, after which the practical evaluation of the Hamiltonian and the canonical coordinates is explained.

5. The Clebsch-Type Coordinate Systems

5.1 Stream Functions

The idea of a stream function (or flow function) is borrowed from hydrodynamic theory (see, e.g., Milne-Thomson (1968)). A stream function is any function that is constant along a stream line. In plasma physics, we are interested in stream functions for the magnetic field. Hence, in our context, a stream function is *any function that is constant along a field line.* In three-dimensional space, the equation $\Phi(u^1, u^2, u^3) \equiv \Phi(\boldsymbol{R}) =$ constant represents a surface. A magnetic-field line can thus be identified with a curve lying at the intersection of two surfaces belonging to two different stream functions.

An interesting property of stream functions $\alpha(\boldsymbol{R})$ and $\beta(\boldsymbol{R})$ to be defined below in (5.2.1), is that the differential magnetic flux through a surface (other than the $\alpha =$ constant and $\beta =$ constant surfaces) is proportional $d\alpha\, d\beta$, or

$$\Psi \propto \int\int d\alpha\, d\beta \ . \tag{5.1.1}$$

We will show below that this property is consistent with the definition of stream functions.

5.2 Generic Clebsch Coordinates

The Clebsch coordinates also have their origin in fluid mechanics, (e.g., Milne-Thomson 1968; Bedeaux 1980). They are related to what is known in hydrodynamics as the Clebsch transformations (Lamb 1945). They were introduced to magnetic-fusion theory by Kruskal and Kulsrud (1958) and by Grad and Rubin (1958). These Clebsch coordinates, commonly denoted by α and β are stream functions of the magnetic field and allow us to write $\boldsymbol{B}$ as the cross product of their gradients

$$\boldsymbol{B} = \nabla\alpha \times \nabla\beta \ . \tag{5.2.1}$$

An intuitive way of obtaining (5.2.1) follows from writing $\boldsymbol{B}$ in terms of a vector potential $\boldsymbol{A}$

$$\boldsymbol{B} = \nabla \times \boldsymbol{A} \ . \tag{5.2.2}$$

Equation (5.2.2) is a consequence of $\nabla \cdot \boldsymbol{B} = 0$. If we now pick two functions α and β such that

$$\boldsymbol{A} = \alpha \nabla \beta \ , \tag{5.2.3}$$

we obtain for (5.2.2)

$$\boldsymbol{B} = \nabla \times (\alpha \nabla \beta) = (\nabla \alpha \times \nabla \beta) + \alpha(\nabla \times \nabla \beta) \ . \tag{5.2.4}$$

Since the last term vanishes identically because of a vector identity, we recover (5.2.1).

A constructive proof of (5.2.1) shows how to find functions α and β (Grad and Rubin 1958; see also, Bateman 1978; Kulsrud 1983; Bernstein 1971). Consider a (transverse) surface S that is intersected by $\boldsymbol{B}$-field lines (Fig. 5.1a) – thus S is nowhere tangent to $\boldsymbol{B}$. In this surface S, we choose a family of lines that we label by $\tilde{\alpha} =$ constant. Any of these lines, take for instance $\tilde{\alpha} = \tilde{\alpha}_0$ (Fig. 5.1b), can be viewed as the intersection of the surface S with a surface traced out

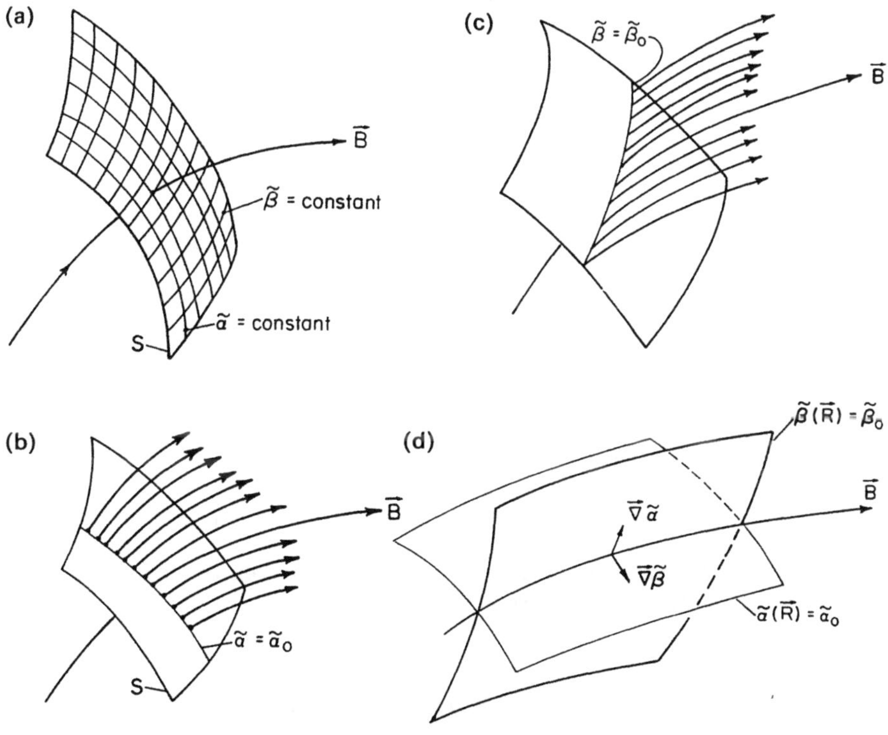

Fig. 5.1a-d. Construction of an (α, β) Clebsch coordinate system. **(a)** $\tilde{\alpha}$, $\tilde{\beta}$ grid laid out on a surface S intersecting a field line $\boldsymbol{B}$. **(b)** $\tilde{\alpha} = \tilde{\alpha}_0$ surface, being constructed from the magnetic-field lines intersecting the curve $\tilde{\alpha} = \tilde{\alpha}_0$ on S. **(c)** $\tilde{\beta} = \tilde{\beta}_0$ surface, constructed from the field lines intersecting the curve $\tilde{\beta} = \tilde{\beta}_0$ on S. **(d)** Representation of the $\tilde{\alpha} = \tilde{\alpha}_0$ and $\tilde{\beta} = \tilde{\beta}_0$ surfaces based on a magnetic-field line. The gradients $\nabla\tilde{\alpha}$ and $\nabla\tilde{\beta}$ to the respective surfaces are also shown

by all the magnetic-field lines passing through $\tilde{\alpha} = \tilde{\alpha}_0$. The latter surface is labeled by carrying the value $\tilde{\alpha} = \tilde{\alpha}_0$ off S on the field lines. Hence, the equation of the surface is $\tilde{\alpha} = \tilde{\alpha}_0$. Analogously, we construct another family of lines $\tilde{\beta}$ = constant in S (Fig. 5.1a) such that the $\tilde{\alpha}$ and $\tilde{\beta}$ lines are nowhere tangent. In other words, the $\tilde{\alpha}$ and $\tilde{\beta}$ lines must form a proper grid in S. Again, as shown in Fig. 5.1c, a particular line $\tilde{\beta} = \tilde{\beta}_0$ is carried off on the field lines, resulting in a surface $\tilde{\beta} = \tilde{\beta}_0$. When we repeat this same procedure for all the field lines, we can thus construct functions $\tilde{\alpha} = \tilde{\alpha}(\boldsymbol{R})$ and $\tilde{\beta} = \tilde{\beta}(\boldsymbol{R})$, such that the geometric loci of points of $\tilde{\alpha}$ = constant and $\tilde{\beta}$ = constant are precisely the types of surfaces discussed above. By construction, the $\boldsymbol{B}$ lines lie in $\tilde{\alpha}$ = constant and $\tilde{\beta}$ = constant surfaces. In that respect, these surfaces can be called "magnetic surfaces". However, in fusion plasma physics, it is customary to reserve this name for toroidal or cylindrical flux surfaces.

A particular magnetic-field line is defined by the intersection of $\tilde{\alpha}$ = constant and $\tilde{\beta}$ = constant surfaces (as is shown in Fig. 5.1d). We can thus interpret the pair of values $\tilde{\alpha}_0$ and $\tilde{\beta}_0$ as the coordinates of the particular field line. A point in space can then be determined by specifying $\tilde{\alpha}$, $\tilde{\beta}$ and a third variable which is related to the length along a field line.

By construction,

$$\boldsymbol{B} \perp \nabla\tilde{\alpha} \quad \text{or} \quad \boldsymbol{B} \cdot \nabla\tilde{\alpha} = 0 \tag{5.2.5a}$$

$$\boldsymbol{B} \perp \nabla\tilde{\beta} \quad \text{or} \quad \boldsymbol{B} \cdot \nabla\tilde{\beta} = 0 \ . \tag{5.2.5b}$$

This states that $\tilde{\alpha}$ and $\tilde{\beta}$ are magnetic stream functions. Another vector that is everywhere orthogonal to both $\nabla\tilde{\alpha}$ and $\nabla\tilde{\beta}$ is their vector product. We must thus have that $\boldsymbol{B}$ is parallel to $\nabla\tilde{\alpha} \times \nabla\tilde{\beta}$ or

$$\boldsymbol{B} = C\nabla\tilde{\alpha} \times \nabla\tilde{\beta} \ , \tag{5.2.6}$$

where C is a proportionality factor. In general, C can vary from point to point. It could thus be a function of $\tilde{\alpha}$, $\tilde{\beta}$ and the third variable which we take for simplicity equal to the length l along a field line. However, from $\nabla \cdot \boldsymbol{B} = 0$, it follows that

$$0 = \nabla \cdot [C\nabla\tilde{\alpha} \times \nabla\tilde{\beta}] = \nabla C \cdot (\nabla\tilde{\alpha} \times \nabla\tilde{\beta}) + C\nabla \cdot (\nabla\tilde{\alpha} \times \nabla\tilde{\beta}) \ , \tag{5.2.7}$$

where the last term vanishes because it involves expressions of the form $\nabla \times \nabla f$ which vanish identically. Thus, we have that

$$\nabla C \cdot (\nabla\tilde{\alpha} \times \nabla\tilde{\beta}) = 0 \ . \tag{5.2.8}$$

With the help of the expression for $\boldsymbol{B}$ in (5.2.6), we obtain

$$\frac{\nabla C}{C} \cdot \boldsymbol{B} = 0 \quad \text{or} \quad \boldsymbol{B} \cdot \nabla C = 0$$

$$\text{or} \quad B\frac{\partial C}{\partial l} = 0 \ . \tag{5.2.9}$$

This states that C is constant along the field lines, and thus is independent of l. Hence, $C = C(\tilde{\alpha}, \tilde{\beta})$ alone.

We now redefine $\tilde{\alpha}$ and $\tilde{\beta}$ such that the factor C is absorbed in the new α and β:

$$\alpha = \tilde{\alpha} \tag{5.2.10a}$$

$$\beta = \int_0^{\tilde{\beta}} d\tilde{\beta}\, C(\tilde{\alpha}, \tilde{\beta}) \qquad \text{or} \qquad \frac{\partial \beta}{\partial \tilde{\beta}} = C(\tilde{\alpha}, \tilde{\beta}) \; . \tag{5.2.10b}$$

If follows from these definitions that

$$\nabla \alpha = \nabla \tilde{\alpha} \tag{5.2.11a}$$

and

$$\nabla \beta = \frac{\partial \beta}{\partial \tilde{\alpha}} \nabla \tilde{\alpha} + \frac{\partial \beta}{\partial \tilde{\beta}} \nabla \tilde{\beta} \; . \tag{5.2.11b}$$

With the replacement (5.2.10b) for $\partial \beta / \partial \tilde{\beta}$, we can thus write for the $\boldsymbol{B}$ expression of (5.2.6):

$$\boldsymbol{B} = C \nabla \tilde{\alpha} \times \nabla \tilde{\beta} = \nabla \tilde{\alpha} \times C \nabla \tilde{\beta} = \nabla \tilde{\alpha} \times \left[\nabla \beta - \frac{\partial \beta}{\partial \tilde{\alpha}} \nabla \tilde{\alpha} \right] = \nabla \alpha \times \nabla \beta \; . \tag{5.2.12}$$

Note that our discussion does not require $\alpha(\boldsymbol{R})$ and $\beta(\boldsymbol{R})$ to be single valued functions. However, $\boldsymbol{B}$, which is a physical quantity, must be single valued. In magnetic-confinement systems, α is usually identified with the flux-surface label ϱ, while β is a magnetic field-line label. β = constant surfaces are shown in Figs. 4.12, 17 for toroidal devices. In mirrors, they are quite similar. The problems encountered with the multiple valuedness of β in tori have been discussed above. In mirrors, β represents an angle-like variable and its use is straightforward.

The functions α and β are not unique for any given magnetic field. For example, we can add to α any arbitrary function of β, or the other way around (but not both at the same time):

$$\nabla \alpha \times \nabla(\beta + g(\alpha)) = (\nabla \alpha \times \nabla \beta) + \left(\nabla \alpha \times \frac{\partial g}{\partial \alpha} \nabla \alpha \right) = \nabla \alpha \times \nabla \beta \; . \tag{5.2.13}$$

5.3 Relationship to the Contra- and Covariant Formalism

When we work with Clebsch coordinates α and β and a variable along a field line, we imply that space has been parameterized by these three variables. In many cases, this last variable is the length l of the field line (from some particular origin). The coordinate surfaces for the coordinates (α, β, l) are thus $\alpha(\boldsymbol{R})$ = constant, $\beta(\boldsymbol{R})$ = constant and $l(\boldsymbol{R})$ = constant. The coordinate lines are intersections of coordinate surfaces as usual. The special feature of the Clebsch representation is that the particular vector we want to describe, namely, $\boldsymbol{B}$, is tangent to one of the coordinate curves: the l curve (since it is by definition the intersection of $\alpha =$ constant and $\beta =$ constant surfaces).

We shall assume that $\hat{\boldsymbol{l}}$ is in the direction of $\boldsymbol{B}$, although, in fact, $\boldsymbol{B} = B\hat{\boldsymbol{B}} = \pm B\hat{\boldsymbol{l}}$.

The covariant-basis vectors in the (α, β, l) coordinate system are denoted by $\boldsymbol{e}_\alpha$, $\boldsymbol{e}_\beta$, and $\boldsymbol{e}_l$, the contravariant-basis vectors by $\boldsymbol{e}^\alpha$, $\boldsymbol{e}^\beta$, $\boldsymbol{e}^l$. The latter are defined by the gradients of the coordinate surfaces

$$\boldsymbol{e}^\alpha \equiv \nabla\alpha \;, \qquad \boldsymbol{e}^\beta \equiv \nabla\beta \;, \qquad \boldsymbol{e}^l \equiv \nabla l \;. \tag{5.3.1}$$

The covariant-basis vectors are tangent to the coordinate curves. Thus, $\boldsymbol{e}_l$ is tangent to the field lines. Moreover, since l measures the length along that coordinate curve, its magnitude is unity:

$$\boldsymbol{e}_l = \frac{\boldsymbol{B}}{B} \equiv \hat{\boldsymbol{B}} \equiv \hat{\boldsymbol{l}} \;. \tag{5.3.2}$$

(Often, our $\hat{\boldsymbol{B}}$ is written as $\hat{\boldsymbol{b}}$; however, we avoid $\hat{\boldsymbol{b}}$ here because we used it to denote the binormal of the Frenet-Serret trihedron.) We shall use the symbols of (5.3.2) interchangeably. The other covariant-basis vectors do not have such a simple form but can be written in terms of the contravariant-basis vectors by means of the Jacobian J:

$$J = \frac{1}{\nabla\alpha\cdot\nabla\beta\times\nabla l} = \frac{1}{\nabla l\cdot\nabla\alpha\times\nabla\beta} \;. \tag{5.3.3}$$

We get

$$\boldsymbol{e}_\alpha = J\boldsymbol{e}^\beta \times \boldsymbol{e}^l = J\nabla\beta\times\nabla l \;, \tag{5.3.4a}$$

$$\boldsymbol{e}_\beta = J\boldsymbol{e}^l \times \boldsymbol{e}^\alpha = J\nabla l\times\nabla\alpha \;. \tag{5.3.4b}$$

Nothing prohibits us from writing the unit vector along $\boldsymbol{B}$ in a similar form:

$$\boldsymbol{e}_l \equiv \hat{\boldsymbol{B}} \equiv \hat{\boldsymbol{l}} = J\nabla\alpha\times\nabla\beta \;. \tag{5.3.4c}$$

The magnetic field can be written in terms of covariant- or contravariant-basis vectors:

$$\boldsymbol{B} = B^\alpha\boldsymbol{e}_\alpha + B^\beta\boldsymbol{e}_\beta + B^l\boldsymbol{e}_l \;, \qquad \text{with} \qquad B^i = \boldsymbol{B}\cdot\boldsymbol{e}^i \;, \tag{5.3.5a}$$

or

$$\boldsymbol{B} = B_\alpha\boldsymbol{e}^\alpha + B_\beta\boldsymbol{e}^\beta + B_l\boldsymbol{e}^l \;, \qquad \text{with} \qquad B_i = \boldsymbol{B}\cdot\boldsymbol{e}_i \;. \tag{5.3.5b}$$

Equation (5.3.5a), the contravariant form of $\boldsymbol{B}$ (since it is written with contravariant components), simplifies to a plausible result. By construction $B^\alpha \equiv$

$\boldsymbol{B}\cdot\nabla\alpha = 0$, and $B^\beta \equiv \boldsymbol{B}\cdot\nabla\beta = 0$, such that the equation in question becomes with the help of (5.3.2):

$$\boldsymbol{B} = B^l \boldsymbol{e}_l = B^l \hat{\boldsymbol{l}} = B^l \hat{\boldsymbol{B}} = \frac{B^l}{B}\boldsymbol{B} \; . \tag{5.3.6}$$

From this we conclude that $B^l \equiv B$ and thus that

$$\boldsymbol{B} = B\hat{\boldsymbol{B}} \tag{5.3.7}$$

is the *contravariant representation* of $\boldsymbol{B}$ (in α, β, l coordinates).

Consider the expression for $\hat{\boldsymbol{B}}$ given in (5.3.4c), but with the Jacobian written explicitly

$$\hat{\boldsymbol{B}} = \frac{\nabla\alpha \times \nabla\beta}{\nabla l\cdot\nabla\alpha \times \nabla\beta} \; . \tag{5.3.8}$$

If we take the dot product of either side with ∇l, we obtain:

$$\hat{\boldsymbol{B}}\cdot\nabla l = 1 \; . \tag{5.3.9}$$

Equation (5.3.9) also follows trivially from the observation that $\hat{\boldsymbol{B}}\cdot\nabla \equiv \partial/\partial l$. It should be noted that (5.3.9) is (in general) the best we can do for a relationship between $\hat{\boldsymbol{B}}$ and ∇l. In general, and unlike $\boldsymbol{e}_l$, $\boldsymbol{e}^l \equiv \nabla l$ is *not* equal to $\hat{\boldsymbol{B}}$; they point in different directions (contrary to what has been assumed frequently in the literature, as e.g., by Lee et al. 1986).

With this expression and the Clebsch representation, $\boldsymbol{B} = \nabla\alpha \times \nabla\beta$, we find the value for the Jacobian of (5.3.3)

$$J = \frac{1}{\nabla l\cdot\nabla\alpha \times \nabla\beta} = \frac{1}{\nabla l\cdot\boldsymbol{B}} = \frac{1}{B\nabla l\cdot\hat{\boldsymbol{B}}} = \frac{1}{B} \; . \tag{5.3.10}$$

This result was also obvious from (5.3.4c), and allows us to identify the contravariant form of (5.3.7) with the Clebsch form:

$$\boldsymbol{B} = B\hat{\boldsymbol{B}} = BJ\,\nabla\alpha \times \nabla\beta = \nabla\alpha \times \nabla\beta \; . \tag{5.3.11}$$

Thus, the "complete" *contravariant* representation of $\boldsymbol{B}$ is

$$\boldsymbol{B} = B\hat{\boldsymbol{B}} = \nabla\alpha \times \nabla\beta \; . \tag{5.3.12}$$

The covariant representation does not provide a simple form because ∇l is in general not parallel to $\boldsymbol{B}$. From (5.3.4) and (5.3.10), we can rewrite the *covariant form* of (5.3.5b) as follows:

$$\begin{aligned}\boldsymbol{B} &= \boldsymbol{B}\cdot\frac{\nabla\beta \times \nabla l}{B}\nabla\alpha + \boldsymbol{B}\cdot\frac{\nabla l \times \nabla\alpha}{B}\nabla\beta + \boldsymbol{B}\cdot\frac{\nabla\alpha \times \nabla\beta}{B}\nabla l \\ &= (\hat{\boldsymbol{B}}\cdot\nabla\beta \times \nabla l)\nabla\alpha + (\hat{\boldsymbol{B}}\cdot\nabla l \times \nabla\alpha)\nabla\beta + B\nabla l \; .\end{aligned} \tag{5.3.13}$$

In the special case where ∇l happens to be equal to $\hat{\boldsymbol{B}}$, this reduces simply to $\boldsymbol{B} = B\hat{\boldsymbol{B}}$. In that case, the co- and contravariant representations coincide.

We shall now determine the proportionality constant in (5.1.1). Consider a surface l = constant. An infinitesimal-area element on this surface is equal to

$$d\boldsymbol{S}(l) = J\,\nabla l\, d\alpha\, d\beta = \frac{\nabla l}{B} d\alpha\, d\beta \ , \tag{5.3.14}$$

where we have set the Jacobian $J = 1/B$. The magnetic flux through an element $d\boldsymbol{S}(l)$ equals

$$d\Psi = \boldsymbol{B}\cdot d\boldsymbol{S}(l) = \frac{\boldsymbol{B}\cdot\nabla l}{B} d\alpha\, d\beta = d\alpha\, d\beta \ . \tag{5.3.15a}$$

This is shown in Fig. 5.2. Every surface S that is intersected by $\boldsymbol{B}$ has a similar (α, β) grid. $\boldsymbol{B}$-lines can never cross α or β surfaces, so they cannot leave a certain volume tube bounded by the surfaces α = constant, $\alpha + d\alpha$ = constant, β = constant and $\beta + d\beta$ = constant. Hence, our conclusion is that the magnetic flux in that tube must remain the same. Thus, the flux through any surface S intersecting $\boldsymbol{B} = \nabla\alpha \times \nabla\beta$ equals

$$\Psi = \iint d\alpha\, d\beta \ . \tag{5.3.15b}$$

In the case where we dealt with the stream functions $\tilde{\alpha}$ and $\tilde{\beta}$, $\boldsymbol{B}$ was written as $C\nabla\tilde{\alpha} \times \nabla\tilde{\beta}$. If we parameterize a field line by $\tilde{\alpha}$, $\tilde{\beta}$, and l, the Jacobian in these coordinates is

$$J = \frac{1}{\nabla l\cdot(\nabla\tilde{\alpha} \times \nabla\tilde{\beta})} = \frac{C}{\nabla l\cdot\boldsymbol{B}} = \frac{C}{B} \ . \tag{5.3.16}$$

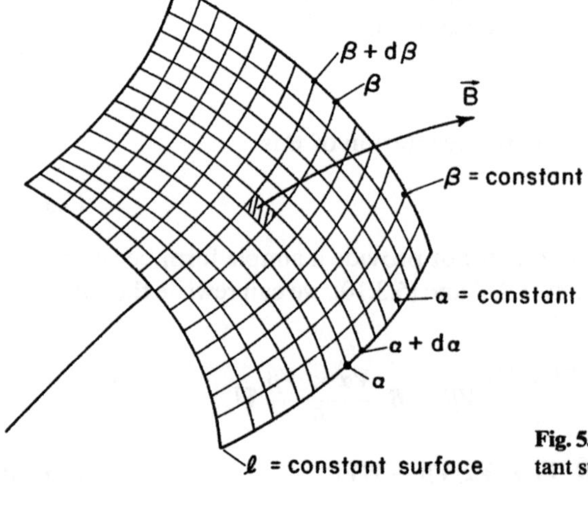

Fig. 5.2. (α, β) grid laid out on an l = constant surface

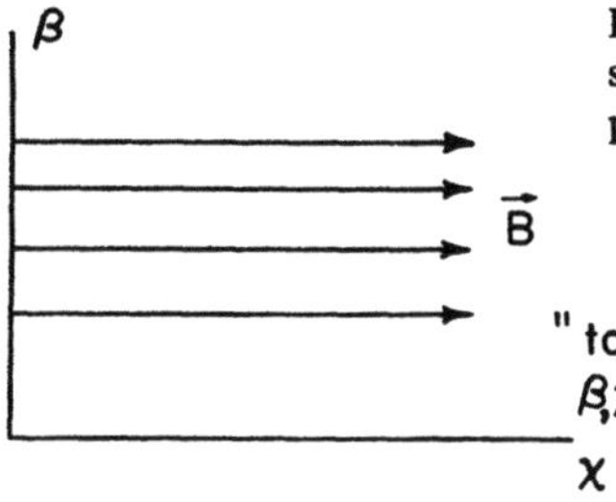

Fig. 5.3. Topological plot of the (β, χ) Boozer-Grad coordinate system showing that magnetic-field lines appear as straight lines, parallel to the χ axis

Consequently, the combination of $d\Psi = \boldsymbol{B} \cdot d\boldsymbol{S}(l)$, with $d\boldsymbol{S}(l)$ now equal to $C \nabla l \, d\tilde{\alpha} \, d\tilde{\beta}/B$ leads to

$$d\Psi = \frac{C}{B} \boldsymbol{B} \cdot \nabla l \, d\tilde{\alpha} \, d\tilde{\beta} = C \, d\tilde{\alpha} \, d\tilde{\beta} \tag{5.3.17a}$$

or

$$\Psi = \int\int C \, d\tilde{\alpha} \, d\tilde{\beta} \ . \tag{5.3.17b}$$

This explains what the proportionality sign in (5.1.1) means. Thus, α, β, $\tilde{\alpha}$ and $\tilde{\beta}$ are all stream functions.

One of the properties of the Clebsch coordinate system is that the field lines are straight lines in the coordinates other than α (the one labeling the flux surface), i.e., the field-line label β and the third coordinate which is used to indicate the "distance" along a field line (which may be but does not have to be the length l). We call this third variable χ. The differential equation for a field line in (β, χ) coordinates is

$$\frac{d\beta}{d\chi} = \frac{B^\beta}{B^\chi} \tag{5.3.18}$$

where $B^\beta = \boldsymbol{B} \cdot \nabla\beta$, $B^\chi = \boldsymbol{B} \cdot \nabla\chi$. With $\boldsymbol{B} = \nabla\alpha \times \nabla\beta$, this equation becomes

$$\frac{d\beta}{d\chi} = 0 \qquad \text{or} \qquad \beta = \text{constant.} \tag{5.3.19}$$

This means that the field line represents a straight line parallel to the χ axis (see Fig. 5.3 – note that this is a "topological" plot, since, in reality, the β and χ axes do not intersect at right angles).

5.4 Boozer-Grad Coordinates

Boozer-Grad coordinates are characterized by a particular choice of the third coordinate in the Clebsch representation. Recently, this coordinate system has been employed frequently by Boozer (1980, 1981) (after discovering it independently) but it was previously used by Grad (1971) (see also, Catto and Hazeltine 1981; White et al. 1983; and Betancourt and Garabedian 1985). We shall discuss the system for toroidal devices that possess flux surfaces, each of which is traced out by one single field line (at least for the irrational flux surfaces). A discussion for mirrors would be similar, except for a few minor points. (One of which is that $\partial \Phi/\partial l = 0$ does not imply that Φ is a flux-surface function in mirrors, but only that Φ is independent of l.)

In this coordinate system, we require that the contravariant representation for $\boldsymbol{B}$ is the usual Clebsch form

$$\boldsymbol{B} = \nabla\varrho \times \nabla\nu \ . \tag{5.4.1}$$

We have made a change of notation: $\alpha \equiv \varrho$ and $\beta \equiv \nu$. Although α and β are normally used as the flux-surface label and field-line label, respectively, several authors have used other notations. Boozer (1980) uses ψ as a flux-surface label and α as a field-line label. He uses β for the Clebsch coordinate of the current – see below. Betancourt and Garabedian (1985) use s as a flux label and ψ as a field-line label. We opted for a different set of variables ϱ and ν, since ϱ normally represents a radial variable, and we used it before as a flux-surface label. ν has been used with the same meaning as here by Solov'ev and Shafranov (1970).

In general, $\nabla\varrho \cdot \nabla\nu \neq 0$ and $\boldsymbol{e}_\varrho \cdot \boldsymbol{e}_\nu \neq 0$. The former implies that the $\nabla\nu$ direction is not tangent to a flux surface (as is often erroneously claimed), although it is perpendicular to $\boldsymbol{B}$. The covariant-basis vector $\boldsymbol{e}_\nu$ is by definition tangent to a flux surface but is not necessarily orthogonal to $\boldsymbol{B}$ (see, e.g., Figs. 4.2 and 5.4).

The idea of the Boozer-Grad coordinate system is to find a covariant representation which reduces to the gradient of a scalar function (the magnetic scalar potential) in a current free region. Rather than taking the arc length l along $\boldsymbol{B}$ as the third coordinate, a function $\tilde{\chi}$ is used such that the third term in the covariant expression, similar to (5.3.13), reads $\nabla\tilde{\chi}$ instead of $B\nabla l$. A second reason for working with $\tilde{\chi}$ rather than with l is that there is a simple relationship between $\tilde{\chi}$ and the angle variables θ and ζ, whereas, no such simple relationship exists between l and (θ, ζ). This issue will be discussed in detail later.

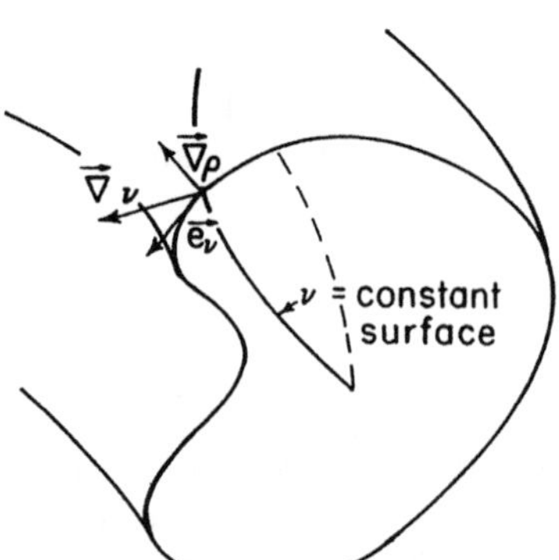

Fig. 5.4. ϱ = constant and ν = constant coordinate surfaces. The contravariant basis vector $\nabla\nu$ is perpendicular to the ν = constant surface, whereas the covariant basis vector $\boldsymbol{e}_\nu$ is parallel to the ϱ = constant surface. $\nabla\nu$ is in general not tangent to the flux surface ϱ = constant

As in the previous section, we assume that $\hat{\boldsymbol{l}} = +\hat{\boldsymbol{B}}$ for simplicity of notation.

Writing $\nabla\tilde{\chi}$ as the third term in the covariant form does *not* imply that $B\nabla l$ equals $\nabla\tilde{\chi}$. The main reason is that our new choice implies a new Jacobian, as well as new covariant-basis vectors. The covariant components B_ϱ and B_ν of the (ϱ, ν, l) system of (5.3.13), have now changed into B'_ϱ and B'_ν, respectively, and consequently, so has the vector $(B_\varrho \nabla\varrho + B_\nu \nabla\nu)$, which now becomes $(B'_\varrho \nabla\varrho + B'_\nu \nabla\nu)$ – which has in general, a different direction. Since both

$$B_\varrho \nabla\varrho + B_\nu \nabla\nu + B\nabla l \qquad \text{and} \qquad B'_\varrho \nabla\varrho + B'_\nu \nabla\nu + \nabla\tilde{\chi}$$

must add up to the same vector $\boldsymbol{B}$, it is clear that in general $\nabla\tilde{\chi}$ cannot be parallel to ∇l. (A rigorous proof of this will be given presently.)

For the covariant representation, we make the Ansatz in the third term

$$\boldsymbol{B} = [\boldsymbol{B}\cdot(\nabla\nu \times \nabla\tilde{\chi})J]\,\nabla\varrho + [\boldsymbol{B}\cdot(\nabla\tilde{\chi} \times \nabla\varrho)J]\,\nabla\nu + \nabla\tilde{\chi} \tag{5.4.2a}$$

or

$$\boldsymbol{B} = \tilde{\lambda}\nabla\varrho + \tilde{\xi}\nabla\nu + \nabla\tilde{\chi} \;. \tag{5.4.2b}$$

We have used the abbreviations

$$\tilde{\lambda} \equiv B'_\varrho \equiv \boldsymbol{B}\cdot(\nabla\nu \times \nabla\tilde{\chi})J$$

and

$$\tilde{\xi} \equiv B'_\nu \equiv \boldsymbol{B}\cdot(\nabla\tilde{\chi} \times \nabla\varrho)J \;.$$

The new Jacobian is easily found by taking the dot product of the contravariant and covariant forms of $\boldsymbol{B}$, (5.4.1) and (5.4.2):

$$J = (\nabla\tilde{\chi}\cdot\nabla\varrho \times \nabla\nu)^{-1} = \frac{1}{(B)^2} \;. \tag{5.4.3}$$

This result is also clear from our requirement, in (5.4.2a), that the third term is $\nabla\tilde{\chi} \equiv [\boldsymbol{B}\cdot(\nabla\varrho \times \nabla\nu)J]\,\nabla\tilde{\chi}$.

This Jacobian allows us to obtain some more insight concerning the function $\tilde{\chi}$. If we take $\hat{\boldsymbol{B}}\cdot\nabla\tilde{\chi}$, (5.4.3) gives

$$\hat{\boldsymbol{B}}\cdot\nabla\tilde{\chi} = \frac{\boldsymbol{B}\cdot\nabla\tilde{\chi}}{B} = \frac{\nabla\varrho \times \nabla\nu}{B}\cdot\nabla\tilde{\chi} = \frac{(B)^2}{B} \tag{5.4.4}$$

or thus

$$\hat{\boldsymbol{B}}\cdot\nabla\tilde{\chi} = \left.\frac{\partial\tilde{\chi}}{\partial l}\right|_{\varrho,\nu} = B \;. \tag{5.4.5}$$

This is an important result. The differential distance along a field line equals $d\tilde{\chi}/B$:

$$dl|_{\text{along } \boldsymbol{B}} = \left.\frac{d\tilde{\chi}}{B}\right|_{\text{along } \boldsymbol{B}} \;. \tag{5.4.6}$$

The tangent $\tilde{\chi}$-basis vector $\boldsymbol{e}_{\tilde{\chi}}$ is by definition equal to

$$\boldsymbol{e}_{\tilde{\chi}} = \frac{\boldsymbol{\nabla}\varrho \times \boldsymbol{\nabla}\nu}{\boldsymbol{\nabla}\tilde{\chi}\cdot(\boldsymbol{\nabla}\varrho \times \boldsymbol{\nabla}\nu)} = \frac{B\hat{\boldsymbol{B}}}{(B)^2} = \frac{\hat{\boldsymbol{B}}}{B} \quad \text{or} \quad \boldsymbol{e}_{\tilde{\chi}} = \frac{\hat{\boldsymbol{l}}}{B} . \tag{5.4.7}$$

From (5.4.7), it also follows that

$$\boldsymbol{e}_{\tilde{\chi}}\cdot \boldsymbol{\nabla} l = \frac{\hat{\boldsymbol{l}}}{B}\cdot \boldsymbol{\nabla} l = \frac{1}{B} . \tag{5.4.8}$$

To show that $B\boldsymbol{\nabla} l \neq \boldsymbol{\nabla}\tilde{\chi}$ for general magnetic-field structures, we apply the chain rule to $\boldsymbol{\nabla}\tilde{\chi}$ in ϱ, ν, l coordinates

$$\boldsymbol{\nabla}\tilde{\chi} = \frac{\partial\tilde{\chi}}{\partial\varrho}\boldsymbol{\nabla}\varrho + \frac{\partial\tilde{\chi}}{\partial\nu}\boldsymbol{\nabla}\nu + \frac{\partial\tilde{\chi}}{\partial l}\boldsymbol{\nabla} l . \tag{5.4.9}$$

Because of (5.4.5), this gives:

$$\boldsymbol{\nabla}\tilde{\chi} = B\boldsymbol{\nabla} l + \left.\frac{\partial\tilde{\chi}}{\partial\varrho}\right|_{\nu,l}\boldsymbol{\nabla}\varrho + \left.\frac{\partial\tilde{\chi}}{\partial\nu}\right|_{\varrho,l}\boldsymbol{\nabla}\nu . \tag{5.4.10}$$

If $B\boldsymbol{\nabla} l$ were equal to $\boldsymbol{\nabla}\tilde{\chi}$, then we would conclude that either $\boldsymbol{\nabla}\varrho = -[(\partial\tilde{\chi}/\partial\nu)/(\partial\tilde{\chi}/\partial\varrho)]\boldsymbol{\nabla}\nu$, or that $\partial\tilde{\chi}/\partial\varrho$ and $\partial\tilde{\chi}/\partial\nu$ are zero. The former is impossible by the definition of the functions ϱ and ν. The latter is not true by the argument which will now be developed. The fact that $\partial\tilde{\chi}/\partial l = B$, (5.4.5), implies that

$$\tilde{\chi}(\varrho,\nu,l) - \tilde{\chi}_0(\varrho,\nu) = \int_0^l B(\varrho,\nu,l)dl . \tag{5.4.11}$$

$\tilde{\chi}_0$ is the value of $\tilde{\chi}$ at $l = 0$. In general, $B = |\boldsymbol{B}|$ depends on all three coordinates, (ϱ, ν, l). Then the following statement must exhibit a contradiction:

$$\left.\frac{\partial\tilde{\chi}}{\partial\varrho}\right|_{\nu,l} \stackrel{?}{=} 0 \stackrel{?}{=} \frac{\partial\tilde{\chi}_0}{\partial\varrho} + \int_0^l \frac{\partial B}{\partial\varrho}(\varrho,\nu,l)dl \tag{5.4.12}$$

Indeed, $\partial\tilde{\chi}_0/\partial\varrho$ is independent of l, while the integral is dependent on l unless the partial derivative $\partial B/\partial\varrho$ vanishes for all l, which is not the case for general B. An analogous argument can be developed for $\partial\tilde{\chi}/\partial\nu$.

Next, we want to redefine $\tilde{\chi}$, $\tilde{\lambda}$, and $\tilde{\xi}$ such that (5.4.2b) simplifies to a form like

$$\boldsymbol{B} = \lambda\boldsymbol{\nabla}\varrho + \boldsymbol{\nabla}\chi . \tag{5.4.13}$$

In other words, we want the term involving $\tilde{\xi}$ to vanish. However, this forces us to change $\tilde{\chi}$ to χ, and $\tilde{\lambda}$ to λ as well. In order to do so, we make use of the ideal MHD force balance

$$\boldsymbol{\nabla} p = \boldsymbol{J} \times \boldsymbol{B} . \tag{5.4.14}$$

($\boldsymbol{J}$ here is the current and not the Jacobian.) From $\boldsymbol{B}\cdot\nabla p = 0$, we conclude that since, in a toroidal system, $\boldsymbol{B}$ traces out flux surfaces ϱ = constant, p is a function of ϱ only, $p \equiv p(\varrho)$. Furthermore, the condition $\boldsymbol{J}\cdot\nabla p = 0$ is equivalent to $\boldsymbol{J}\cdot\nabla\varrho\, dp/d\varrho = 0$ or

$$\boldsymbol{J}\cdot\nabla\varrho = 0 \tag{5.4.15}$$

(except possibly at points where $dp/d\varrho = 0$). This says that the current-density lines lie in constant ϱ surfaces.

Ampère's law permits us to find an expression for $\boldsymbol{J}$. From $\boldsymbol{J} = (1/\mu_0)\nabla\times\boldsymbol{B}$, we obtain with the help of (5.4.2b)

$$\boldsymbol{J} = \frac{1}{\mu_0}\nabla\times[\tilde{\lambda}\nabla\varrho + \tilde{\xi}\nabla v + \nabla\tilde{\chi}] = \frac{1}{\mu_0}[(\nabla\tilde{\lambda}\times\nabla\varrho) + (\nabla\tilde{\xi}\times\nabla v)] \ . \tag{5.4.16}$$

The condition $\boldsymbol{J}\cdot\nabla\varrho = 0$ implies that

$$\nabla\varrho\cdot(\nabla\tilde{\xi}\times\nabla v) = 0$$

$$(\nabla\varrho\times\nabla v)\cdot\nabla\tilde{\xi} = 0$$

so

$$\boldsymbol{B}\cdot\nabla\tilde{\xi} = 0 \tag{5.4.17}$$

or

$$\frac{\partial\tilde{\xi}}{\partial l} = 0 \ . \tag{5.4.18}$$

In a toroidal device with flux surfaces traced out by single field lines, this states that $\tilde{\xi}$ is a flux-surface quantity

$$\tilde{\xi} \equiv \tilde{\xi}(\varrho) \ . \tag{5.4.19}$$

If we look back at our original covariant form for $\boldsymbol{B}$, $\boldsymbol{B} = \tilde{\lambda}\nabla\varrho + \tilde{\xi}\nabla v + \nabla\tilde{\chi}$, we observe that the term $\tilde{\xi}(\varrho)\nabla v$ pushes the field line in the ∇v direction (which would potentially cause a secular behavior) (see Fig. 5.4). A field line does not move in the ∇v direction, however. Because $\boldsymbol{B}\cdot\nabla v = 0$, field lines lie in constant v surfaces. From (5.4.10) and the discussion below (5.4.12), we see that $\nabla\tilde{\chi}$ does have components in the $\nabla\varrho$ and ∇v directions. The terms $\nabla\tilde{\chi}$ and $\tilde{\lambda}\nabla\varrho$ will cancel the "secular" $\tilde{\xi}\nabla v$ term. By looking for a new χ, we shall try to absorb the component of $\nabla\tilde{\chi}$ which lies in the ∇v direction into the new $\nabla\chi$. (Equation (5.4.2b) does not imply that $\boldsymbol{B}$ has a component along $\nabla\varrho$ either. Since $\boldsymbol{B}\cdot\nabla\varrho = 0$, the two other terms $\tilde{\xi}\nabla v$ and $\nabla\tilde{\chi}$ must have components in the $\nabla\varrho$ direction to make the net component of $\boldsymbol{B}$ along $\nabla\varrho$ vanish.)

Because $\tilde{\xi}$ is a flux function, its gradient is along the $\nabla\varrho$ direction: $\nabla\tilde{\xi} = \dot{\tilde{\xi}}\nabla\varrho$, where the dot represents the derivative with respect to ϱ, $\dot{\tilde{\xi}} \equiv d\tilde{\xi}/d\varrho$. The expression for the current density in (5.4.16) can be written successively as follows:

$$\begin{aligned}\mu_0 \boldsymbol{J} &= (\nabla\tilde{\lambda} \times \nabla\varrho) + (\dot{\tilde{\xi}}\nabla\varrho \times \nabla\nu)\\ &= \nabla\varrho \times [\dot{\tilde{\xi}}\nabla\nu - \nabla\tilde{\lambda}]\\ &= \nabla\varrho \times [\nabla(\dot{\tilde{\xi}}\nu) - \nabla\tilde{\lambda}]\\ &= \nabla\varrho \times \nabla(\dot{\tilde{\xi}}\nu - \tilde{\lambda}) \ . \end{aligned} \tag{5.4.20}$$

Introducing a new function λ, such that the current has a simple Clebsch representation, gives

$$\mu_0 \boldsymbol{J} = \nabla\lambda \times \nabla\varrho \tag{5.4.21}$$

with

$$\lambda \equiv (\tilde{\lambda} - \dot{\tilde{\xi}}\nu) \ . \tag{5.4.22}$$

Thus, λ is a Clebsch (stream) function for the current density $\boldsymbol{J}$.

We now substitute the $\tilde{\lambda}$ from (5.4.22) in the covariant $\boldsymbol{B}$ for (5.4.2b), $\boldsymbol{B} = \tilde{\lambda}\nabla\varrho + \tilde{\xi}\nabla\nu + \nabla\tilde{\chi}$, and obtain

$$\begin{aligned}\boldsymbol{B} &= (\lambda + \dot{\tilde{\xi}}\nu)\nabla\varrho + \tilde{\xi}\nabla\nu + \nabla\tilde{\chi}\\ &= \lambda\nabla\varrho + (\dot{\tilde{\xi}}\nu\nabla\varrho + \tilde{\xi}\nabla\nu) + \nabla\tilde{\chi}\\ &= \lambda\nabla\varrho + \nabla(\tilde{\xi}\nu) + \nabla\tilde{\chi}\\ &= \lambda\nabla\varrho + \nabla(\tilde{\xi}\nu + \tilde{\chi}) \ . \end{aligned} \tag{5.4.23}$$

When we define the new function χ as

$$\chi \equiv \tilde{\chi} + \tilde{\xi}\nu \ , \tag{5.4.24}$$

the magnetic field becomes in *covariant form*:

$$\boldsymbol{B} = \lambda\nabla\varrho + \nabla\chi \ . \tag{5.4.25}$$

Since, in general, $\nabla\nu \cdot \nabla\varrho \neq 0 \neq \nabla\nu \cdot \nabla\chi$, both $\nabla\varrho$ and $\nabla\chi$ have components along $\nabla\nu$. However, they offset each other so that no secular term remains.

In summary, to arrive at the simple covariant form given in (5.4.25), we have made two substitutions:

$$\begin{aligned}\chi &= \tilde{\chi} + \tilde{\xi}\nu\\ \lambda &= \tilde{\lambda} - \dot{\tilde{\xi}}\nu \ . \end{aligned}$$

Neither $\tilde{\xi}$ nor ν varies with l. Consequently,

$$\hat{\boldsymbol{B}} \cdot \nabla\chi = \frac{\partial\chi}{\partial l} = \frac{\partial\tilde{\chi}}{\partial l} = B \ .$$

Because of this result, the relationships expressed in (5.4.3), (5.4.5), and (5.4.7) are still valid for χ as well as $\tilde{\chi}$:

$$\hat{\boldsymbol{B}} \cdot \nabla \chi = B \tag{5.4.26}$$

$$J^{-1} = \nabla \chi \cdot (\nabla \varrho \times \nabla v) = (B)^2 \tag{5.4.27}$$

$$\boldsymbol{e}_\chi = \frac{\hat{\boldsymbol{l}}}{B} = \frac{\hat{\boldsymbol{B}}}{B} \, . \tag{5.4.28}$$

Having established the identity of the new function χ, we can rewrite the "*complete*" *contravariant form* of $\boldsymbol{B}$. Multiplying the Clebsch representation in (5.4.1) by J in the numerator and the denominator, and using (5.4.27), we obtain

$$\boldsymbol{B} = (B)^2 \boldsymbol{e}_\chi = \nabla \varrho \times \nabla v \, . \tag{5.4.29}$$

These are the two equivalent contravariant representations of $\boldsymbol{B}$.

It is not easy to visualize the $\chi =$ constant surfaces. In Chap. 6, we shall show that the $v =$ constant surfaces are ruled surfaces in the toroidal coordinates (θ_f, ζ_f) in which the magnetic field is "straight", with an equation of the form $v = A(\varrho)\theta_f - D(\varrho)\zeta_f$, (6.6.14b) (see also the discussion related to Fig. 4.17, below (4.9.15)). The v surfaces are traced out by "straight" field lines. Similarly, we shall show there that the $\chi =$ constant surfaces are ruled surfaces traced out by a "straight" line, obeying an equation of the form $\chi = a(\varrho)\theta_f + d(\varrho)\zeta_f$, (6.6.14c) (which intersects a field line).

The (ϱ, v, χ) Boozer-Grad coordinate system suffers from the same shortcomings when it comes to single valuedness in toroidal devices, as those encountered with the (α, β, l) coordinates. Again, this system is useful for formal derivations, but practical calculations are better done after conversion to an angle system. This transformation will be discussed in an upcoming section.

Finally, we show that the λ Clebsch coordinate of the current density can be related to the pressure gradient and the Pfirsch-Schlüter current. Following Boozer (1980), we obtain from (5.4.21) with $\nabla \lambda = (\partial \lambda / \partial \varrho) \nabla \varrho + (\partial \lambda / \partial v) \nabla v + (\partial \lambda / \partial \chi) \nabla \chi$

$$\mu_0 \boldsymbol{J} = \frac{\partial \lambda}{\partial v} \nabla v \times \nabla \varrho + \frac{\partial \lambda}{\partial \chi} \nabla \chi \times \nabla \varrho \tag{5.4.30}$$

or

$$\mu_0 \boldsymbol{J} = -\boldsymbol{B} \frac{\partial \lambda}{\partial v} + \nabla \chi \times \nabla \varrho \frac{\partial \lambda}{\partial \chi} \, . \tag{5.4.31}$$

Since $\hat{\boldsymbol{B}} \cdot \nabla \varrho \times \nabla \chi = \hat{\boldsymbol{B}} \cdot \nabla \varrho \times (\boldsymbol{B} - \lambda \nabla \varrho) = 0$, the parallel component of (5.4.30) leads to

$$\frac{\partial \lambda}{\partial v} = -\mu_0 \frac{J_\parallel}{B} \, . \tag{5.4.32}$$

To find an interesting relationship for λ from the perpendicular component, we substitute $\boldsymbol{J}$ in $\nabla p = \boldsymbol{J} \times \boldsymbol{B}$. Recalling that $p \equiv p(\varrho)$, we get from (5.4.30)

$$\dot{p}\nabla\varrho = \frac{1}{\mu_0}\left(\nabla\chi \times \nabla\varrho\frac{\partial\lambda}{\partial\chi}\right) \times \boldsymbol{B} \tag{5.4.33}$$

which, with the *bac-cab* rule, becomes

$$\begin{aligned}\dot{p}\nabla\varrho &= \frac{1}{\mu_0}\frac{\partial\lambda}{\partial\chi}[\nabla\varrho(\boldsymbol{B}\cdot\nabla\chi) - \nabla\chi(\boldsymbol{B}\cdot\nabla\varrho)]\\ &= \frac{1}{\mu_0}\frac{\partial\lambda}{\partial\chi}[\nabla\varrho(\nabla\varrho \times \nabla v\cdot\nabla\chi)]\\ &= \frac{(B)^2}{\mu_0}\frac{\partial\lambda}{\partial\chi}\nabla\varrho\ .\end{aligned}$$

Equating the coefficients of both sides, we obtain an expression for $\partial\lambda/\partial\chi$:

$$\frac{\partial\lambda}{\partial\chi} = \frac{\mu_0}{(B)^2}\frac{dp}{d\varrho}\ . \tag{5.4.34}$$

The function λ is determined via the differential equations, (5.4.32) and (5.4.34), and is thus determined up to an additive constant (constant in v and χ). To take away some of the arbitrariness of λ (and χ), we impose the initial condition at $\chi = v = 0$,

$$\lambda(\varrho, v = 0, \chi = 0) = 0\ . \tag{5.4.35}$$

This condition implies that $\lambda = 0$ when $\boldsymbol{J} = 0$, as we shall show momentarily. Indeed, $\boldsymbol{J} = 0$ makes $J_\parallel$ and $\dot{p}$ vanish so that we obtain from (5.4.32) and (5.4.34),

$$\lambda \equiv \lambda(\varrho) \quad \text{when} \quad \boldsymbol{J} = 0\ . \tag{5.4.36}$$

This can also be seen from (5.4.21). The "initial" condition (5.4.36) then sets

$$\lambda = 0 \quad \text{when} \quad \boldsymbol{J} = 0\ . \tag{5.4.37}$$

In vacuum, the covariant form of $\boldsymbol{B}$ becomes

$$\boldsymbol{B} = \nabla\chi, \qquad \boldsymbol{J} = 0\ . \tag{5.4.38}$$

When $\boldsymbol{J} = 0$, the function χ is the magnetic scalar potential.

However, this still does not remove all the arbitrariness of the (ϱ, v, χ) coordinate system. As with any Clebsch coordinate, v is determined only to within an additive flux function (cf. (5.2.13)). If $v^* = v + g(\varrho)$, $\boldsymbol{B}$ is unchanged: $\boldsymbol{B} = \nabla\varrho \times \nabla v = \nabla\varrho \times \nabla v^*$. Since v does not appear in the covariant form, this change in v does not force any other quantities to change. The magnetic field and the properties of χ are also left the same if we construct a new χ by adding a flux function. That the properties of a new $\chi^* = \chi + f(\varrho)$ do not change has already been made clear when we substituted χ for $\tilde{\chi}$, using (5.4.24). We saw that χ retains its properties as long as the additive term is independent of l. This is certainly true for a flux function. However, changing χ to χ^* forces λ to change to a new λ^* as follows:

$$\chi^* = \chi + f(\varrho) \Rightarrow \lambda^* = \lambda - \frac{df}{d\varrho} . \tag{5.4.39}$$

6. Toroidal Flux Coordinates

6.1 Straight Field-Line Coordinates

Toroidal flux coordinates are a set of poloidal and toroidal "angles" θ_f and ζ_f chosen such that the equation of a field line is that of a straight line in those coordinates. It is common to say that the magnetic-field lines appear as straight lines, or that the magnetic field *is* straight in (θ_f, ζ_f).

To construct functions $\theta_f(\boldsymbol{R})$ and $\zeta_f(\boldsymbol{R})$ it is necessary that the *magnetic field* $\boldsymbol{B}$ *is known* completely. In this section, we assume the vector $\boldsymbol{B}$ to be known everywhere as a function of arbitrary angle coordinates θ and ζ. (These angles could, for example, be the "elementary" angles θ_e and ζ_e.) In a later section, we shall see how to transform from Cartesian components of $\boldsymbol{B}$ to the desired flux coordinates (θ_f, ζ_f).

The first coordinate of the system will be the flux-surface label ϱ. Surfaces of constant ϱ are the flux surfaces. ϱ stands for any flux-surface function and can be "replaced" anywhere by a quantity such as the poloidal or toroidal flux enclosed by (or outside of) the flux surface Ψ_{tor}, Ψ_{pol}, the volume of a flux surface V, the pressure on that flux surface p, the rotational transform or its inverse, ι, $q, \ldots$ (as long as $dp/d\varrho \neq 0$, $d\iota/d\varrho \neq 0$, etc. at places other than the magnetic axis). Recall that we only accept radially increasing ϱ's. Thus, $\varrho \equiv -p$; $\varrho \equiv -\Psi^{\text{d}}_{\text{pol}}$; $\varrho \equiv \iota$ and $\varrho \equiv -q$ in most stellarators; $\varrho \equiv q$ and $\varrho \equiv -\iota$ in tokamaks.

The angles θ and ζ on the surface are defined by *multiple-valued functions* $\theta(\boldsymbol{R})$ and $\zeta(\boldsymbol{R})$, in the sense that a point on the flux surface has (θ, ζ) coordinates equal to $\theta + 2\pi m$ and $\zeta + 2\pi n$, where m and n are arbitrary integers. Their *gradients must be single valued*, however (i.e., periodic functions of θ and ζ).

The magnetic field $\boldsymbol{B}$ can be written in terms of its contravariant components in the (ϱ, θ, ζ) coordinate system as usual:

$$\boldsymbol{B} = B^{\varrho} \boldsymbol{e}_{\varrho} + B^{\theta} \boldsymbol{e}_{\theta} + B^{\zeta} \boldsymbol{e}_{\zeta} \ , \tag{6.1.1}$$

where $B^i = \boldsymbol{B} \cdot \nabla u^i$. From the definition of a flux surface,

$$\boldsymbol{B} \cdot \nabla \varrho = 0 \ , \tag{6.1.2}$$

we conclude immediately that

$$B^{\varrho} = 0 \ . \tag{6.1.3}$$

Thus $\boldsymbol{B}$ equals:

$$\boldsymbol{B} = B^\theta \boldsymbol{e}_\theta + B^\zeta \boldsymbol{e}_\zeta = B^\theta \sqrt{g}(\nabla\zeta \times \nabla\varrho) + B^\zeta \sqrt{g}(\nabla\varrho \times \nabla\theta) \ . \tag{6.1.4}$$

Here we used $\boldsymbol{e}_i = (\nabla u^j \times \nabla u^k)\sqrt{g}$, where $\sqrt{g} = [\nabla\varrho \cdot (\nabla\theta \times \nabla\zeta)]^{-1}$ is the Jacobian. Recall that $g = (J)^2$, with g the determinant of the metric coefficient matrix. Instead of using J for the Jacobian, we shall henceforth use the symbol $\sqrt{g}$. (In the flux-coordinate context, the symbol J is often used for a current-density flux, which is a "current", not a current-density. However, we shall not use the symbol J at all to avoid confusion with the Jacobian or the magnitude of the current-density.) Note that g, in general, is a function of (ϱ, θ, ζ): $g \equiv g(\varrho, \theta, \zeta)$; the Jacobian is a known single valued (i.e., periodic) function in (θ, ζ).

The two components B^θ and B^ζ are not independent of each other as can be seen from $\nabla \cdot \boldsymbol{B} = 0$,

$$\frac{\partial}{\partial\theta}(\sqrt{g}B^\theta) + \frac{\partial}{\partial\zeta}(\sqrt{g}B^\zeta) = 0 \ . \tag{6.1.5}$$

This relationship suggests that B^θ and B^ζ are derived from a single function $v \equiv v(\varrho, \theta, \zeta)$ as follows:

$$B^\theta = -\frac{1}{\sqrt{g}}\frac{\partial v}{\partial\zeta} \ , \qquad B^\zeta = \frac{1}{\sqrt{g}}\frac{\partial v}{\partial\theta} \ . \tag{6.1.6a, b}$$

These B^θ and B^ζ satisfy (6.1.5), as can be checked by substitution. We can now rewrite (6.1.4):

$$\boldsymbol{B} = \nabla\varrho \times \left(\frac{\partial v}{\partial\zeta}\nabla\zeta + \frac{\partial v}{\partial\theta}\nabla\theta\right) \ . \tag{6.1.7}$$

Adding a term $(\partial v/\partial\varrho)\nabla\varrho$ between the brackets leaves $\boldsymbol{B}$ unchanged, since $\nabla\varrho \times \nabla\varrho = 0$. Then we can use $\nabla v = (\partial v/\partial\varrho)\nabla\varrho + (\partial v/\partial\theta)\nabla\theta + (\partial v/\partial\zeta)\nabla\zeta$, allowing us to write the simple Clebsch form for $\boldsymbol{B}$:

$$\boldsymbol{B} = \nabla\varrho \times \nabla v \ . \tag{6.1.8}$$

This equation states that the $\boldsymbol{B}$ lines lie on constant v surfaces. The same conclusion can be drawn from the equation of a field line on a $\varrho = \text{constant}$ surface. If we substitute (6.1.6) in $d\theta/B^\theta = d\zeta/B^\zeta$, we have that $dv = 0$, since $v \equiv v(\varrho, \theta, \zeta)$ and $d\varrho = 0$ on a flux surface. Hence, the equation of a field line lying in a flux surface $\varrho = \varrho_0$ is

$$v(\varrho_0, \theta, \zeta) = \text{constant} \ . \tag{6.1.9}$$

From the expression in (6.1.7), we can infer what form the function $v(\varrho, \theta, \zeta)$ must have. $\boldsymbol{B}$ is a physical quantity and is thus single valued, i.e., its components must be periodic in (θ, ζ). We need not require the gradient of v to be single valued (periodic), but $\nabla\varrho \times \nabla v$ must be. Because of the cross product with $\nabla\varrho$, the ϱ

dependence of ν is immaterial as far as $\nabla\varrho \times \nabla\nu$ is concerned. If ν is a periodic function in θ and ζ, then $\nabla\nu$ is periodic and so evidently is $\nabla\varrho \times \nabla\nu$. Hence with regard to the requirement that $\nabla\varrho \times \nabla\nu$ be single valued, a function $\nu(\varrho, \theta, \zeta)$, with an arbitrary ϱ dependence but which is periodic in (θ, ζ) is permitted. Any non-periodic function which is not a simple polynomial in θ or ζ must be rejected, since $\nabla\varrho \times \nabla\nu$ would then have non-periodic terms containing θ or ζ, implying multiple valuedness (the reader may wish to try functions like $\exp(\theta)$, $\ln(\zeta)$, $1/(1+\zeta^2), \ldots$). A function ν having the form of a polynomial in, say θ,

$$a(\varrho) + b(\varrho)\theta + c(\varrho)\theta^2 + \ldots \tag{6.1.10}$$

is acceptable provided we keep only the first two terms

$$a(\varrho) + b(\varrho)\theta \ . \tag{6.1.11}$$

Otherwise, $\nabla\nu$ would contain terms like $(\partial(c\theta^2)/\partial\theta)\,\nabla\theta = 2c\theta\,\nabla\theta$, which is not acceptable because of the multiple valuedness in θ which remains after the cross product with $\nabla\varrho$. Note that with (6.1.11), we would have that $\nabla\nu = (\partial a/\partial\varrho)\,\nabla\varrho + (\partial b/\partial\varrho)\theta\,\nabla\varrho + b\,\nabla\theta$, which contains a secular term involving θ as well. However, this is permissible, since it contains a $\nabla\varrho$, and $\nabla\varrho \times (\ )$ eliminates that term.

Similar arguments hold for polynomials in ζ. Our conclusion is that the most general form of $\nu(\varrho, \theta, \zeta)$, consistent with the periodicity of $\nabla\varrho \times \nabla\nu$ is

$$\nu(\varrho, \theta, \zeta) = A(\varrho) + B(\varrho)\theta + C(\varrho)\zeta + \tilde{\nu}(\varrho, \theta, \zeta) \ . \tag{6.1.12}$$

Here, $\tilde{\nu}$ represents a periodic function in θ and ζ. $A(\varrho)$ can be absorbed into $\tilde{\nu}$ so that we write:

$$\nu(\varrho, \theta, \zeta) = B(\varrho)\theta + C(\varrho)\zeta + \tilde{\nu}(\varrho, \theta, \zeta) \ . \tag{6.1.13}$$

One could have come to the same conclusion by requiring that the components B^θ and B^ζ in (6.1.6) are periodic. The form of (6.1.13) can rigorously be obtained if one expands both $\sqrt{g}B^\theta$ and $\sqrt{g}B^\zeta$ in Fourier series in θ and ζ. Then one could separate the terms $n = 0$ and $m = 0$; $n = 0$ and $m \neq 0$; $n \neq 0$ and $m = 0$; and m and n different from zero and manipulate these forms according to the prescriptions expressed by the differential equations, (6.1.6). To actually solve for ν, one could integrate (6.1.6a) with respect to ζ such that the result contains an integration constant which is only a function of θ. Then that form of ν is differentiated with respect to θ and is set equal to $\sqrt{g}B^\zeta$ according to (6.1.6b). This would provide an equation for the unknown function of θ, which could be solved for, after integration with respect to θ (up to an additive constant).

B^θ, B^ζ and $\sqrt{g}$ are known functions of ϱ; so the ϱ dependence of ν is fixed. The ϱ-dependent functions $B(\varrho)$ and $C(\varrho)$ are found from (6.1.6) as follows. Recall the form-invariant expression for Ψ_{tor} (4.7.10)

$$\Psi_{\text{tor}} = \frac{1}{2\pi} \iiint\limits_V \boldsymbol{B} \cdot \nabla\zeta \, d^3R \ . \tag{6.1.14}$$

The volume element d^3R stands for $\sqrt{g}\,d\varrho\,d\theta\,d\zeta$. The derivative of Ψ_{tor} with respect to ϱ, is equal to:

$$\dot{\Psi}_{tor} \equiv \frac{d\Psi_{tor}}{d\varrho} = \frac{1}{2\pi}\int_0^{2\pi}\int_0^{2\pi} d\theta\,d\zeta\,\sqrt{g}\boldsymbol{B}\cdot\nabla\zeta\ . \tag{6.1.15}$$

In the literature, this quantity, $\dot{\Psi}_{tor}$ is sometimes referred to as a magnetic flux density (see, e.g., Hirshman and Sigmar 1981). We do not use this name here, since a flux density in physics is commonly identified with the magnetic induction (field) $\boldsymbol{B}$. We can substitute $\sqrt{g}\boldsymbol{B}\cdot\nabla\zeta = \sqrt{g}B^\zeta$ from (6.1.6b) in order to obtain

$$\dot{\Psi}_{tor} = \frac{1}{2\pi}\int_0^{2\pi}\int_0^{2\pi} d\theta\,d\zeta\,\frac{\partial v}{\partial\theta}\ . \tag{6.1.16}$$

From the general form of v in (6.1.13), we find that:

$$\frac{\partial v}{\partial\theta} = B(\varrho) + \frac{\partial\tilde{v}}{\partial\theta}\ . \tag{6.1.17}$$

The integral over θ of the function $\partial\tilde{v}/\partial\theta$ vanishes because $\tilde{v}$ is periodic in θ. The other integration is trivial:

$$\dot{\Psi}_{tor} = 2\pi B(\varrho)\ . \tag{6.1.18}$$

Analogous manipulations for the flux Ψ^{T}_{pol} lead to a similar result:

$$\dot{\Psi}^{T}_{pol} \equiv \frac{d\Psi^{T}_{pol}}{d\varrho} = \frac{1}{2\pi}\int_0^{2\pi}\int_0^{2\pi} d\theta\,d\zeta\,\sqrt{g}\boldsymbol{B}\cdot\nabla\theta$$

$$= \frac{-1}{2\pi}\int_0^{2\pi}\int_0^{2\pi} d\theta\,d\zeta\,\frac{\partial v}{\partial\zeta} = -2\pi C(\varrho)\ . \tag{6.1.19}$$

As a result of (6.1.18) and (6.1.19), the function v becomes:

$$v(\varrho,\theta,\zeta) = \frac{1}{2\pi}(\dot{\Psi}_{tor}\theta - \dot{\Psi}^{T}_{pol}\zeta) + \tilde{v}(\varrho,\theta,\zeta)\ . \tag{6.1.20}$$

If $\tilde{v}$ in (6.1.20) happens to be zero or a flux-surface function, the θ and ζ coordinates we started from are already flux coordinates. Indeed, because the field lines lie on constant v surfaces (6.1.9) and (6.1.20) would give for the equation of a field line

$$\dot{\Psi}_{tor}\theta - \dot{\Psi}^{T}_{pol}\zeta = \text{constant}\ . \tag{6.1.21}$$

This is the equation of a straight line in (θ,ζ) coordinates. Thus, $\zeta_f = \zeta$ and $\theta_f = \theta$.

If $\tilde{v}$ is a function of θ and ζ, we perform a change of variables in order to eliminate $\tilde{v}$:

$$\theta_f = \theta + 2\pi \frac{\tilde{\nu}}{\dot{\Psi}_{tor}} \qquad \text{and} \qquad \zeta_f = \zeta \tag{6.1.22a}$$

or

$$\theta_f = \theta \qquad \text{and} \qquad \zeta_f = \zeta - 2\pi \frac{\tilde{\nu}}{\dot{\Psi}^{T}_{pol}} \ . \tag{6.1.22b}$$

Note that these transformations are consistent with (4.6.2). The function ν in terms of θ_f and ζ_f becomes:

$$\nu(\varrho, \theta_f, \zeta_f) = \frac{1}{2\pi}\dot{\Psi}_{tor}\theta_f - \frac{1}{2\pi}\dot{\Psi}^{T}_{pol}\zeta_f \ . \tag{6.1.23}$$

Again, $\nu =$ constant is the equation of a straight line

$$\dot{\Psi}_{tor}\theta_f - \dot{\Psi}^{T}_{pol}\zeta_f = \text{constant} \ . \tag{6.1.24}$$

The contravariant components of $\boldsymbol{B}$ are still defined by (6.1.6) but with θ replaced by θ_f and ζ by ζ_f:

$$B^{\varrho} = 0 \tag{6.1.25a}$$

$$B^{\theta_f} = \frac{1}{2\pi\sqrt{g_f}}\dot{\Psi}^{T}_{pol} \tag{6.1.25b}$$

$$B^{\zeta_f} = \frac{1}{2\pi\sqrt{g_f}}\dot{\Psi}_{tor} \ . \tag{6.1.25c}$$

From these equations for B^{θ_f} and B^{ζ_f}, we observe that

$$\frac{B^{\theta_f}}{B^{\zeta_f}} = \frac{\dot{\Psi}^{T}_{pol}}{\dot{\Psi}_{tor}} = \text{flux function} \ , \tag{6.1.26}$$

thus proving a result that we anticipated before, in (4.8.10), during the "derivation" of the expression for the rotational transform. The fact that B^{θ_f}/B^{ζ_f} is a flux function is crucial for the straightness of $\boldsymbol{B}$. From the equation for a field line,

$$\frac{d\theta_f}{d\zeta_f} = \frac{B^{\theta_f}}{B^{\zeta_f}} = \frac{\dot{\Psi}^{T}_{pol}}{\dot{\Psi}_{tor}} = \iota\!\!\!-(\varrho) \ , \tag{6.1.27}$$

we see that the slope of the line is indeed constant on the flux surface, and equal to the rotational transform $\iota\!\!\!-$.

In summary, a transformation of one of the coordinates is sufficient to make the field lines straight. In tokamaks, it is customary to retain the symmetry angle (which measures rotation about the major axis) $\zeta = \zeta_f$. Consequently, θ then must be changed.

We should remark that the angles θ_f and ζ_f are not the only ones in which the magnetic field appears straight. Indeed, for example, if we further transform the θ_f and ζ_f coordinates into new coordinates θ_F and ζ_F, as follows:

$$\theta_F = \theta_f + \dot{\Psi}^r_{pol} G(\varrho, \theta_f, \zeta_f) \tag{6.1.28a}$$

$$\zeta_F = \zeta_f + \dot{\Psi}_{tor} G(\varrho, \theta_f, \zeta_f) \ , \tag{6.1.28b}$$

where G is an arbitrary periodic function, we see that upon substitution of these equations into (6.1.23) for v, we obtain

$$v(\varrho, \theta_F, \zeta_F) = \frac{1}{2\pi} \dot{\Psi}_{tor} \theta_F - \frac{1}{2\pi} \dot{\Psi}^r_{pol} \zeta_F \ , \tag{6.1.29}$$

showing that θ_F and ζ_F are flux coordinates as well. One can use this degree of freedom to deform the coordinates further to make other quantities look simpler. In Hamada coordinates, the system is deformed until the current-density lines become straight in addition to the magnetic-field lines. Boozer uses this freedom to free the magnetic scalar potential (that would exist in vacuum) from a periodic part. In axisymmetric devices, it is customary to choose θ_f such that $\nabla\theta_f \perp \nabla\zeta_o$, where ζ_o is the ignorable coordinate. These three choices have been discussed qualitatively above, but will be considered in detail below.

Using the flux coordinates θ_f and ζ_f, the magnetic field in Clebsch form is equal to:

$$\boldsymbol{B} = \nabla\varrho \times \nabla\left(\frac{\dot{\Psi}_{tor}}{2\pi}\theta_f - \frac{\dot{\Psi}^r_{pol}}{2\pi}\zeta_f\right) . \tag{6.1.30}$$

In terms of the poloidal flux through a disk, Ψ^d_{pol}, the expression for $\boldsymbol{B}$ is

$$\boldsymbol{B} = \nabla\varrho \times \nabla\left(\frac{\dot{\Psi}_{tor}}{2\pi}\theta_f + \frac{\dot{\Psi}^d_{pol}}{2\pi}\zeta_f\right) . \tag{6.1.31}$$

Here we have used the fact that $d\Psi^r_{pol} = -d\Psi^d_{pol}$ (see (4.7.4)).

If the flux surface volume V is chosen as a flux-surface label, (6.1.30) becomes

$$\boldsymbol{B} = \nabla V \times \nabla\left(\frac{1}{2\pi}\frac{d\Psi_{tor}}{dV}\theta_f - \frac{1}{2\pi}\frac{d\Psi^r_{pol}}{dV}\zeta_f\right) . \tag{6.1.32}$$

In tokamaks, it is customary to use the poloidal flux as a flux label, and to use the safety factor q. If one takes $\varrho \equiv \Psi^r_{pol}/2\pi \equiv \Psi$, one obtains from (6.1.30)

$$\boldsymbol{B} = \nabla\Psi \times \nabla(q\theta_f - \zeta_f) . \tag{6.1.33}$$

For the poloidal disk flux as a flux label, the expression is exactly the same as in (6.1.33), but with $\varrho \equiv \Psi \equiv -\Psi^d_{pol}/2\pi$.

In stellarators, the choice of flux label ϱ is commonly Ψ_{tor}, such that with $\varrho \equiv \psi \equiv \Psi_{tor}/2\pi$, we obtain

$$\boldsymbol{B} = \nabla\psi \times \nabla(\theta_f - \iota\zeta_f) . \tag{6.1.34}$$

The general Clebsch form of (6.1.30) can be written as a sum of two terms

$$\boldsymbol{B} = \frac{\dot{\Psi}^{\mathrm{r}}_{\mathrm{pol}}}{2\pi}(\boldsymbol{\nabla}\zeta_{\mathrm{f}} \times \boldsymbol{\nabla}\varrho) + \frac{\dot{\Psi}_{\mathrm{tor}}}{2\pi}(\boldsymbol{\nabla}\varrho \times \boldsymbol{\nabla}\theta_{\mathrm{f}}) \equiv \boldsymbol{B}_{\mathrm{P}} + \boldsymbol{B}_{\mathrm{T}} \ . \tag{6.1.35}$$

$\boldsymbol{B}_{\mathrm{P}}$ is called the poloidal field, and $\boldsymbol{B}_{\mathrm{T}}$ the toroidal field. Note that $\boldsymbol{B}_{\mathrm{P}}$ does not lie along $\boldsymbol{\nabla}\theta_{\mathrm{f}}$ but along $\boldsymbol{e}_{\theta_{\mathrm{f}}}$. Analogously, $\boldsymbol{B}_{\mathrm{T}}$ is not directed along $\boldsymbol{\nabla}\zeta_{\mathrm{f}}$ but along $\boldsymbol{e}_{\zeta_{\mathrm{f}}}$. Only in axisymmetric devices is $\boldsymbol{\nabla}\zeta_{\mathrm{f}} \propto \boldsymbol{e}_{\zeta_{\mathrm{f}}} \propto \boldsymbol{\zeta}_{\mathrm{f}}$, so, $\boldsymbol{B}_{\mathrm{T}} \propto \boldsymbol{\zeta}_{\mathrm{f}}$ if ζ_{f} is chosen as symmetry angle.

The covariant components of $\boldsymbol{B}$, $B_i = \boldsymbol{B}\cdot\boldsymbol{e}_i$ can readily be found from the contravariant components B^i via the use of the metric coefficients:

$$B_i = (g_{ij})_{\mathrm{f}}B^j = g_{i\theta_{\mathrm{f}}}B^{\theta_{\mathrm{f}}} + g_{i\zeta_{\mathrm{f}}}B^{\zeta_{\mathrm{f}}} \ . \tag{6.1.36}$$

The metric coefficients are those of the flux-coordinate system $(\theta_{\mathrm{f}}, \zeta_{\mathrm{f}})$.

With the form of $\boldsymbol{B}$ given in (6.1.30), the symbol $\sqrt{g_{\mathrm{f}}}$ could be replaced in all of the above as follows:

$$\sqrt{g_{\mathrm{f}}} = \sqrt{g_{\mathrm{f}}(\varrho, \theta_{\mathrm{f}}, \zeta_{\mathrm{f}})} = \frac{\dot{\Psi}^{\mathrm{r}}_{\mathrm{pol}}}{2\pi}(\boldsymbol{B}\cdot\boldsymbol{\nabla}\theta_{\mathrm{f}})^{-1} \ , \tag{6.1.37}$$

or, for example, in the tokamak case (6.1.33):

$$\sqrt{g_{\mathrm{f}}} = \sqrt{g_{\mathrm{f}}(\Psi, \theta_{\mathrm{f}}, \zeta_{\mathrm{f}})} = (\boldsymbol{B}\cdot\boldsymbol{\nabla}\theta_{\mathrm{f}})^{-1} \ . \tag{6.1.38}$$

The straight $\boldsymbol{B}$ coordinate systems are discussed in many references. They basically all follow the treatment discussed in Appendix I of the classic paper by Greene and Johnson (1962). The possibility of writing $\boldsymbol{B}$ as $\boldsymbol{\nabla}\varrho \times \boldsymbol{\nabla}v$ was suggested by Kruskal and Kulsrud (1958), and the transformation to angles in which $\boldsymbol{B}$ lines are straight, was, according to Greene and Johnson, first used by Kadomtsev (1960) and Mercier (1961) – Refs. 3 and 8 in the Greene and Johnson paper. Discussions similar to those in Greene and Johnson (1962) can be found in Bateman (1978), p. 125 (note a disturbing misprint on page 126: the designations contravariant and covariant basis vectors must be reversed); Solov'ev and Shafranov (1970); Miyamoto (1980); Hazeltine and Meiss (1985); Kieras (1982). Important also are the contributions by Shafranov (1968), and Solov'ev (1975). For an alternative approach to straight $\boldsymbol{B}$ toroidal-flux cooordinates, see the sequence of Boozer's papers (1980–1983).

6.2 Symmetry Flux Coordinates in a Tokamak

In an ideal axisymmetric tokamak, common sense suggests that we take the symmetry angle ζ_{o} as the toroidal angle. The ζ_{o}-coordinate surfaces are vertical planes so that $\boldsymbol{\nabla}\zeta_{\mathrm{o}}$ points in the symmetry direction. In axisymmetric devices the flux surfaces are such that

$$\nabla\Psi \cdot \nabla\zeta_o = 0 \ . \tag{6.2.1}$$

To have $\partial/\partial\zeta_o|_{\theta_f,\varrho} \equiv 0$ for every physical quantity, the $\theta_f =$ constant surfaces must be chosen so that

$$\nabla\zeta_o \cdot \nabla\theta_o = 0 \ . \tag{6.2.2}$$

We use a subscript "o" for angles (flux coordinates or not) that satisfy (6.2.1) and (6.2.2) to indicate their symmetry aspects. (Thus we drop the subscript "f" in favor of "o".) These statements are equivalent to requiring that the ζ_o-coordinate curves are circles, whose tangents point in the symmetry direction:

$$\boldsymbol{e}_{\zeta_o} \propto \nabla\zeta_o \ . \tag{6.2.3}$$

In general, we *cannot* take $\nabla\varrho \cdot \nabla\theta_o = 0$. If we insist on doing so, we have a completely orthogonal system, and this spoils the straightness of $\boldsymbol{B}$ lines; because of the Shafranov shift amongst other effects, the $\theta =$ constant and $\varrho =$ constant surfaces are not orthogonal. Because of (6.2.1) and (6.2.2), we can say that we work in a partially orthogonal system.

The fact that ζ_o is an ignorable coordinate allows us to write $\boldsymbol{B}$ in an alternative form. The $\boldsymbol{B}$-field representation in Clebsch form is still valid in (θ_o, ζ_o) *flux coordinates*

$$\boldsymbol{B} = \nabla\Psi \times \nabla(q\theta_o - \zeta_o) = q(\nabla\Psi \times \nabla\theta_o) + (\nabla\zeta_o \times \nabla\Psi) \ . \tag{6.2.4}$$

We have chosen the poloidal flux as a flux label for simplicity (see (6.1.33)). We represent the distance from the major (symmetry) axis to a particular point on the surface by R; it represents the radial coordinate in a cylindrical coordinate system $(R, \zeta_c = \pi/2 - \zeta_o, z)$. It is then clear that

$$|\nabla\zeta_o| = \frac{1}{R} \ . \tag{6.2.5}$$

From orthogonality, $\nabla\Psi \cdot \nabla\zeta_o = 0 = \nabla\theta_o \cdot \nabla\zeta_o$, it follows that $\nabla\Psi \times \nabla\theta_o$ is parallel to $\nabla\zeta_o$:

$$\nabla\Psi \times \nabla\theta_o = k\nabla\zeta_o \ . \tag{6.2.6}$$

k is the proportionality factor. The Jacobian $\sqrt{g_o}$ is defined as usual by $(\nabla\zeta_o \cdot \nabla\Psi \times \nabla\theta_o)^{-1}$, so that from (6.2.6)

$$\sqrt{g_o} = \frac{1}{k|\nabla\zeta_o|^2} \ . \tag{6.2.7}$$

This can be solved for k, as g_o is known,

$$k = \frac{1}{\sqrt{g_o}\,|\nabla\zeta_o|^2} \ . \tag{6.2.8}$$

The toroidal field $\boldsymbol{B}_{\mathrm{T}} = q\nabla\Psi \times \nabla\theta_{\mathrm{o}}$ can thus be written upon combining (6.2.6) and (6.2.8)

$$\boldsymbol{B}_{\mathrm{T}} = q(\nabla\Psi \times \nabla\theta_{\mathrm{o}}) = \frac{q\nabla\zeta_{\mathrm{o}}}{\sqrt{g_{\mathrm{o}}}\,|\nabla\zeta_{\mathrm{o}}|^2} \tag{6.2.9}$$

$$= \frac{qR^2}{\sqrt{g_{\mathrm{o}}}}\nabla\zeta_{\mathrm{o}} \ . \tag{6.2.10}$$

It is customary to introduce a separate symbol for the factor in front of the $\nabla\zeta_{\mathrm{o}}$:

$$I \equiv \frac{qR^2}{\sqrt{g_{\mathrm{o}}}} \ . \tag{6.2.11}$$

This I can also be written as

$$I = \sqrt{g_{\mathrm{o}}}(\nabla\Psi \times \nabla\theta_{\mathrm{o}})\cdot\boldsymbol{B} = \frac{\sqrt{g_{\mathrm{o}}}}{q}\boldsymbol{B}_{\mathrm{T}}\cdot\boldsymbol{B} = \frac{\sqrt{g_{\mathrm{o}}}}{q}\boldsymbol{B}_{\mathrm{T}}\cdot\boldsymbol{B}_{\mathrm{T}} \ . \tag{6.2.12}$$

The magnetic field in a tokamak then becomes

$$\boldsymbol{B} = I\nabla\zeta_{\mathrm{o}} + (\nabla\zeta_{\mathrm{o}} \times \nabla\Psi) \ , \tag{6.2.13}$$

which is the form used most often in the tokamak literature.

An important feature of the "function" I is that it is a flux-surface function: $I \equiv I(\Psi)$ for an isotropic pressure plasma equilibrium. This can be shown by considering Ampère's law:

$$\nabla \times \boldsymbol{B} = \mu_0\boldsymbol{J} \tag{6.2.14}$$

$$(\nabla \times \boldsymbol{B})^i = \frac{1}{\sqrt{g}}\left(\frac{\partial}{\partial u^j}B_k - \frac{\partial}{\partial u^k}B_j\right) , \qquad i, j, k \text{ cyc } 1, 2, 3 \ . \tag{6.2.15}$$

For the radial (Ψ) component, this becomes

$$J^{\Psi} = \mu_0^{-1}(\nabla \times \boldsymbol{B})^{\Psi} = \frac{1}{\mu_0\sqrt{g_{\mathrm{o}}}}\left(\frac{\partial}{\partial\theta_{\mathrm{o}}}B_{\zeta_{\mathrm{o}}} - \frac{\partial}{\partial\zeta_{\mathrm{o}}}B_{\theta_{\mathrm{o}}}\right) . \tag{6.2.16}$$

Because, on the one hand, axisymmetry implies that $\partial/\partial\zeta_{\mathrm{o}} \equiv 0$, and on the other hand $\boldsymbol{J}\cdot\nabla\Psi = 0$ requires that $J^{\Psi} = 0$, we have

$$\frac{\partial}{\partial\theta_{\mathrm{o}}}B_{\zeta_{\mathrm{o}}} = 0 \ . \tag{6.2.17}$$

The third covariant component of $\boldsymbol{B}$, $B_3 = \boldsymbol{B}\cdot\boldsymbol{e}_3$ turns out to be just equal to I:

$$B_{\zeta_{\mathrm{o}}} \equiv \boldsymbol{B}\cdot\boldsymbol{e}_{\zeta_{\mathrm{o}}} = \boldsymbol{B}\cdot(\nabla\Psi \times \nabla\theta_{\mathrm{o}})\sqrt{g_{\mathrm{o}}} = I \ . \tag{6.2.18}$$

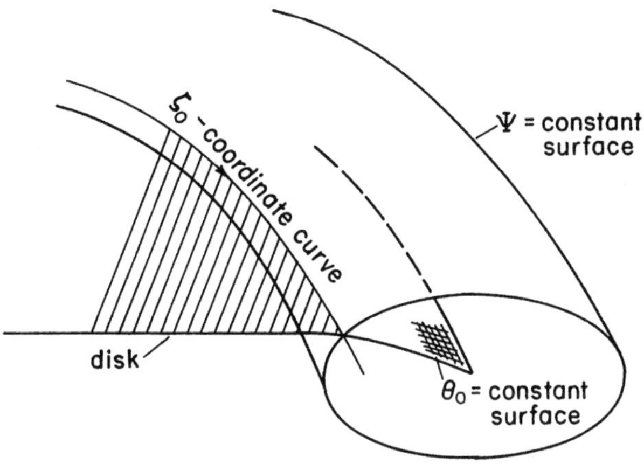

Fig. 6.1. A ζ_o-coordinate curve intersecting a constant Ψ and θ_o surface in an axisymmetric tokamak

Here we used (6.2.12). Thus,

$$\frac{\partial}{\partial \theta_o} I = 0 \ , \tag{6.2.19}$$

and because of axisymmetry,

$$\frac{\partial}{\partial \zeta_o} I = 0 \ ,$$

hence,

$$I \equiv I(\Psi) \ . \tag{6.2.20}$$

(When viscosity is included in the force balance, I is a function of θ as well.) The standard form for $\boldsymbol{B}$ is thus

$$\boldsymbol{B} = I(\Psi)\,\nabla\zeta_o + (\nabla\zeta_o \times \nabla\Psi) \ . \tag{6.2.21}$$

To clarify the physical meaning of the function $I \equiv I(\Psi)$, we consider the integral form of Ampère's law:

$$\oint \boldsymbol{B}\cdot d\boldsymbol{l} = \mu_0 \times (\text{enclosed current}) \ . \tag{6.2.22}$$

We apply this loop integral along a ζ_o-coordinate curve for θ_o values on the inside of a toroidal surface as shown in Fig. 6.1. In Chap. 2, we found that the differential arc length along a coordinate curve u^i equals $dl(i) = |\boldsymbol{e}_i|\,du^i$ (without summation over i, because along the u^i curve, $du^j = du^k = 0$ for $j \neq i \neq k$). We then have that $d\boldsymbol{l}_{\zeta_o} = \boldsymbol{e}_{\zeta_o} d\zeta_o$. From (6.2.3), it is easily established that $\boldsymbol{e}_{\theta_o}\cdot\boldsymbol{e}_{\zeta_o} = 0$. Then with $\boldsymbol{B} = I\,\nabla\zeta_o + \nabla\zeta_o \times \nabla\Psi$ and $\nabla\zeta_o \times \nabla\Psi \propto \boldsymbol{e}_{\theta_o}$, we have from the LHS of (6.2.22)

$$\oint (I\nabla\zeta_o + \nabla\zeta_o \times \nabla\Psi)\cdot e_{\zeta_o} d\zeta_o = \oint I\nabla\zeta_o \cdot e_{\zeta_o} d\zeta_o = 2\pi I \ . \tag{6.2.23}$$

The "enclosed current" is the total poloidal current outside the flux surface which flows through the disk in Fig. 6.1. We represent this enclosed current by $I^{\mathrm{d}}_{\mathrm{pol}}$ in analogy with our notations for magnetic fluxes. In conclusion, we have

$$I \equiv \frac{\mu_0}{2\pi} I^{\mathrm{d}}_{\mathrm{pol}} \ . \tag{6.2.24}$$

Thus the function $I \equiv I(\Psi)$ effectively measures the total poloidal current outside the flux surface $\Psi =$ constant. $I^{\mathrm{d}}_{\mathrm{pol}}$ contains both plasma currents and the toroidal-field-coil currents.

6.3 Interlude: Non-Flux Coordinates in Tokamaks

In the previous section we saw that given a proper choice of the *flux coordinates* θ_{f} and ζ_{f} in tokamaks (denoted here by θ_o and ζ_o because they satisfy (6.2.1) and (6.2.2)), the Clebsch form of (6.1.33) can be written in the alternative form of (6.2.21):

$$\boldsymbol{B} = \nabla\Psi \times \nabla(q\theta_o - \zeta_o) \rightarrow \boldsymbol{B} = I(\Psi)\nabla\zeta_o + (\nabla\zeta_o \times \nabla\Psi) \ . \tag{6.3.1}$$

In this section we shall show that axisymmetry, which permits us to choose angles θ_o and ζ_o satisfying (6.2.1) and (6.2.2), always allows us to write $\boldsymbol{B}$ in a form like (6.2.21), even when θ_o and ζ_o are not flux coordinates. Only when θ_o and ζ_o are flux coordinates in addition to satisfying (6.2.1) and (6.2.2), can the arrow in (6.3.1) be reversed. Below we shall discuss the requirement which must be imposed on θ_o in order for it to become a flux-coordinate angle. Furthermore, we shall show how to construct a flux-coordinate angle $\theta_{o,\mathrm{f}} \equiv \Theta_o$ given the non-flux coordinate θ_o.

To see that a form like (6.2.21) is possible even in a non-flux coordinate system, we argue as follows. Consider the poloidal flux through a disk,

$$\Psi^{\mathrm{d}}_{\mathrm{pol}} \equiv \iint_{S_{\mathrm{disk}}} \boldsymbol{B}\cdot d\boldsymbol{S} = \iint_{S_{\mathrm{disk}}} d\boldsymbol{S}\cdot \nabla\times\boldsymbol{A} = \oint_{\Gamma} d\boldsymbol{l}\cdot\boldsymbol{A} \ . \tag{6.3.2}$$

The contour Γ along which the loop integral must be taken is the boundary curve of the disk (where the disk touches the axisymmetric toroidal magnetic surface). Because of axisymmetry, Γ is a circle with circumference $2\pi R$, and $\boldsymbol{A}$ must be constant along that loop: the magnetic vector potential $\boldsymbol{A}$ is a physical quantity, not merely a mathematical construct. Hence, we obtain the simple result:

$$\Psi^{\mathrm{d}}_{\mathrm{pol}} = 2\pi A_{\zeta_o} = 2\pi R\tilde{A}_{\zeta_o} \ . \tag{6.3.3}$$

Here $\boldsymbol{A}_{\mathrm{T}} \equiv A_{\zeta_o}\nabla\zeta_o = \tilde{A}_{\zeta_o}\hat{\zeta}$ so $\tilde{A}_{\zeta_o} = A_{\zeta_o}/R$. ($\tilde{A}_{\zeta_o}$ is the *physical* toroidal component, as opposed to the contravariant or covariant components.) We define the function Ψ as

$$\Psi \equiv -\frac{\Psi_{\text{pol}}^{\text{d}}}{2\pi} \,. \tag{6.3.4}$$

The minus sign is to make Ψ increase radially. Then,

$$\Psi = -A_{\zeta_o} = -R\tilde{A}_{\zeta_o} \,. \tag{6.3.5}$$

We can now define the poloidal magnetic field $\boldsymbol{B}_\text{P}$. It is the field that is responsible for the flux through the disk; it is thus everything except the toroidal field $\boldsymbol{B}_\text{T}$, which is along $\nabla\zeta_o$. We have

$$\begin{aligned}\boldsymbol{B}_\text{P} \equiv \nabla \times \boldsymbol{A}_\text{T} &= \nabla \times (A_{\zeta_o} \nabla\zeta_o) \\ &= A_{\zeta_o}(\nabla \times \nabla\zeta_o) + (\nabla A_{\zeta_o} \times \nabla\zeta_o) \,.\end{aligned} \tag{6.3.6}$$

The first term vanishes and $A_{\zeta_o} = -\Psi$, so

$$\boldsymbol{B}_\text{P} = -\nabla\Psi \times \nabla\zeta_o \,. \tag{6.3.7}$$

Ψ is thus a stream function for the poloidal field $\boldsymbol{B}_\text{P}$.

To find the field $\boldsymbol{B} = \boldsymbol{B}_\text{P} + \boldsymbol{B}_\text{T}$, we write $\boldsymbol{B}_\text{T}$ (the field along $\nabla\zeta_o$) as

$$\boldsymbol{B}_\text{T} = B_{\zeta_o} \nabla\zeta_o \,. \tag{6.3.8}$$

If we now use the symbol I to denote $B_{\zeta_o} = R\tilde{B}_{\zeta_o}$ we obtain

$$\boldsymbol{B} = I\,\nabla\zeta_o - (\nabla\Psi \times \nabla\zeta_o) \,. \tag{6.3.9}$$

$B_{\zeta_o} \equiv I$ is a flux function, the proof of which (see (6.2.14–20)) applies also to the present situation. In conclusion, we find

$$\boldsymbol{B} = I(\Psi)\,\nabla\zeta_o + (\nabla\zeta_o \times \nabla\Psi) \,, \tag{6.3.10}$$

which is the required expression. The flux function $I = I(\Psi)$ is

$$I = RB_\text{T} \,. \tag{6.3.11}$$

Because of (6.2.1), expressing orthogonality between $\nabla\zeta_o$ and $\nabla\Psi$, it follows from (6.3.10) that

$$|\nabla\Psi| = RB_\text{P} \,. \tag{6.3.12}$$

So far we have not mentioned the angle θ_o. Let us find out which θ_o we must choose to convert (6.3.9) into a Clebsch form. The conditions in (6.2.1) and (6.2.2) permit us to write $\nabla\zeta_o$ in (6.3.9) as a cross product, $\nabla\zeta_o = c(\nabla\Psi \times \nabla\theta_o)$. To find c, we dot multiply this last expression by $\nabla\zeta_o$ and find:

$$\nabla\zeta_o = \frac{\sqrt{g}}{R^2}(\nabla\Psi \times \nabla\theta_o) \,. \tag{6.3.13}$$

Here $g \equiv g(\Psi, \theta_o, \zeta_o)$. Equation (6.3.9) for $\boldsymbol{B}$ is now

$$\boldsymbol{B} = \nabla\Psi \times \left[\frac{I(\Psi)\sqrt{g}}{R^2}\,\nabla\theta_o - \nabla\zeta_o\right] . \tag{6.3.14}$$

If $\sqrt{g}/R^2$ is a flux-surface function (and only then), we can move the ∇ operator in front of the square brackets, leading to a Clebsch form (the $\nabla\Psi \times (\ )$ will eliminate the secular $\theta_o[d(I\sqrt{g}/R^2)/d\Psi]\,\nabla\Psi$ term). If θ_o were a flux coordinate, the Jacobian would have been such that (6.2.11) was satisfied: $I\sqrt{g}/R^2 = q(\Psi)$. Thus, we conclude that it is possible to go from a form like (6.3.9) to a Clebsch form provided θ_o is chosen such that (6.2.11) is satisfied and thus

$$\sqrt{g} \propto R^2 \tag{6.3.15}$$

in order to make $I\sqrt{g}/R^2$ a flux surface quantity.

Note that even in the "elementary" toroidal system (θ_e, ζ_e) (but based on the magnetic axis), we can write $\boldsymbol{B}$ as in (6.3.9) regardless of the shape of the tokamak's flux surfaces. However, it is not possible to write $\boldsymbol{B}$ in a Clebsch form in terms of θ_e.

Because $\boldsymbol{B} = I\,\nabla\zeta_o + \nabla\zeta_o \times \nabla\Psi$ in an axisymmetric device, we can easily write an explicit expression for a newly constructed flux-coordinate angle, $\theta_{o,f} \equiv \Theta_o$, in terms of the old poloidal angle θ_o.

The stream function ν of (6.1.20) expressed in the non-flux coordinates $(\varrho = \Psi, \theta = \theta_o, \zeta = \zeta_o)$ is

$$\nu(\Psi,\theta_o,\zeta_o) = (q\theta_o - \zeta_o) + \tilde{\nu}(\Psi,\theta_o)\ . \tag{6.3.16}$$

Because of axisymmetry, $\tilde{\nu}$ is independent of ζ_o; q is the safety factor defined in (4.8.3) and (4.8.4): $q \equiv d\Psi_{\rm tor}/d\Psi^{\rm r}_{\rm pol} = -d\Psi_{\rm tor}/d\Psi^{\rm d}_{\rm pol}$. For this $\nu(\Psi,\theta_o,\zeta_o)$, we can compute the B^{ζ_o} component from (6.1.6b):

$$\boldsymbol{B}\cdot\nabla\zeta_o = B^{\zeta_o} = \frac{1}{\sqrt{g}}\frac{\partial\nu}{\partial\theta_o} = \frac{1}{\sqrt{g}}\left(q + \frac{\partial\tilde{\nu}}{\partial\theta_o}\right)\ . \tag{6.3.17}$$

Integration with respect to θ_o gives:

$$\tilde{\nu}(\Psi,\theta_o) = -q\theta_o + \int_0^{\theta_o} \sqrt{g}(\boldsymbol{B}\cdot\nabla\zeta_o)\,d\theta_o\ . \tag{6.3.18}$$

From $\boldsymbol{B}$ in (6.3.10) we have that $\boldsymbol{B}\cdot\nabla\zeta_o = I/R^2$, leading to

$$\tilde{\nu}(\Psi,\theta_o) = -q\theta_o + \int_0^{\theta_o} \frac{I\sqrt{g}}{R^2}\,d\theta_o\ . \tag{6.3.19}$$

To find a flux-coordinate angle $\theta_{o,f} \equiv \Theta_o$, we apply the transformation in (6.1.22a) for $\varrho \equiv \Psi \equiv -\Psi^{\rm d}_{\rm pol}/2\pi$ in order to find $\theta_{o,f} = \theta_o + \tilde{\nu}/q$, or

$$\theta_{o,f} \equiv \Theta_o = \frac{1}{q}\int_0^{\theta_o} \sqrt{g}\,\frac{I}{R^2}\,d\theta_o\ . \tag{6.3.20}$$

This expression shows explicitly that θ_o is already a flux coordinate, equal to Θ_o, if (6.2.11) is satisfied.

6.4 Straight Current-Density-Line Coordinates

The current density $\boldsymbol{J}$ satisfies two important relationships:

$$\boldsymbol{J}\cdot\nabla\varrho = 0 \tag{6.4.1}$$

$$\nabla\cdot\boldsymbol{J} = 0\ . \tag{6.4.2}$$

Similar equations are obeyed by the magnetic field $\boldsymbol{B}$. This means that we can use the same procedure for the current density $\boldsymbol{J}$ as we did for $\boldsymbol{B}$ for we looked for coordinates in which $\boldsymbol{B}$ is straight. Our goal this time is to start from arbitrary angle coordinates θ and ζ and to find coordinates in which the current-density lines are straight (regardless of what the $\boldsymbol{B}$ lines look like). We will call these coordinates *current coordinates*, and represent them by θ_J and ζ_J. The treatment is analogous to that for $\boldsymbol{B}$ in Sect. 6.1., so we omit details.

The current density $\boldsymbol{J}$ is presumed to be known. From (6.4.2), we have:

$$\frac{\partial}{\partial\theta}(\sqrt{g}J^{\theta}) + \frac{\partial}{\partial\zeta}(\sqrt{g}J^{\zeta}) = 0\ , \tag{6.4.3}$$

implying that we can write J^{θ} and J^{ζ} as derivatives of a Clebsch (stream) function $\eta(\varrho,\theta,\zeta)$:

$$J^{\theta} = -\frac{1}{\sqrt{g}}\frac{\partial\eta}{\partial\zeta}\ ,\qquad J^{\zeta} = \frac{1}{\sqrt{g}}\frac{\partial\eta}{\partial\theta}\ . \tag{6.4.4}$$

This means that $\boldsymbol{J}$ is expressible in a Clebsch form:

$$\boldsymbol{J} = \nabla\varrho\times\nabla\eta \tag{6.4.5}$$

such that the current-density lines on a flux surface (ϱ = constant) are defined by the equation

$$\eta(\varrho,\theta,\zeta) = \text{constant}\ . \tag{6.4.6}$$

η is of the form

$$\eta(\varrho,\theta,\zeta) = \frac{1}{2\pi}\dot{I}_{\text{tor}}(\varrho)\theta - \frac{1}{2\pi}\dot{I}^{\text{r}}_{\text{pol}}(\varrho)\zeta + \tilde{\eta}(\varrho,\theta,\zeta)\ , \tag{6.4.7}$$

Here, $\dot{I}^{\text{r}}_{\text{pol}}(\varrho) \equiv dI^{\text{r}}_{\text{pol}}/d\varrho$, where $I^{\text{r}}_{\text{pol}}(\varrho)$ represents the poloidal current flowing *within* the flux surface. $I^{\text{r}}_{\text{pol}}$ is the poloidal current-density flux (just as $\Psi^{\text{r}}_{\text{pol}}$ is the poloidal magnetic-density flux):

$$I^{\text{r}}_{\text{pol}} = \frac{1}{2\pi}\iiint_V (\boldsymbol{J}\cdot\nabla\theta)\,d^3R = \iint_{S_{\text{pol}}} \boldsymbol{J}\cdot d\boldsymbol{S}\ . \tag{6.4.8}$$

One could have used the poloidal current *outside* a flux surface instead, by analogy with the disk flux $\Psi^{\text{d}}_{\text{pol}}$. In that case, the current would include the coil current. The flux function $I_{\mathrm{tor}}(\varrho)$ represents the total toroidal current flowing within the volume enclosed by the flux surface.

$$I_{\mathrm{tor}}(\varrho) = \frac{1}{2\pi} \iiint_V (\boldsymbol{J} \cdot \boldsymbol{\nabla}\zeta)\, d^3R = \iint_{S_{\mathrm{tor}}} \boldsymbol{J} \cdot d\boldsymbol{S} \ . \tag{6.4.9}$$

$\tilde{\eta}$ is a periodic function in θ and ζ. To make $\boldsymbol{J}$-lines straight, we must get rid of $\tilde{\eta}$ or make it a flux function. A transformation similar to (6.1.22a, b) will suffice:

$$\theta_J = \theta + \frac{2\pi\tilde{\eta}}{\dot{I}_{\mathrm{tor}}} \quad \text{and} \quad \zeta_J = \zeta \tag{6.4.10a}$$

or

$$\theta_J = \theta \quad \text{and} \quad \zeta_J = \zeta - \frac{2\pi\tilde{\eta}}{\dot{I}^{\mathrm{r}}_{\mathrm{pol}}} \ . \tag{6.4.10b}$$

The equation of a current-density line in (θ_J, ζ_J) coordinates is

$$\frac{\dot{I}_{\mathrm{tor}}}{2\pi}\theta_J - \frac{\dot{I}^{\mathrm{r}}_{\mathrm{pol}}}{2\pi}\zeta_J = \text{constant} \ , \tag{6.4.11}$$

and the Clebsch form for $\boldsymbol{J}$ reads

$$\boldsymbol{J} = \boldsymbol{\nabla}\varrho \times \boldsymbol{\nabla}\left(\frac{\dot{I}_{\mathrm{tor}}}{2\pi}\theta_J - \frac{\dot{I}^{\mathrm{r}}_{\mathrm{pol}}}{2\pi}\zeta_J\right) \ . \tag{6.4.12}$$

The contravariant components of $\boldsymbol{J}$ are

$$J^{\varrho} = 0 \tag{6.4.13a}$$

$$J^{\theta_J} = \frac{\dot{I}^{\mathrm{r}}_{\mathrm{pol}}}{2\pi\sqrt{g_J}} \tag{6.4.13b}$$

$$J^{\zeta_J} = \frac{\dot{I}_{\mathrm{tor}}}{2\pi\sqrt{g_J}} \ . \tag{6.4.13c}$$

Usually, straight $\boldsymbol{J}$-lines are not used if the magnetic field is not straight. It would be interesting, however, if we could start with flux coordinates θ_{f} and ζ_{f} in which $\boldsymbol{B}$ is straight, and deform these so that in addition to $\boldsymbol{B}$, $\boldsymbol{J}$ also becomes straight. If we replaced the θ and ζ by θ_{f} and ζ_{f} in the development of this section, everything would remain correct until (6.4.9). However, the transformation equations, (6.4.10), would indeed make $\boldsymbol{J}$ straight, but they would spoil the straightness of $\boldsymbol{B}$. Consequently, we have to invent new transformation relations that make $\boldsymbol{J}$ straight but keep $\boldsymbol{B}$ straight. A transformation of the form given in (6.1.28) is adequate (i.e., make $\tilde{\eta}$ vanish or make it a flux function, while keeping the $\tilde{\nu}$ of $\boldsymbol{B}$ equal to zero or a flux function). We must deform *both* θ_{f} and ζ_{f}. The coordinates created in this way are known as Hamada coordinates. They are characterized by a very simple Jacobian. This system will be discussed in Sect. 6.8.

6.5 Covariant B Components and Their Relationship to the Boozer-Grad Form

In this section, we further investigate a flux coordinate system in which $\boldsymbol{B}$ is straight regardless of the form of the $\boldsymbol{J}$ lines. This means that we set $\theta = \theta_f$ and $\zeta = \zeta_f$ in the previous section. We also keep $\tilde{\eta}$ non-zero in (6.4.7).

6.5.1 The Vector Potential

The contravariant components of $\boldsymbol{B}$, expressed in flux coordinates, are according to (6.1.25):

$$B^i = \frac{1}{2\pi\sqrt{g_f}}(0, \dot{\Psi}^{\mathrm{T}}_{\mathrm{pol}}, \dot{\Psi}_{\mathrm{tor}}) \; . \tag{6.5.1}$$

These components were derived from a function $\nu(\varrho, \theta_f, \zeta_f)$ in (6.1.23)

$$\nu(\varrho, \theta_f, \zeta_f) = \frac{\dot{\Psi}_{\mathrm{tor}}}{2\pi}\theta_f - \frac{\dot{\Psi}^{\mathrm{T}}_{\mathrm{pol}}}{2\pi}\zeta_f \tag{6.5.2}$$

via (6.1.6)

$$B^{\theta_f} = -\frac{1}{\sqrt{g_f}}\frac{\partial \nu}{\partial \zeta_f} \; ; \qquad B^{\zeta_f} = \frac{1}{\sqrt{g_f}}\frac{\partial \nu}{\partial \theta_f} \; . \tag{6.5.3}$$

We now want to give physical meaning to the function ν. It is clear that it acts as a "potential" function for $\boldsymbol{B}$. The expressions for the $\boldsymbol{B}$ components in (6.5.3) suggest that there might be a "curl" operation involved. The formula for the curl relates covariant and contravariant components as follows:

$$(\nabla \times \boldsymbol{A})^i = \frac{1}{\sqrt{g}}\left(\frac{\partial A_k}{\partial u^j} - \frac{\partial A_j}{\partial u^k}\right) , \qquad i, j, k \text{ cyc } 1, 2, 3 \; . \tag{6.5.4}$$

If we identify $\boldsymbol{A}$ in the above formula with a "vector potential", which we will denote $\boldsymbol{A}^*$, so that $\boldsymbol{B} = \nabla \times \boldsymbol{A}^*$, we observe that (6.5.4) is a relationship between the contravariant components of $\boldsymbol{B}$, B^i, and the covariant components of $\boldsymbol{A}^*$, A_j^*. The meaning of the asterisk will be explained below. For now, it can be disregarded. This suggests that according to (6.5.3), $\boldsymbol{B}$ is derivable from a "vector potential" $\boldsymbol{A}^*$ in which ν is the first and only non-vanishing component:

$$A_i^* = (-\nu, 0, 0) \tag{6.5.5a}$$

or

$$A_i^* = \left(-\frac{\dot{\Psi}_{\text{tor}}}{2\pi}\theta_f + \frac{\dot{\Psi}^{\text{r}}_{\text{pol}}}{2\pi}\zeta_f, 0, 0 \right) . \tag{6.5.5b}$$

Thus $\nu(\varrho, \theta_f, \zeta_f)$ is a "potential" function in the sense that it is the only component of a "vector potential" $\boldsymbol{A}^*$.

The quotation marks and the asterisk in $\boldsymbol{A}^*$ indicate that although the ν and $\boldsymbol{A}^*$ produce the correct magnetic field $\boldsymbol{B}$, they are *not* acceptable as physical quantities themselves because of their non-periodicity, i.e., non-single valuedness in θ_f and ζ_f. This problem is not insurmountable, since it follows from vector identities that as far as $\boldsymbol{B}$ is concerned, a magnetic vector potential is only defined up to an additive term which is the gradient of any "arbitrary" function. Since $\boldsymbol{B} = \nabla \times \boldsymbol{A}$, we have that $\nabla \times (\boldsymbol{A} + \nabla f) = \nabla \times \boldsymbol{A} + \nabla \times \nabla f = \nabla \times \boldsymbol{A} = \boldsymbol{B}$. Both $\boldsymbol{A}$ and $\boldsymbol{A} + \nabla f$ are legitimate mathematical vector potentials for $\boldsymbol{B}$; f is completely arbitrary.

Although it was thought for a long time that a magnetic vector potential is merely a mathematical tool, it has been established that a magnetic vector potential does have a physical meaning (see, e.g., Feynman, Leighton and Sands 1963). In order for $\boldsymbol{A}$ to be a physical quantity, we must require that it be single valued, i.e., periodic in θ_f and ζ_f. Thus $f(\varrho, \theta_f, \zeta_f)$ is arbitrary with the only limitation that $\boldsymbol{A} + \nabla f$ must be periodic.

Our strategy is straightforward; we choose a particular form of f such that $\boldsymbol{A}^*$ of (6.5.5) will be transformed into an acceptable vector potential $\boldsymbol{A} = \boldsymbol{A}^* + \nabla f$. To convert $\boldsymbol{A}^*$ into a periodic vector potential, we must split up f into a periodic part $\tilde{f}$ and a secular part $\acute{f}$. The secular part must be chosen such that $\nabla \acute{f}$ cancels the linear terms in θ_f and ζ_f of $\boldsymbol{A}^*$ in (6.5.5). The components of $\boldsymbol{A}^*$ appearing in (6.5.5) are covariant components, so that $\boldsymbol{A}^*$ itself equals

$$\begin{aligned} \boldsymbol{A}^* &= A_\varrho^* \nabla\varrho + A_{\theta_f}^* \nabla\theta_f + A_{\zeta_f}^* \nabla\zeta_f \\ &= \left(-\frac{\dot{\Psi}_{\text{tor}}}{2\pi}\theta_f + \frac{\dot{\Psi}^{\text{r}}_{\text{pol}}}{2\pi}\zeta_f \right) \nabla\varrho . \end{aligned} \tag{6.5.6}$$

The most general form of $\acute{f}$ for our purpose is

$$\acute{f} = \left[a + \frac{\Psi_{\text{tor}}(\varrho)}{2\pi} \right] \theta_f + \left[b - \frac{\Psi^{\text{r}}_{\text{pol}}(\varrho)}{2\pi} \right] \zeta_f , \tag{6.5.7}$$

where a and b are constants, independent of ϱ. With this form of $\acute{f}$, and with $f = \acute{f} + \tilde{f}$, ∇f becomes

$$\nabla f = \frac{\dot{\Psi}_{\text{tor}}}{2\pi}\theta_f \nabla\varrho - \frac{\dot{\Psi}^{\text{r}}_{\text{pol}}}{2\pi}\zeta_f \nabla\varrho + \left(a + \frac{\Psi_{\text{tor}}}{2\pi} \right) \nabla\theta_f + \left(b - \frac{\Psi^{\text{r}}_{\text{pol}}}{2\pi} \right) \nabla\zeta_f + \nabla\tilde{f}. \tag{6.5.8}$$

The first two terms of ∇f in (6.5.8), which are secular terms, cancel the secular first covariant component of $\boldsymbol{A}^*$ in (6.5.5). The new and acceptable vector potential $\boldsymbol{A}$ can thus be written as

$$\boldsymbol{A} = \left(a + \frac{\Psi_{\text{tor}}}{2\pi}\right)\nabla\theta_{\text{f}} + \left(b - \frac{\Psi^{\text{r}}_{\text{pol}}}{2\pi}\right)\nabla\zeta_{\text{f}} + \nabla\tilde{f} \qquad (6.5.9)$$

or

$$A_i = \left(0,\, a + \frac{\Psi_{\text{tor}}}{2\pi},\, b - \frac{\Psi^{\text{r}}_{\text{pol}}}{2\pi}\right) + \frac{\partial\tilde{f}}{\partial u^i}\,, \qquad (6.5.10)$$

where i and u^i stand for ϱ, θ_{f}, ζ_{f}. $\tilde{f}$ is a completely arbitrary periodic function. The first component of $\boldsymbol{A}$ is zero, whereas in $\boldsymbol{A}^*$ the first component was the only nonzero component. The crucial point here is that any physically acceptable vector potential must have non-vanishing second and third components in addition to those involving $\tilde{f}$. The vector potential given by Solov'ev (1975) is equivalent to (6.5.10) but with the constants $a = b = 0$ and $\tilde{f} = 0$.

We could have written down the "bracketed" components of $\boldsymbol{A}$ in (6.5.10) right away from a term by term identification of (6.5.4) with (6.5.3) after replacing the dot by $d/d\varrho$ or $\partial/\partial\varrho$.

6.5.2 Covariant *B* Components

It is now appropriate to discuss some aspects of Ampère's law

$$\mu_0\boldsymbol{J} = \nabla\times\boldsymbol{B}\ . \qquad (6.5.11)$$

This represents an inhomogeneous vector differential equation for the magnetic field with a source term equal to $\mu_0\boldsymbol{J}$. (In component form, it represents a system of three coupled inhomogeneous partial differential equations.) According to the theory of differential equations, if $\boldsymbol{J}$ is given, the equation is linear and the solution consists of two parts, $\boldsymbol{B}_0$, the solution of the homogeneous equation (also called the "complementary solution"), and $\boldsymbol{B}_J$, a particular solution of the inhomogeneous equation, i.e., the driven equation with source term $\mu_0\boldsymbol{J}$:

$$\boldsymbol{B} = \boldsymbol{B}_0 + \boldsymbol{B}_J\ . \qquad (6.5.12)$$

$\boldsymbol{B}_0$ satisfies

$$\nabla\times\boldsymbol{B}_0 = 0\ , \qquad (6.5.13)$$

while $\boldsymbol{B}_J$ obeys

$$\nabla\times\boldsymbol{B}_J = \mu_0\boldsymbol{J}\ . \qquad (6.5.14)$$

Equation (6.5.13) represents the equation for a magnetic field in vacuum. It implies that $\boldsymbol{B}_0$ can be derived from a *magnetic scalar potential* Φ via the gradient

$$\boldsymbol{B}_0 = \nabla\Phi\ . \qquad (6.5.15)$$

Then the total $\boldsymbol{B}$ can be written as

$$\boldsymbol{B} = \nabla\Phi + \boldsymbol{B}_J \ . \tag{6.5.16}$$

This problem is somewhat different from those encountered in the theory of simple differential equations because $\nabla \times \nabla h = 0$ for every arbitrary h. If we want to give Φ and $\boldsymbol{B}_J$ their usual meanings, we must make sure that $\boldsymbol{B}_J$ is not merely a particular solution of (6.5.14), but that it is the correct *physical* solution with all the integration constants included, whereas in simple differential equation theory, we need not worry about integration constants for a particular solution. An easy way to find the proper solution for $\boldsymbol{B}_J$ is to find the simplest solution of (6.5.14) by observation or manipulation. We then check whether that solution is in agreement with physical common sense by means of taking "Ampère loop integrals." Then, to obtain the proper $\boldsymbol{B}_J$ we add the gradient of an appropriate scalar function.

Once $\boldsymbol{B}_J$ is established correctly, we can say that the Φ solution of (6.5.13) and (6.5.15) is the magnetic scalar potential in vacuum. This potential Φ is determined by currents *exterior* to the region of interest; in our applications, it is related to the coil currents. However, Φ is not always single valued. Since $\boldsymbol{B}_0 = \nabla\Phi$ is a single-valued function, Φ can have at most linear terms in its secular part. This means that a closed loop integral can give a non-zero result if Φ has such a secular part Φ'. For example, for $\Phi = \alpha(\varrho)\theta$:

$$\oint_{\substack{\text{closed } \theta \text{ coor-} \\ \text{dinate curve}}} \boldsymbol{B}\cdot d\boldsymbol{l} = \oint \nabla\Phi\cdot d\boldsymbol{l} = \oint d\Phi = \alpha(\varrho)2\pi = \Delta\Phi \ . \tag{6.5.17}$$

For a single-valued (i.e., periodic) Φ, this integral would have vanished. A closed integration loop *not* encircling the $\varrho = 0$ axis—in the above example—also vanishes.

The loop integral $\oint \boldsymbol{B}\cdot d\boldsymbol{l}$ around a current-carrying conductor is constant and equal to the current enclosed. We can thus say that the magnetic field $\boldsymbol{B}$ produced by such a conductor can be derived from a secular (non-periodic) potential. If there exist poloidal or toroidal integration loops that do not encircle current conductors and stay entirely inside the torus, the magnetic field potential due to these conductors is single valued inside the torus. (A *poloidal* (*toroidal*) loop is any closed curve that encircles the *magnetic* (*major*) axis of the toroid.)

As a simple example, consider Figs. 6.2, 3. In Fig. 6.2, a circular loop of wire and a straight wire are shown. Both carry steady currents as indicated. In a (r, θ, ζ) coordinate system, with $R = R_0 + r\cos\theta$, the magnetic scalar potential is $\Phi = (\mu_0/2\pi)\{I_t\zeta + I_p\theta\}$; $\boldsymbol{B}$ is computed via $\boldsymbol{B} = \nabla\Phi = (1/r)(\partial\Phi/\partial\theta)\hat{\theta} + (1/R)(\partial\Phi/\partial\zeta)\hat{\zeta}$. In Fig. 6.3, we show a torus and several sets of coils. There is no plasma present. Every closed contour that remains inside the torus will never enclose the helical windings or the equilibrium coils. Hence, these coils "give rise to" a periodic potential. The coil in the middle (on the minor axis) can easily be encircled and consequently, it will "generate" a potential that is secular along the path of the loop.

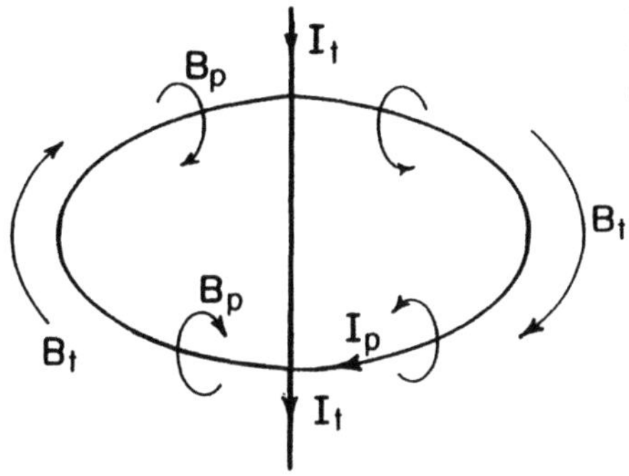

Fig. 6.2. "Toroidal" and "poloidal" magnetic-field behavior for a straight wire in the center of a circular loop. The current I_p generates B_p and B_t is due to I_t

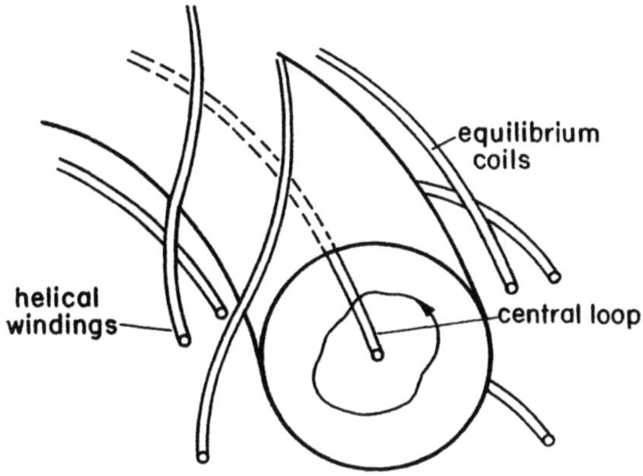

Fig. 6.3. A closed integration curve which encircles a central current carrying loop picks up a secular term in the scalar magnetic potential. None of the helical windings is encircled by an integration curve lying entirely inside the torus, and hence, these windings can only generate a single-valued (or periodic) magnetic potential, inside the torus

Scalar magnetic "potentials" with a secular part as discussed above are only mathematical potentials because they are not single valued. In the latter part of this section, we will consider only physical or single valued (i.e., periodic) potentials.

Now we find the covariant components of $\boldsymbol{B}_J$ in terms of the contravariant current components. Ampere's law for $\boldsymbol{B}_J$ given in (6.5.14) reads in component form:

$$J^i = \frac{1}{\mu_0}(\nabla \times \boldsymbol{B})^i = \frac{1}{\mu_0}\frac{1}{\sqrt{g}}\left(\frac{\partial B_k}{\partial u^j} - \frac{\partial B_j}{\partial u^k}\right), \qquad i, j, k \text{ cyc } 1, 2, 3 \ . \tag{6.5.18}$$

We have omitted the index J on the $\boldsymbol{B}$-components for clarity. Here we are dealing with $\boldsymbol{B}$ due to currents unless stated otherwise. The contravariant current-density components were given in (6.4.4), with η given by (6.4.7); however, here we are interested in $\theta = \theta_f$ and $\zeta = \zeta_f$. With these components on the left side, (6.5.18) leads to three coupled equations:

$$J^\varrho = 0 = \left(\frac{\partial B_{\zeta_f}}{\partial \theta_f} - \frac{\partial B_{\theta_f}}{\partial \zeta_f}\right) \tag{6.5.19a}$$

$$J^{\theta_f} = \frac{1}{\sqrt{g_f}}\left[\frac{1}{2\pi}\frac{\partial I^r_{pol}}{\partial \varrho} - \frac{\partial \tilde{\eta}}{\partial \zeta_f}\right] = \frac{1}{\mu_0}\frac{1}{\sqrt{g_f}}\left(\frac{\partial B_\varrho}{\partial \zeta_f} - \frac{\partial B_{\zeta_f}}{\partial \varrho}\right) \tag{6.5.19b}$$

$$J^{\zeta_f} = \frac{1}{\sqrt{g_f}}\left[\frac{1}{2\pi}\frac{\partial I_{tor}}{\partial \varrho} + \frac{\partial \tilde{\eta}}{\partial \theta_f}\right] = \frac{1}{\mu_0}\frac{1}{\sqrt{g_f}}\left(\frac{\partial B_{\theta_f}}{\partial \varrho} - \frac{\partial B_\varrho}{\partial \theta_f}\right) . \tag{6.5.19c}$$

Here we have set $\dot{I}(\varrho) \equiv dI/d\varrho \equiv \partial I/\partial \varrho$, because $I \equiv I(\varrho)$. A glance at the second and third equations suggests the identification

$$B^*_\varrho = -\mu_0 \tilde{\eta}(\varrho, \theta_f, \zeta_f) \tag{6.5.20a}$$

$$B^*_{\theta_f} = \frac{\mu_0}{2\pi} I_{tor}(\varrho) \tag{6.5.20b}$$

$$B^*_{\zeta_f} = -\frac{\mu_0}{2\pi} I^r_{pol}(\varrho) . \tag{6.5.20c}$$

The solutions are consistent with the first equation (6.5.19a) as can be checked by substitution.

To check whether the components in (6.5.20) make sense physically, we consider the loop integrals along the θ_f and ζ_f coordinate curves, respectively.

The geometry for the θ_f loop integral is depicted in Fig. 6.4. As at the end of Sect. 6.2., we borrow from Chap. 2 the result that a differential arc length vector along the u^i coordinate curve equals $d\boldsymbol{l}(i) = \boldsymbol{e}_i du^i$. There is *no* summation over i here, since along the u^i curve $du^j = 0 = du^k$, where $j \neq i \neq k$. With $\boldsymbol{B} = \sum_j B_j \boldsymbol{e}^j$, we have then the θ_f-loop integral:

$$\begin{aligned} \oint_{\Gamma_{\theta_f}} \boldsymbol{B}\cdot d\boldsymbol{l} &= \oint_{\Gamma_{\theta_f}} \sum_j B_j \boldsymbol{e}^j \cdot \boldsymbol{e}_i du^i \\ &= \oint_{\Gamma_{\theta_f}} B_i du^i \qquad \text{(no summation over } i) \\ &= \int_0^{2\pi} B_{\theta_f} d\theta_f . \end{aligned} \tag{6.5.21}$$

From Ampère's law:

$$\oint \boldsymbol{B}\cdot d\boldsymbol{l} = \mu_0 \times \text{(enclosed current)}$$

$$\int_0^{2\pi} B_{\theta_f} d\theta_f = \mu_0 I_{tor}(\varrho) . \tag{6.5.22}$$

This relationship is in agreement with the B_{θ_f} component suggested in (6.5.20). (Boozer (1981) uses an alternative form of $d\boldsymbol{l}(i)$ to arrive at the same conclusion. As shown in Chap. 2, $d\boldsymbol{l}(i) = \sqrt{g}(\nabla u^j \times \nabla u^k) du^i$, with i, j, k cyc 1, 2, 3. To find (6.5.22), the definition of $(\sqrt{g})^{-1} = \nabla u^i \cdot (\nabla u^j \times \nabla u^k)$ must be used.)

For the toroidal component B_{ζ_f}, we consider a contour integral along the ζ_f curve (Fig. 6.5). Performing this integral, we see that we pick up the contribution from the poloidal current I^{out}_{pol} which represents the plasma current outside the

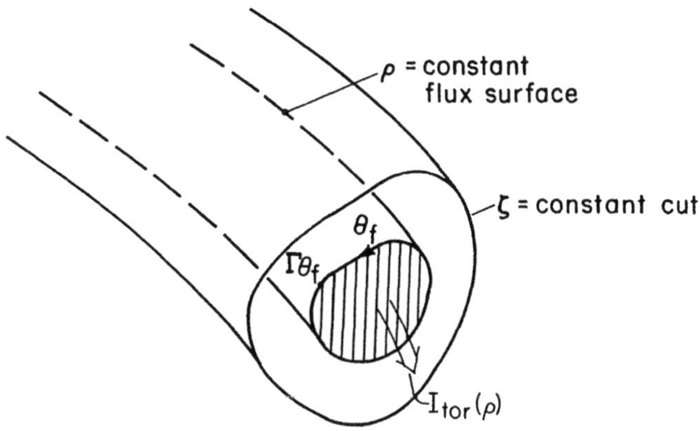

Fig. 6.4. Geometry for integration along a poloidal coordinate curve Γ_{θ_f}, lying at the intersection of a ϱ = constant flux surface and a ζ = constant cut

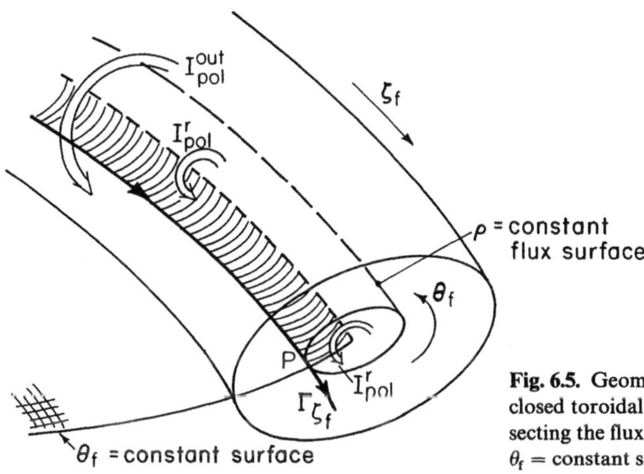

Fig. 6.5. Geometry for integration along a closed toroidal coordinate curve Γ_{ζ_f}, intersecting the flux surface ϱ = constant and a θ_f = constant surface

flux surface ϱ = constant:

$$\int_0^{2\pi} B_{\zeta_f} d\zeta_f = \mu_0 I_{\rm pol}^{\rm out}(\varrho) = \mu_0 (I_{\rm pol}^{\rm plas} - I_{\rm pol}^{\rm r}) \ . \tag{6.5.23}$$

$I_{\rm pol}^{\rm plas}$ is the total poloidal current inside the plasma. (Note that the current inside the ribbon $I_{\rm pol}^{\rm r}$ does not contribute to the field at point P (Fig. 6.5).) Coil currents are not included in this part of the discussion because we are still solving for the components of $\boldsymbol{B}_J$, which is the magnetic field component due to only the plasma currents. Equation (6.5.23) requires that the solution for B_{ζ_f} should include an integration constant $\mu_0 I_{\rm pol}^{\rm plas}/2\pi$. Instead of (6.5.20), we have

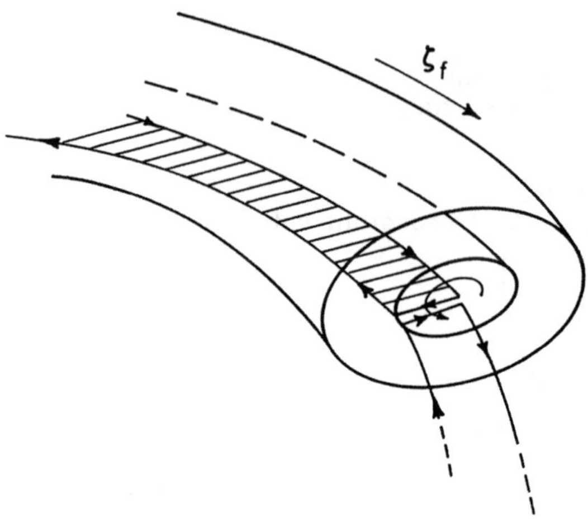

Fig. 6.6. Integration along a closed curve made up of a toroidal coordinate curve, the magnetic axis and two infinitesimally close connection pieces

$$(B_\varrho)_J = -\mu_0 \tilde{\eta}(\varrho, \theta_f, \zeta_f) \tag{6.5.24a}$$

$$(B_{\theta_f})_J = \frac{\mu_0}{2\pi} I_{tor}(\varrho) \tag{6.5.24b}$$

$$(B_{\zeta_f})_J = \frac{\mu_0}{2\pi}(I_{pol}^{plas} - I_{pol}^{r}(\varrho)) \ . \tag{6.5.24c}$$

Equation (6.5.24c) is consistent with (6.5.23).

An alternative way to look at this result for $(B_{\zeta_f})_J$ is to consider the loop along a ζ_f-curve that encloses the poloidal ribbon current $I_{pol}^r(\varrho)$ as shown in Fig. 6.6. Then Ampère's loop law becomes

$$\oint \boldsymbol{B} \cdot d\boldsymbol{l} = -\int_0^{2\pi} B_{\zeta_f}(\varrho)\, d\zeta_f + \int_0^{2\pi} B_{\zeta_f}(0)\, d\zeta_f = \mu_0 I_{pol}^r(\varrho) \ , \tag{6.5.25}$$

with

$$\int_0^{2\pi} B_{\zeta_f}(0)\, d\zeta_f = \mu_0 I_{pol}^{plas} \ . \tag{6.5.26}$$

Thus the loop integral for $B_{\zeta_f}(\varrho)$ equals (6.5.23).

The argument developed above is equivalent to saying that the assumed particular solution $\boldsymbol{B}_J^*$ must be augmented by the gradient of a function $h = (\mu_0/2\pi) I_{pol}^{plas} \zeta_f$ such that $\boldsymbol{B}_J = \boldsymbol{B}_J^* + \nabla h$.

Next we want to find the structure of the vacuum scalar potential Φ. It is the vacuum field component $\boldsymbol{B}_0$ or scalar magnetic potential Φ that will bring the coil-current contributions into the picture. Referring to Fig. 6.3, we see that there are no closed poloidal loops that enclose helical or equilibrium coils as long as we stay inside the torus. The central conductor parallel to the minor axis in Fig. 6.3 is absent except in a levitron or levitated octupole producing a contribution to the potential $(\mu_0/2\pi)I_{\text{coil}}\theta_f$. We assume it is absent from here on. Usually, however, the scalar potential cannot have a term linear in θ_f (cf. our discussion near (6.5.17)).

For toroidal integration loops, remaining inside the torus, it should be clear that the coils crossing the disk in the hole of the doughnut that carry a unidirectional current will contribute to a net toroidal field and $\oint \boldsymbol{B}\cdot d\boldsymbol{l} \sim \oint \nabla\Phi\cdot d\boldsymbol{l} \sim \oint d\Phi \neq 0$. Consequently, the magnetic scalar potential must contain a linear (secular) term in ζ_f: $\Phi \sim (\mu_0/2\pi)(\sum I_{\text{coils}})\zeta_f$. Except for this term, the potential must be single valued. For example, the part of the scalar potential due to the helical coils in a stellarator—which carry currents in opposide directions—is periodic. Thus, we can represent the magnetic scalar potential as follows:

$$\Phi = \Phi' + \tilde{\Phi} = \frac{\mu_0}{2\pi}\left(\sum I_{\text{coils}}\right)\zeta_f + \tilde{\Phi}(\varrho, \theta_f, \zeta_f) \ . \tag{6.5.27}$$

Note that there exists a hierarchy in the contributions of the coils. The coils that wrap around in the toroidal direction (but which stay outside the torus) only give rise to a single valued potential. No secular part exists. The coils going around the short way can provide a secular part to the potential, since the toroidal loop integrals do not always vanish. Of course, the latter coils can also provide a periodic part. In a stellarator, the helical coils will also intersect the disk in the doughnut hole. However, in a stellarator the current is of opposite sign in neighboring wires so the net poloidal coil current due to these coils is zero. In a torsatron all helical coils conduct current in the same direction; these coils thus provide a net toroidal field and a secular term $\propto \zeta_f$ in the potential.

The result for the covariant $\boldsymbol{B}$ components is, with $\boldsymbol{B} = \boldsymbol{B}_J + \nabla\Phi = \boldsymbol{B}_J + \nabla\Phi' + \nabla\tilde{\Phi}$:

$$B_\varrho = -\mu_0\tilde{\eta}(\varrho, \theta_f, \zeta_f) + \frac{\partial\tilde{\Phi}(\varrho, \theta_f, \zeta_f)}{\partial\varrho} \tag{6.5.28a}$$

$$B_{\theta_f} = \frac{\mu_0}{2\pi}I_{\text{tor}}(\varrho) + \frac{\partial\tilde{\Phi}(\varrho, \theta_f, \zeta_f)}{\partial\theta_f} \tag{6.5.28b}$$

$$B_{\zeta_f} = \frac{\mu_0}{2\pi}\left(I_{\text{pol}}^{\text{plas}} - I_{\text{pol}}^{\text{r}}(\varrho)\right) + \frac{\mu_0}{2\pi}\left(\sum I_{\text{coils}}\right) + \frac{\partial\tilde{\Phi}(\varrho, \theta_f, \zeta_f)}{\partial\zeta_f} \ . \tag{6.5.28c}$$

A more rigorous school of thought requires the magnetic scalar potential to be a physical quantity. This implies that Φ cannot have secular terms and must be periodic. This viewpoint requires that a scalar magnetic potential cannot be used if one considers loop integrals encircling currents (Panofsky and Phillips 1962, p. 127). In our case, it means that the coil-currents which are encircled and consequently contribute to secular terms cannot be described in terms of a physical scalar potential. As a result, (6.5.13) and (6.5.15) would not be equivalent for a *physical* potential Φ. The implication for the covariant $\boldsymbol{B}$-components is that we only deal with a periodic potential $\tilde{\Phi}$ and that we absorb $\sum I_{\text{coils}}$ into the integration constant of (6.5.28c) as follows:

$$I_{\text{pol}}^{\text{tot}} = I_{\text{pol}}^{\text{plas}} + \sum I_{\text{coils}} \quad . \tag{6.5.29}$$

$I_{\text{pol}}^{\text{tot}}$ is the total poloidal current crossing a disk touching the magnetic axis. The poloidal plasma current and the coil currents crossing that disk are included. It is convenient to introduce a symbol for the poloidal current (including the coil currents) outside a flux surface:

$$I_{\text{pol}}^{\text{d}}(\varrho) = I_{\text{pol}}^{\text{tot}} - I_{\text{pol}}^{\text{r}}(\varrho) \quad . \tag{6.5.30}$$

The index "d" stands for *d*isk.

The covariant **B**-components are thus alternatively

$$B_\varrho = -\mu_0 \tilde{\eta}(\varrho, \theta_{\text{f}}, \zeta_{\text{f}}) + \frac{\partial \tilde{\Phi}(\varrho, \theta_{\text{f}}, \zeta_{\text{f}})}{\partial \varrho} \tag{6.5.31a}$$

$$B_{\theta_{\text{f}}} = \frac{\mu_0}{2\pi} I_{\text{tor}}(\varrho) + \frac{\partial \tilde{\Phi}(\varrho, \theta_{\text{f}}, \zeta_{\text{f}})}{\partial \theta_{\text{f}}} \tag{6.5.31b}$$

$$B_{\zeta_{\text{f}}} = \frac{\mu_0}{2\pi} I_{\text{pol}}^{\text{d}}(\varrho) + \frac{\partial \tilde{\Phi}(\varrho, \theta_{\text{f}}, \zeta_{\text{f}})}{\partial \zeta_{\text{f}}} \quad . \tag{6.5.31c}$$

Similar expressions have been given by Shafranov (1968). Although not explicitly stated, it must be understood that our integration constants, in (6.5.28c), have been absorbed in Shafranov's I. See also Hirshman (1982).

6.5.3 Relationship to the Boozer-Grad Form

Now that we know the covariant magnetic-field components, we convert the usual covariant form to the covariant Boozer-Grad form of (5.4.25). The generic covariant representation of **B** is

$$\boldsymbol{B} = B_\varrho \boldsymbol{\nabla}\varrho + B_{\theta_{\text{f}}} \boldsymbol{\nabla}\theta_{\text{f}} + B_{\zeta_{\text{f}}} \boldsymbol{\nabla}\zeta_{\text{f}} \tag{6.5.32}$$

with the covariant components given in (6.5.31). After substitution of these components, we can rearrange (6.5.32) as follows:

$$\begin{aligned}
\boldsymbol{B} &= -\mu_0 \tilde{\eta} \boldsymbol{\nabla}\varrho + \frac{\partial \tilde{\Phi}}{\partial \varrho} \boldsymbol{\nabla}\varrho + \frac{\mu_0}{2\pi} I_{\text{tor}} \boldsymbol{\nabla}\theta_{\text{f}} + \frac{\partial \tilde{\Phi}}{\partial \theta_{\text{f}}} \boldsymbol{\nabla}\theta_{\text{f}} + \frac{\mu_0}{2\pi} I_{\text{pol}}^{\text{d}} \boldsymbol{\nabla}\zeta_{\text{f}} + \frac{\partial \tilde{\Phi}}{\partial \zeta_{\text{f}}} \boldsymbol{\nabla}\zeta_{\text{f}} \\
&= \left(\frac{\partial \tilde{\Phi}}{\partial \varrho} \boldsymbol{\nabla}\varrho + \frac{\partial \tilde{\Phi}}{\partial \theta_{\text{f}}} \boldsymbol{\nabla}\theta_{\text{f}} + \frac{\partial \tilde{\Phi}}{\partial \zeta_{\text{f}}} \boldsymbol{\nabla}\zeta_{\text{f}} \right) - \mu_0 \tilde{\eta} \boldsymbol{\nabla}\varrho + \frac{\mu_0}{2\pi} I_{\text{tor}} \boldsymbol{\nabla}\theta_{\text{f}} + \frac{\mu_0}{2\pi} I_{\text{pol}}^{\text{d}} \boldsymbol{\nabla}\zeta_{\text{f}} \\
&= \boldsymbol{\nabla}\tilde{\Phi} - \mu_0 \tilde{\eta} \boldsymbol{\nabla}\varrho + \frac{\mu_0}{2\pi} \boldsymbol{\nabla}(I_{\text{tor}} \theta_{\text{f}}) - \frac{\mu_0}{2\pi} \theta_{\text{f}} \dot{I}_{\text{tor}} \boldsymbol{\nabla}\varrho \\
&\quad + \frac{\mu_0}{2\pi} \boldsymbol{\nabla}(I_{\text{pol}}^{\text{d}} \zeta_{\text{f}}) - \frac{\mu_0}{2\pi} \zeta_{\text{f}} \dot{I}_{\text{pol}}^{\text{d}} \boldsymbol{\nabla}\varrho
\end{aligned} \tag{6.5.33a}$$

$$= \boldsymbol{\nabla}\tilde{\Phi} + \boldsymbol{\nabla}\left(\frac{\mu_0}{2\pi} I_{\text{tor}} \theta_{\text{f}}\right) + \boldsymbol{\nabla}\left(\frac{\mu_0}{2\pi} I_{\text{pol}}^{\text{d}} \zeta_{\text{f}}\right)$$

$$+ \boldsymbol{\nabla}\varrho\left(-\mu_0 \tilde{\eta} - \frac{\mu_0}{2\pi} \dot{I}_{\text{tor}} \theta_{\text{f}} - \frac{\mu_0}{2\pi} \dot{I}_{\text{pol}}^{\text{d}} \zeta_{\text{f}}\right)$$

$$= \boldsymbol{\nabla}\left(\tilde{\Phi} + \frac{\mu_0}{2\pi} I_{\text{tor}} \theta_{\text{f}} + \frac{\mu_0}{2\pi} I_{\text{pol}}^{\text{d}} \zeta_{\text{f}}\right) + \lambda \boldsymbol{\nabla}\varrho \ . \tag{6.5.33b}$$

Here we have defined the following "Boozer-Grad quantity"

$$\lambda \equiv -\mu_0\left(\tilde{\eta} + \frac{\dot{I}_{\text{tor}}}{2\pi}\theta_{\text{f}} + \frac{\dot{I}_{\text{pol}}^{\text{d}}}{2\pi}\zeta_{\text{f}}\right) . \tag{6.5.34}$$

In terms of the poloidal ribbon current, $I_{\text{pol}}^{\text{r}}$, for which we have

$$\dot{I}_{\text{pol}}^{\text{d}} = -\dot{I}_{\text{pol}}^{\text{r}} \ , \tag{6.5.35}$$

we obtain

$$\lambda \equiv -\mu_0\left(\tilde{\eta} + \frac{\dot{I}_{\text{tor}}}{2\pi}\theta_{\text{f}} - \frac{\dot{I}_{\text{pol}}^{\text{r}}}{2\pi}\zeta_{\text{f}}\right) \tag{6.5.36}$$

or with (6.4.7)

$$\lambda = -\mu_0 \eta \ . \tag{6.5.37}$$

With this λ, we obtain for the current density $\boldsymbol{J}$ which was originally given as $\boldsymbol{J} = \boldsymbol{\nabla}\varrho \times \boldsymbol{\nabla}\eta$ in (6.4.5)

$$\boldsymbol{J} = \frac{1}{\mu_0} \boldsymbol{\nabla}\lambda \times \boldsymbol{\nabla}\varrho \ . \tag{6.5.38}$$

This is exactly the Boozer-Grad form obtained earlier in (5.4.21).

If we introduce the "Boozer-Grad quantity" χ in (6.5.33),

$$\chi = \tilde{\Phi} + \frac{\mu_0}{2\pi}(I_{\text{tor}}\theta_{\text{f}} + I_{\text{pol}}^{\text{d}}\zeta_{\text{f}}) \ , \tag{6.5.39}$$

the covariant form of $\boldsymbol{B}$ reduces to precisely the Boozer-Grad form

$$\boldsymbol{B} = \boldsymbol{\nabla}\chi + \lambda \boldsymbol{\nabla}\varrho \ , \tag{6.5.40}$$

or with (6.5.37),

$$\boldsymbol{B} = \boldsymbol{\nabla}\chi - \mu_0 \eta \boldsymbol{\nabla}\varrho \ . \tag{6.5.41}$$

To make the link with Boozer's work somewhat clearer, we give below a table of our symbols and his (omitting c's, 4π's, $\mu_0, \ldots$). For completeness, we also give the symbols used by Solov'ev and Shafranov.

Ours	Boozer (1981)	Solov'ev-Shafranov (1970)
$\tilde{\eta}$	$-\beta^*$	$\tilde{\nu}$
λ	β	ν
$I^{\mathrm{d}}_{\mathrm{pol}}$	g	constant-I
I_{tor}	I	J
θ_{f}	θ	θ
ζ_{f}	ϕ	ζ
ν	θ_0	ϑ

6.6 Boozer's Toroidal Flux Coordinates

We have already mentioned that the flux coordinates θ_{f} and ζ_{f} are not unique. In other words, there is more than one pair of angle functions $\theta_{\mathrm{f}}(\boldsymbol{R})$ and $\zeta_{\mathrm{f}}(\boldsymbol{R})$ in which the magnetic field is straight. Boozer makes that particular choice of flux coordinates, which we denote by θ_{B} and ζ_{B}, such that the periodic part of the scalar magnetic potential vanishes. Thus, he deforms the $(\theta_{\mathrm{f}}, \zeta_{\mathrm{f}})$ coordinate curves so that $\tilde{\Phi} \equiv 0$. This gives his Jacobian a rather simple form. Following Boozer's (1981) paper, we compute the Jacobian from the contravariant and covariant representations of $\boldsymbol{B}$, (6.1.30) and (6.5.33a), respectively:

$$\boldsymbol{B} = \frac{\dot{\Psi}_{\mathrm{tor}}}{2\pi} \nabla\varrho \times \nabla\theta_{\mathrm{f}} - \frac{\dot{\Psi}^{\mathrm{r}}_{\mathrm{pol}}}{2\pi} \nabla\varrho \times \nabla\zeta_{\mathrm{f}} = \frac{\dot{\Psi}^{\mathrm{r}}_{\mathrm{pol}}}{2\pi\sqrt{g_{\mathrm{f}}}} \boldsymbol{e}_{\theta_{\mathrm{f}}} + \frac{\dot{\Psi}_{\mathrm{tor}}}{2\pi\sqrt{g_{\mathrm{f}}}} \boldsymbol{e}_{\zeta_{\mathrm{f}}} \tag{6.6.1}$$

and

$$\boldsymbol{B} = \left(-\mu_0\tilde{\eta} + \frac{\partial\tilde{\Phi}}{\partial\varrho}\right)\nabla\varrho + \left(\frac{\mu_0}{2\pi} I_{\mathrm{tor}} + \frac{\partial\tilde{\Phi}}{\partial\theta_{\mathrm{f}}}\right)\nabla\theta_{\mathrm{f}} + \left(\frac{\mu_0}{2\pi} I^{\mathrm{d}}_{\mathrm{pol}} + \frac{\partial\tilde{\Phi}}{\partial\zeta_{\mathrm{f}}}\right)\nabla\zeta_{\mathrm{f}} \; . \tag{6.6.2}$$

Since $\boldsymbol{B}\cdot\boldsymbol{B} = (B)^2 = B^i B_i$, we obtain from the dot product of these equations:

$$\begin{aligned}(B)^2 &= \frac{\dot{\Psi}^{\mathrm{r}}_{\mathrm{pol}}}{2\pi\sqrt{g_{\mathrm{f}}}}\left(\frac{\mu_0}{2\pi} I_{\mathrm{tor}} + \frac{\partial\tilde{\Phi}}{\partial\theta_{\mathrm{f}}}\right) + \frac{\dot{\Psi}_{\mathrm{tor}}}{2\pi\sqrt{g_{\mathrm{f}}}}\left(\frac{\mu_0}{2\pi} I^{\mathrm{d}}_{\mathrm{pol}} + \frac{\partial\tilde{\Phi}}{\partial\zeta_{\mathrm{f}}}\right) \\ &= \frac{1}{\sqrt{g_{\mathrm{f}}}}\frac{\mu_0}{4\pi^2}(\dot{\Psi}^{\mathrm{r}}_{\mathrm{pol}} I_{\mathrm{tor}} + \dot{\Psi}_{\mathrm{tor}} I^{\mathrm{d}}_{\mathrm{pol}}) + \frac{\dot{\Psi}^{\mathrm{r}}_{\mathrm{pol}}}{2\pi\sqrt{g_{\mathrm{f}}}}\frac{\partial\tilde{\Phi}}{\partial\theta_{\mathrm{f}}} + \frac{\dot{\Psi}_{\mathrm{tor}}}{2\pi\sqrt{g_{\mathrm{f}}}}\frac{\partial\tilde{\Phi}}{\partial\zeta_{\mathrm{f}}}\end{aligned}$$

or

$$(B)^2 = \frac{1}{\sqrt{g_{\mathrm{f}}}}\frac{\mu_0}{4\pi^2}(\dot{\Psi}^{\mathrm{r}}_{\mathrm{pol}} I_{\mathrm{tor}} + \dot{\Psi}_{\mathrm{tor}} I^{\mathrm{d}}_{\mathrm{pol}}) + \boldsymbol{B}\cdot\nabla\tilde{\Phi} \; . \tag{6.6.3}$$

Recall that $\boldsymbol{B}\cdot\nabla = B^i\partial/\partial u^i$, the B^i being given in (6.1.25). After solving this equation for $\sqrt{g_{\mathrm{f}}}$ we find:

$$\sqrt{g_{\mathrm{f}}} = \frac{\mu_0}{4\pi^2}\frac{(\dot{\Psi}^{\mathrm{r}}_{\mathrm{pol}} I_{\mathrm{tor}} + \dot{\Psi}_{\mathrm{tor}} I^{\mathrm{d}}_{\mathrm{pol}})}{(B)^2 - \boldsymbol{B}\cdot\nabla\tilde{\Phi}} \; . \tag{6.6.4a}$$

Somewhat more symmetrically, this is

$$\sqrt{g_f} = \frac{\mu_0}{4\pi^2} \frac{(\dot{\Psi}_{tor} I^d_{pol} - \dot{\Psi}^d_{pol} I_{tor})}{(B)^2 - B \cdot \nabla \tilde{\Phi}} . \tag{6.6.4b}$$

Boozer takes $\tilde{\Phi} \equiv 0$. Therefore, Boozer's Jacobian, with ϱ still used as a radial coordinate, is

$$\sqrt{g_B} = \frac{\mu_0}{4\pi^2} \frac{(\dot{\Psi}^T_{pol} I_{tor} + \dot{\Psi}_{tor} I^d_{pol})}{(B)^2} . \tag{6.6.5}$$

Boozer (1981) uses the toroidal flux as a flux label, $\varrho \equiv \Psi_{tor}$. The Jacobian in this case is denoted by $(\sqrt{g})_B$ and is given by

$$(\sqrt{g})_B^{-1} \equiv \nabla \Psi_{tor} \cdot \nabla \theta_B \times \nabla \zeta_B . \tag{6.6.6}$$

It relates to the Jacobian $\sqrt{g_B}$ with ϱ as a coordinate, as follows:

$$(\sqrt{g})_B^{-1} = \dot{\Psi}_{tor} (\nabla \varrho \cdot \nabla \theta_B \times \nabla \zeta_B) = \dot{\Psi}_{tor} (\sqrt{g_B})^{-1} . \tag{6.6.7}$$

Making use of the definition of iota-bar, $\iota\!\!- = \dot{\Psi}^T_{pol} / \dot{\Psi}_{tor}$, we obtain the Jacobian as used by Boozer:

$$(\sqrt{g})_B = \dot{\Psi}_{tor}^{-1} \frac{\mu_0}{4\pi^2} \left(\frac{\dot{\Psi}^T_{pol} I_{tor} + \dot{\Psi}_{tor} I^d_{pol}}{(B)^2} \right) \tag{6.6.8a}$$

or

$$(\sqrt{g})_B = \frac{\mu_0}{4\pi^2} \frac{\iota\!\!- I_{tor} + I^d_{pol}}{B^2} ; \qquad \varrho \equiv \Psi_{tor} . \tag{6.6.8b}$$

To find the toroidal Boozer flux coordinates θ_B and ζ_B (in which $\tilde{\Phi} \equiv 0$) in terms of other flux coordinates θ_f and ζ_f (in which, in general, $\tilde{\Phi} \neq 0$), we use the freedom provided by (6.1.28):

$$\theta_B = \theta_f + \dot{\Psi}^T_{pol} G_B(\varrho, \theta_f, \zeta_f) \tag{6.6.9a}$$

$$\zeta_B = \zeta_f + \dot{\Psi}_{tor} G_B(\varrho, \theta_f, \zeta_f) . \tag{6.6.9b}$$

Recall that this transformation preserves the straightness of the magnetic field lines. A proper choice of G_B gives the coordinate system extra properties such as a zero periodic scalar magnetic potential in Boozer's case.

The new Jacobian $\sqrt{g_B}$ was calculated in (6.6.5), while the old one is written in (6.6.4). Now we look for an alternative relationship between $\sqrt{g_f}$ and $\sqrt{g_B}$ in order to determine G_B. By definition, the inverse Jacobian is

$$(\sqrt{g_B})^{-1} = \nabla \varrho \cdot (\nabla \theta_B \times \nabla \zeta_B) . \tag{6.6.10}$$

Upon substitution of the Boozer angles θ_B and ζ_B from (6.6.9), this becomes

$$
\begin{aligned}
(\sqrt{g_\mathrm{B}})^{-1} &= \nabla\varrho\cdot[\nabla(\theta_\mathrm{f} + \dot{\Psi}^\mathrm{r}_\mathrm{pol} G_\mathrm{B}) \times \nabla(\zeta_\mathrm{f} + \dot{\Psi}_\mathrm{tor} G_\mathrm{B})] \\
&= \nabla\varrho\cdot(\nabla\theta_\mathrm{f} \times \nabla\zeta_\mathrm{f}) + \dot{\Psi}^\mathrm{r}_\mathrm{pol}\,\nabla\varrho\cdot(\nabla G_\mathrm{B} \times \nabla\zeta_\mathrm{f}) \\
&\quad + \dot{\Psi}_\mathrm{tor}\,\nabla\varrho\cdot(\nabla\theta_\mathrm{f} \times \nabla G_\mathrm{B}) \\
&= (\sqrt{g_\mathrm{f}})^{-1} + \dot{\Psi}^\mathrm{r}_\mathrm{pol}(\nabla\zeta_\mathrm{f} \times \nabla\varrho)\cdot\nabla G_\mathrm{B} + \dot{\Psi}_\mathrm{tor}(\nabla\varrho \times \nabla\theta_\mathrm{f})\cdot\nabla G_\mathrm{B} \\
&= (\sqrt{g_\mathrm{f}})^{-1} + \frac{\dot{\Psi}^\mathrm{r}_\mathrm{pol}}{2\pi\sqrt{g_\mathrm{f}}}(2\pi)\boldsymbol{e}_{\theta_\mathrm{f}}\cdot\nabla G_\mathrm{B} + \frac{\dot{\Psi}_\mathrm{tor}}{2\pi\sqrt{g_\mathrm{f}}}(2\pi)\boldsymbol{e}_{\zeta_\mathrm{f}}\cdot\nabla G_\mathrm{B}
\end{aligned}
$$

or

$$
(\sqrt{g_\mathrm{B}})^{-1} = (\sqrt{g_\mathrm{f}})^{-1} + 2\pi\boldsymbol{B}\cdot\nabla G_\mathrm{B}\ . \tag{6.6.11}
$$

Here we used (6.1.25) for B^i.

This is a magnetic differential equation for G_B. If we replace the Jacobians in (6.6.11) by their computed values given in (6.6.4) and (6.6.5), we obtain

$$
\boldsymbol{B}\cdot\nabla G_\mathrm{B} = \frac{2\pi}{\mu_0}\frac{(B)^2}{\dot{\Psi}^\mathrm{r}_\mathrm{pol} I_\mathrm{tor} + \dot{\Psi}_\mathrm{tor} I^\mathrm{d}_\mathrm{pol}} - \frac{2\pi}{\mu_0}\frac{(B)^2 - \boldsymbol{B}\cdot\nabla\tilde{\Phi}}{\dot{\Psi}^\mathrm{r}_\mathrm{pol} I_\mathrm{tor} + \dot{\Psi}_\mathrm{tor} I^\mathrm{d}_\mathrm{pol}}
$$

or

$$
\boldsymbol{B}\cdot\nabla G_\mathrm{B} = \frac{2\pi}{\mu_0}\frac{\boldsymbol{B}\cdot\nabla\tilde{\Phi}}{\dot{\Psi}^\mathrm{r}_\mathrm{pol} I_\mathrm{tor} + \dot{\Psi}_\mathrm{tor} I^\mathrm{d}_\mathrm{pol}} \tag{6.6.12}
$$

We find that the simple choice

$$
G_\mathrm{B} = \frac{2\pi}{\mu_0}\frac{\tilde{\Phi}}{\dot{\Psi}^\mathrm{r}_\mathrm{pol} I_\mathrm{tor} + \dot{\Psi}_\mathrm{tor} I^\mathrm{d}_\mathrm{pol}} \tag{6.6.13}
$$

allows the unique determination of the toroidal Boozer flux coordinates θ_B and ζ_B via (6.6.9).

We can now relate the Boozer toroidal flux coordinates, $(\varrho, \theta_\mathrm{B}, \zeta_\mathrm{B})$, to the Boozer-Grad Clebsch coordinates, (ϱ, ν, χ), given above in (6.1.23) and (6.5.39) with $\tilde{\Phi} \equiv 0$:

$$
\varrho = \varrho \tag{6.6.14a}
$$

$$
\nu = \frac{\dot{\Psi}_\mathrm{tor}}{2\pi}\theta_\mathrm{B} - \frac{\dot{\Psi}^\mathrm{r}_\mathrm{pol}}{2\pi}\zeta_\mathrm{B} \tag{6.6.14b}
$$

$$
\chi = \frac{\mu_0}{2\pi} I_\mathrm{tor}\theta_\mathrm{B} + \frac{\mu_0}{2\pi} I^\mathrm{d}_\mathrm{pol}\zeta_\mathrm{B}\ . \tag{6.6.14c}
$$

The Jacobian $\sqrt{g_\mathrm{B}}$ is given in (6.6.5). Note that we still use ϱ as a flux-surface label, whereas Boozer normally uses Ψ_tor. The only difference is that $\nabla\Psi_\mathrm{tor} =$

$\dot{\Psi}_{tor}\,\nabla\varrho$ and that the $\dot{\Psi}_{tor}$ is absorbed in Boozer's ν; it is for this reason that his ν contains an ι.

It is interesting to investigate what happens to Boozer's covariant form for $\boldsymbol{B}$ in vacuum. To do this, we start from the Boozer-Grad form given in (6.5.40), $\boldsymbol{B} = \nabla\chi + \lambda\nabla\varrho$ with χ and λ given by (6.6.14c) and (6.5.36), respectively. From our discussion in Sect. 5.4., we recall that $\lambda = 0$ when $\boldsymbol{J} = 0$ (see (5.4.37)). With the replacement $I^{d}_{pol} = I^{plas}_{pol} + \sum I_{coils} - I^{r}_{pol}(\varrho) \rightarrow \sum I_{coils}$ in vacuum, we obtain a simple form for the vacuum field:

$$\boldsymbol{B} = \sum(I_{coils})\,\nabla\zeta_B\ . \tag{6.6.15}$$

The interpretation of this equation is important. It *looks like* the magnetic field has only a "toroidal" component in vacuum; however, this is not necessarily the case. For a *tokamak*, $\sum I_{coils}$ stands for the toroidal-field coil currents that encircle the torus poloidally. In the vacuum case, the poloidal and toroidal coordinate curves intersect at right angles (see Fig. 6.7a). In an axisymmetric device, the tangent-basis vector $\boldsymbol{e}_\zeta$ and the reciprocal basis vector $\nabla\zeta$ are parallel such that indeed the vacuum field is in the toroidal direction. In a *stellarator*, $\sum(I_{coils})$ also consists only of the toroidal-field coil currents, since the currents in the helical windings add up to zero. However, we know that stellarator field lines

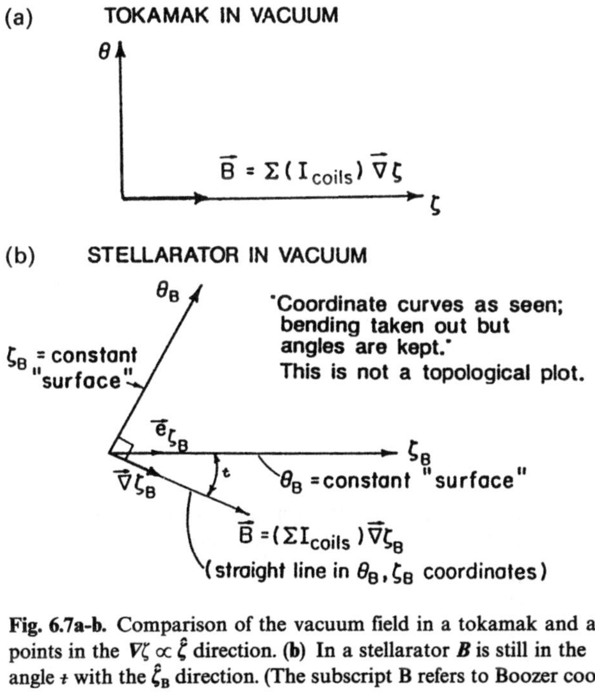

Fig. 6.7a-b. Comparison of the vacuum field in a tokamak and a stellarator. **(a)** In a tokamak $\boldsymbol{B}$ points in the $\nabla\zeta \propto \hat{\zeta}$ direction. **(b)** In a stellarator $\boldsymbol{B}$ is still in the $\nabla\zeta_B$ direction, but $\nabla\zeta_B$ makes an angle ι with the $\hat{\zeta}_B$ direction. (The subscript B refers to Boozer coordinates.)

follow spiralling paths in vacuum and are thus not in the usual toroidal direction. Yet Boozer's expression in (6.6.15) is correct. The explanation lies in Fig. 6.7b. Boozer has adjusted his coordinates so that the periodic scalar potential $\tilde{\Phi}$ vanishes while the $\boldsymbol{B}$ lines remain straight. As a consequence, the coordinate surfaces θ_B = constant and ζ_B = constant do not intersect at right angles nor do the coordinate curves. This has the important consequence that the contravariant-basis vector $\nabla\zeta_B$ does not lie in the $\boldsymbol{e}_{\zeta_B}$ direction, the latter normally being called the toroidal direction. As shown in Fig. 6.7b, the field line lies in the $\nabla\zeta_B$ direction which makes an angle with the ζ_B coordinate curve. This angle is related to the rotational transform.

6.7 Ideal-MHD-Equilibrium Conditions for Toroidally Confined Plasmas

To find an equation for $\tilde{\eta}$, the periodic part of the current stream function η, we substitute the contravariant components of $\boldsymbol{B}$ and $\boldsymbol{J}$, given in (6.1.25) and the LHS of (6.5.19) in the MHD equilibrium equation $\nabla p = \boldsymbol{J} \times \boldsymbol{B}$. The first component of this force balance is, in vector form,

$$\dot{p}\,\nabla\varrho = J^{\theta_f} B^{\zeta_f} \boldsymbol{e}_{\theta_f} \times \boldsymbol{e}_{\zeta_f} + J^{\zeta_f} B^{\theta_f} \boldsymbol{e}_{\zeta_f} \times \boldsymbol{e}_{\theta_f}$$

or

$$\dot{p}\,\nabla\varrho = J^{\theta_f} B^{\zeta_f}\, \nabla\varrho \sqrt{g_f} - J^{\zeta_f} B^{\theta_f}\, \nabla\varrho \sqrt{g_f}\ . \tag{6.7.1}$$

Using

$$B^i = \frac{1}{2\pi\sqrt{g_f}}(0,\, \dot{\Psi}^{\mathrm{r}}_{\mathrm{pol}},\, \dot{\Psi}_{\mathrm{tor}}) \tag{6.7.2}$$

$$J^j = \frac{1}{2\pi\sqrt{g_f}}\left(0,\, \dot{I}^{\mathrm{r}}_{\mathrm{pol}} - 2\pi\frac{\partial\tilde{\eta}}{\partial\zeta_f},\, \dot{I}_{\mathrm{tor}} + 2\pi\frac{\partial\tilde{\eta}}{\partial\theta_f}\right), \tag{6.7.3}$$

we obtain

$$\frac{1}{4\pi^2\sqrt{g_f}}\left\{-\dot{I}_{\mathrm{tor}}\dot{\Psi}^{\mathrm{r}}_{\mathrm{pol}} + \dot{I}^{\mathrm{r}}_{\mathrm{pol}}\dot{\Psi}_{\mathrm{tor}} - 2\pi\dot{\Psi}^{\mathrm{r}}_{\mathrm{pol}}\frac{\partial\tilde{\eta}}{\partial\theta_f} - 2\pi\dot{\Psi}_{\mathrm{tor}}\frac{\partial\tilde{\eta}}{\partial\zeta_f}\right\} = \dot{p}\ . \tag{6.7.4}$$

Since $\dot{\Psi}^{\mathrm{r}}_{\mathrm{pol}} \sim B^{\theta_f}$, $\dot{\Psi}_{\mathrm{tor}} \sim B^{\zeta_f}$ and $\partial\tilde{\eta}/\partial\theta_f = (\nabla\tilde{\eta})_{\theta_f}$, $\partial\tilde{\eta}/\partial\zeta_f = (\nabla\tilde{\eta})_{\zeta_f}$ and $\mathrm{B}^{\varrho} \equiv 0$, the last two terms inside the curly brackets of (6.7.4) represent the dot product of $(4\pi^2\sqrt{g_f}\boldsymbol{B})$ with $\nabla\tilde{\eta}$. We obtain

$$\frac{-\dot{I}_{\mathrm{tor}}\dot{\Psi}^{\mathrm{r}}_{\mathrm{pol}} + \dot{I}^{\mathrm{r}}_{\mathrm{pol}}\dot{\Psi}_{\mathrm{tor}}}{4\pi^2\sqrt{g_f}} - \boldsymbol{B}\cdot\nabla\tilde{\eta} = \dot{p}\ . \tag{6.7.5}$$

Thus, we find a magnetic differential equation for $\tilde{\eta}$:

$$\boldsymbol{B}\cdot\nabla\tilde{\eta} = \left\{\frac{\dot{I}^{\mathrm{r}}_{\mathrm{pol}}\dot{\Psi}_{\mathrm{tor}} - \dot{I}_{\mathrm{tor}}\dot{\Psi}^{\mathrm{r}}_{\mathrm{pol}}}{4\pi^2\sqrt{g_{\mathrm{f}}}} - \dot{p}\right\} . \tag{6.7.6}$$

The second and third components of $\nabla p = \boldsymbol{J}\times\boldsymbol{B}$ give no useful information because $J^{\varrho} = B^{\varrho} = 0$.

In order for (6.7.6) to have a solution, the source term must satisfy two constraint conditions ((4.10.7) and (4.10.4))

$$\iiint_{dV}\{\quad\}\,d^3R = 0 \tag{6.7.7}$$

and

$$\oint\frac{dl}{B}\{\quad\} = 0 . \tag{6.7.8}$$

To impose the first condition, we note that $d^3R = \sqrt{g_{\mathrm{f}}}\,d\varrho\,d\theta_{\mathrm{f}}\,d\zeta_{\mathrm{f}}$. The volume integral must be taken over the infinitesimal volume dV between two flux surfaces. Hence, we obtain for the first condition:

$$d\varrho\int_0^{2\pi}\int_0^{2\pi}\sqrt{g_{\mathrm{f}}}\,d\theta_{\mathrm{f}}\,d\zeta_{\mathrm{f}}\left\{\frac{\dot{I}^{\mathrm{r}}_{\mathrm{pol}}\dot{\Psi}_{\mathrm{tor}} - \dot{I}_{\mathrm{tor}}\dot{\Psi}^{\mathrm{r}}_{\mathrm{pol}}}{4\pi^2\sqrt{g_{\mathrm{f}}}} - \dot{p}\right\} = 0 , \tag{6.7.9}$$

or, since all the quantities involved are flux-surface quantities,

$$\dot{I}^{\mathrm{r}}_{\mathrm{pol}}\dot{\Psi}_{\mathrm{tor}} - \dot{I}_{\mathrm{tor}}\dot{\Psi}^{\mathrm{r}}_{\mathrm{pol}} = \dot{p}\int_0^{2\pi}\int_0^{2\pi}\sqrt{g_{\mathrm{f}}}\,d\theta_{\mathrm{f}}\,d\zeta_{\mathrm{f}} . \tag{6.7.10}$$

The double integral is proportional to the volume of the thin shell we integrated over

$$dV = d\varrho\int_0^{2\pi}\int_0^{2\pi}\sqrt{g_{\mathrm{f}}}\,d\theta_{\mathrm{f}}\,d\zeta_{\mathrm{f}} . \tag{6.7.11}$$

We thus have for this solvability condition:

$$\dot{I}^{\mathrm{r}}_{\mathrm{pol}}\dot{\Psi}_{\mathrm{tor}} - \dot{I}_{\mathrm{tor}}\dot{\Psi}^{\mathrm{r}}_{\mathrm{pol}} = \dot{p}\frac{dV}{d\varrho} . \tag{6.7.12}$$

This is a frequently used form of the ideal MHD equilibrium condition in toroidally confined plasmas.

The second solvability condition must be taken over a closed field line. From the equation of a field line, we have that $dl/B = d\zeta_{\mathrm{f}}/B^{\zeta_{\mathrm{f}}}$ which with $B^{\zeta_{\mathrm{f}}} = \dot{\Psi}_{\mathrm{tor}}/(2\pi\sqrt{g_{\mathrm{f}}})$ equals $2\pi\,d\zeta_{\mathrm{f}}\sqrt{g_{\mathrm{f}}}/\dot{\Psi}_{\mathrm{tor}}$. If we integrate over a field line that closes upon itself after N toroidal transits, we can write

$$\dot{p}\oint\frac{dl}{B} = \frac{\dot{I}^{\mathrm{r}}_{\mathrm{pol}}\dot{\Psi}_{\mathrm{tor}} - \dot{I}_{\mathrm{tor}}\dot{\Psi}^{\mathrm{r}}_{\mathrm{pol}}}{2\pi\dot{\Psi}_{\mathrm{tor}}}\int_0^{2\pi N}d\zeta = \frac{(\dot{I}^{\mathrm{r}}_{\mathrm{pol}}\dot{\Psi}_{\mathrm{tor}} - \dot{I}_{\mathrm{tor}}\dot{\Psi}^{\mathrm{r}}_{\mathrm{pol}})N}{\dot{\Psi}_{\mathrm{tor}}} .$$

If we now make use of the first constraint in (6.7.12), we obtain

$$\dot{p}\oint\frac{dl}{B}=N\frac{dV}{d\varrho}\frac{\dot{p}}{\dot{\Psi}_{\rm tor}}$$

or with $\dot{p}\equiv dp/d\varrho$ and $\dot{\Psi}_{\rm tor}\equiv d\Psi_{\rm tor}/d\varrho$

$$\frac{dp}{d\varrho}\left[\frac{dV}{d\Psi_{\rm tor}}-\frac{1}{N}\oint\frac{dl}{B}\right]=0\ . \tag{6.7.13}$$

For $\dot{p}\equiv dp/d\varrho\neq 0$ this condition states that the loop integral of dl/B must be constant on rational surfaces:

$$\frac{1}{N}\oint\frac{dl}{B}=\frac{dV}{d\Psi_{\rm tor}}\ . \tag{6.7.14}$$

The loop integral on the left-hand side is called the "proper length of a field line". This condition requires that the proper lengths of all the field lines on a rational surface must be the same. This condition is often referred to as the *current closure condition*. If it is satisfied, the device is said to possess true flux surfaces. In axisymmetric devices, this condition is satisfied. To see this, consider a particular field line at a point A as shown in Fig. 6.8 and compute its $(1/N)\oint dl/B$. Do the same for a different field line in the same poloidal plane starting at a point B. Now move a toroidal distance $\Delta\zeta_{\rm f}=q\Delta\theta_{\rm f}$ backwards, where q is the safety factor. The field line arriving at point C is equivalent to that arriving at point B because of axisymmetry. However, the former field line goes to point A because it has a rotational transform $\iota=1/q$. Hence, the proper length of both field lines must be the same. In nonaxisymmetric systems such as stellarators, the argument does not hold, and condition (6.7.14) is *not* satisfied. If this condition is violated, the Fourier coefficients of $\tilde{\eta}$ will exhibit resonant denominators of the form $(m\iota-n)$ leading to "infinite" Pfirsch-Schlüter currents and magnetic islands having fiber-like structures. We will discuss in Chap. 12 in what sense the two quantities on either side of (6.7.14) differ from each other in general.

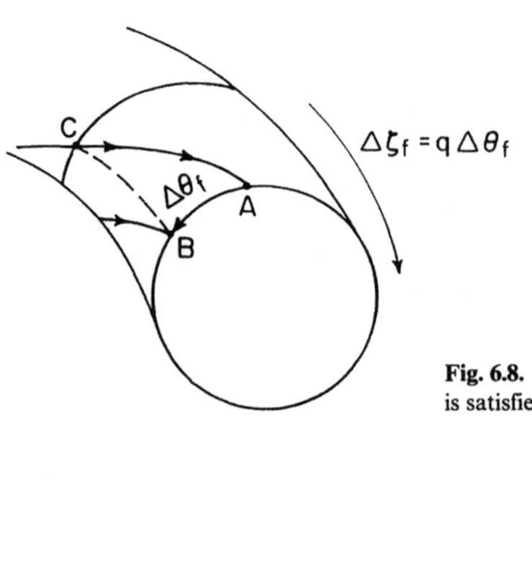

Fig. 6.8. Illustrating that the current-closure condition is satisfied in an axisymmetric tokamak. See text

It is customary to put the magnetic differential equation for $\tilde{\eta}$ in a simpler form by substituting the equilibrium condition given in (6.7.12) into (6.7.6). We then have

$$\boldsymbol{B}\cdot\boldsymbol{\nabla}\tilde{\eta} = \dot{p}\left[\frac{1}{4\pi^2\sqrt{g_{\mathrm{f}}}}\frac{dV}{d\varrho} - 1\right] \tag{6.7.15}$$

or with $dV/d\varrho \equiv \dot{V}$

$$\boldsymbol{B}\cdot\boldsymbol{\nabla}\tilde{\eta} = \dot{p}\left[\frac{\dot{V}}{4\pi^2\sqrt{g_{\mathrm{f}}}} - 1\right] . \tag{6.7.16}$$

In $(\theta_{\mathrm{f}}, \zeta_{\mathrm{f}})$ coordinates, the LHS can be found from $B^i(\boldsymbol{\nabla}\tilde{\eta})_i$ with the B^i given in (6.1.25). Equation (6.7.16) then becomes

$$\dot{\Psi}_{\mathrm{pol}}\frac{\partial\tilde{\eta}}{\partial\theta_{\mathrm{f}}} + \dot{\Psi}_{\mathrm{tor}}\frac{\partial\tilde{\eta}}{\partial\zeta_{\mathrm{f}}} = \frac{\dot{p}}{2\pi}(\dot{V} - 4\pi^2\sqrt{g_{\mathrm{f}}}) . \tag{6.7.17}$$

From our expression for the flux-surface average given in (4.9.11) and using (6.7.11), we can write the flux-surface average of $1/\sqrt{g_{\mathrm{f}}}$

$$\langle(\sqrt{g_{\mathrm{f}}})^{-1}\rangle = \frac{\int_0^{2\pi}\int_0^{2\pi} d\theta_{\mathrm{f}}\, d\zeta_{\mathrm{f}}}{\int_0^{2\pi}\int_0^{2\pi} d\theta_{\mathrm{f}}\, d\zeta_{\mathrm{f}}\sqrt{g_{\mathrm{f}}}} = \frac{4\pi^2}{dV/d\varrho} = \frac{4\pi^2}{\dot{V}} . \tag{6.7.18}$$

(Note that the average of $\sqrt{g_{\mathrm{f}}}$ that Shafranov (1968) uses is not the same as a flux-surface average. For Shafranov, $(\sqrt{g_{\mathrm{f}}}) = (1/4\pi^2)\int\int d\theta_{\mathrm{f}}\, d\zeta_{\mathrm{f}}\sqrt{g_{\mathrm{f}}}$.) The specific flux volume can thus alternatively be written as

$$\dot{V} = \frac{4\pi^2}{\langle(\sqrt{g_{\mathrm{f}}})^{-1}\rangle} = 4\pi^2\langle(\sqrt{g_{\mathrm{f}}})^{-1}\rangle^{-1} \tag{6.7.19}$$

and consequently, the magnetic differential equation for $\tilde{\eta}$ of (6.7.16) reads

$$\boldsymbol{B}\cdot\boldsymbol{\nabla}\tilde{\eta} = \dot{p}\left[\frac{\langle(\sqrt{g_{\mathrm{f}}})^{-1}\rangle^{-1}}{\sqrt{g_{\mathrm{f}}}} - 1\right] . \tag{6.7.20}$$

It is interesting to consider the alternative path that Boozer (1981) uses to arrive at the same differential equation for $\tilde{\eta}$. Starting from his Clebsch form for the current $\mu_0\boldsymbol{J} = \boldsymbol{\nabla}\lambda \times \boldsymbol{\nabla}\varrho$ and expanding $\boldsymbol{\nabla}\lambda$ in its covariant components as $\boldsymbol{\nabla}\lambda = (\partial\lambda/\partial\varrho)\,\boldsymbol{\nabla}\varrho + (\partial\lambda/\partial\theta_{\mathrm{f}})\,\boldsymbol{\nabla}\theta_{\mathrm{f}} + (\partial\lambda/\partial\zeta_{\mathrm{f}})\,\boldsymbol{\nabla}\zeta_{\mathrm{f}}$, he writes an expression for the current density

$$\boldsymbol{J} = \frac{1}{\mu_0}\left[(\boldsymbol{\nabla}\theta_{\mathrm{f}} \times \boldsymbol{\nabla}\varrho)\frac{\partial\lambda}{\partial\theta_{\mathrm{f}}} + (\boldsymbol{\nabla}\zeta_{\mathrm{f}} \times \boldsymbol{\nabla}\varrho)\frac{\partial\lambda}{\partial\zeta_{\mathrm{f}}}\right] . \tag{6.7.21}$$

Substitution of this form for $\boldsymbol{J}$ in the MHD equilibrium equation $\nabla p = \boldsymbol{J} \times \boldsymbol{B}$ results in a differential equation for λ as will be shown presently. We use the contravariant form for $\boldsymbol{B}$

$$\boldsymbol{B} = \frac{\dot{\Psi}_{\text{tor}}}{2\pi}(\nabla\varrho \times \nabla\theta_{\text{f}}) - \frac{\dot{\Psi}^{\text{r}}_{\text{pol}}}{2\pi}(\nabla\varrho \times \nabla\zeta_{\text{f}}) \tag{6.7.22}$$

and obtain for the MHD force balance

$$\mu_0 \dot{p}\,\nabla\varrho = \left[-\frac{\partial\lambda}{\partial\theta_{\text{f}}}(\nabla\theta_{\text{f}} \times \nabla\varrho) \times (\nabla\varrho \times \nabla\zeta_{\text{f}})\frac{\dot{\Psi}^{\text{r}}_{\text{pol}}}{2\pi} + \frac{\partial\lambda}{\partial\zeta_{\text{f}}}(\nabla\zeta_{\text{f}} \times \nabla\varrho) \times (\nabla\varrho \times \nabla\theta_{\text{f}})\frac{\dot{\Psi}_{\text{tor}}}{2\pi} \right] . \tag{6.7.23}$$

Upon application of the *bac-cab* rule, this becomes

$$\mu_0 \dot{p}\,\nabla\varrho = \left(\frac{\dot{\Psi}^{\text{r}}_{\text{pol}}}{2\pi}\frac{\partial\lambda}{\partial\theta_{\text{f}}} + \frac{\dot{\Psi}_{\text{tor}}}{2\pi}\frac{\partial\lambda}{\partial\zeta_{\text{f}}} \right)[-\nabla\varrho(\nabla\zeta_{\text{f}} \cdot \nabla\theta_{\text{f}} \times \nabla\varrho)] . \tag{6.7.24}$$

With the definition of the Jacobian $(\sqrt{g_{\text{f}}})^{-1} = \nabla\varrho \cdot (\nabla\theta_{\text{f}} \times \nabla\zeta_{\text{f}}) = -\nabla\zeta_{\text{f}} \cdot (\nabla\theta_{\text{f}} \times \nabla\varrho)$, we obtain

$$\dot{p} = \frac{1}{2\pi\mu_0\sqrt{g_{\text{f}}}}\left(\dot{\Psi}^{\text{r}}_{\text{pol}}\frac{\partial\lambda}{\partial\theta_{\text{f}}} + \dot{\Psi}_{\text{tor}}\frac{\partial\lambda}{\partial\zeta_{\text{f}}} \right) . \tag{6.7.25}$$

With λ given in (6.5.36) and (6.5.37),

$$\lambda = -\mu_0\eta = -\mu_0\tilde{\eta} - \frac{\mu_0}{2\pi}\dot{I}_{\text{tor}}\theta_{\text{f}} + \frac{\mu_0}{2\pi}\dot{I}^{\text{r}}_{\text{pol}}\zeta_{\text{f}} ,$$

the above differential equation for λ reduces to (6.7.6).

To solve for the unknown function $\tilde{\eta}$, we must solve its magnetic differential equation, following the usual procedure (see Sect. 4.10). The functions $\tilde{\eta}$ and $\sqrt{g_{\text{f}}}$ appearing in (6.7.17) are expanded in a double Fourier series in θ_{f} and ζ_{f} and use is made of the equation of a field line $\theta_{\text{f0}} = \theta_{\text{f}} - \iota\zeta_{\text{f}}$ to eliminate one of the variables. The two solubility conditions put constraints on the Fourier expansion coefficients and allow proper treatment of resonant denominators of the form $(m\iota - n)$ on rational surfaces.

6.8 Hamada Coordinates

In Sect. 6.4., we discussed the condition for straight current lines. The current is represented by $\boldsymbol{J} = \nabla\varrho \times \nabla\eta$, where $\eta = (1/2\pi)\dot{I}_{\text{tor}}\theta - (1/2\pi)\dot{I}^{\text{r}}_{\text{pol}}\zeta + \tilde{\eta}$, which implies that they are straight only when $\tilde{\eta}$ is a flux function or simply zero.

We wish to use the freedom of (6.1.28) to find a new set of flux coordinates in which the $\boldsymbol{J}$ lines are straight in addition to the $\boldsymbol{B}$ lines being straight. This will result in a coordinate system with a Jacobian equal to a constant (which can be made unity if desired). These new coordinates are known as *Hamada coordinates or natural coordinates* and are represented by θ_{H} and ζ_{H}.

The existence of Hamada coordinates was first proved by Hamada (1962). An independent proof was later established by Greene and Johnson (1962). Most treatments (and ours) follow this latter reference. See also Solov'ev and Shafranov (1970); and Boozer (1981).

The requirement that $\tilde{\eta}$ be a flux function or zero implies that $\boldsymbol{B}\cdot\nabla\tilde{\eta}$ must vanish. From (6.7.20), we observe that this is the case (for nonzero $\dot{p}$) if the inverse of the flux-surface average of the inverse Jacobian equals the Jacobian itself. In other words, the *Jacobian* must be a flux function. (There is a confusing misprint in Solov'ev and Shafranov, above (2.87). D is the total Jacobian from Cartesian coordinates (x, y, z) to Hamada coordinates, $(\varrho, \theta_{\mathrm{H}}, \zeta_{\mathrm{H}})$. It is not the Jacobian from θ, ζ to θ'(our θ_{H}) and ζ'(our ζ_{H}) as stated above (2.87).)

Equation (6.7.19) suggests that the choice of flux label,

$$\varrho \equiv V\,, \tag{6.8.1}$$

provides a very simple expression for the flux-surface average of the inverse Jacobian. Since then $\dot{V} = 1$, we have

$$\langle(\sqrt{g_{\mathrm{f}}})^{-1}\rangle^{-1} = \frac{1}{4\pi^2}\,. \tag{6.8.2}$$

Hence, to make $\boldsymbol{B}$ and $\boldsymbol{J}$ lines straight, we must take the inverse Jacobian equal to $4\pi^2$ when $\varrho \equiv V$. Thus

$$(\sqrt{g})_{\mathrm{H}} \equiv \frac{1}{4\pi^2}\,, \qquad \varrho_{\mathrm{H}} \equiv V\,. \tag{6.8.3}$$

$\varrho = V$ is not necessary for the $\boldsymbol{J}$ lines to be straight, but it is very convenient because the ubiquitous $\dot{V}$ equals one. If $\varrho \neq V$, then $\sqrt{g_{\mathrm{H}}} = f(\varrho)$ is needed. (As before, we denote the Jacobian with V as flux label by $(\sqrt{g})_{\mathrm{H}}$; the symbol $\sqrt{g_{\mathrm{H}}}$ is the "Hamada Jacobian" for the flux label ϱ.)

Since we began from the magnetic differential equation for $\tilde{\eta}$, in (6.7.20), we may conclude that a Hamada coordinate system exists when the solvability conditions of the $\boldsymbol{B}\cdot\nabla\tilde{\eta}$ equation are satisfied. The first implies $\langle\mathrm{RHS}\rangle \equiv 0$ as stated in (4.10.5). This is equivalent to the volume integral condition; cf. (4.10.7). It is clear that with $\sqrt{g_{\mathrm{f}}} \equiv \sqrt{g_{\mathrm{H}}}$ equal to a flux function, the flux-surface average of the RHS of (6.7.20) is identically zero. Thus this condition reduces to an identity. This is not surprising because (6.7.20) is a rewritten form of (6.7.15), which was, in turn, obtained by *subtracting* the solvability constraint (6.7.12) from the original $\boldsymbol{B}\cdot\nabla\tilde{\eta}$ equation (6.7.6).

The other condition is the "current closure condition"

$$\frac{1}{N}\oint\frac{dl}{B} = \frac{dV}{d\Psi_{\mathrm{tor}}}\,. \tag{6.8.4}$$

Thus, Hamada coordinates exist when the proper equilibrium conditions are satisfied.

In terms of Hamada coordinates θ_H and ζ_H, the contravariant components of $\boldsymbol{B}$ are flux functions themselves. With $\sqrt{g_f} \to (\sqrt{g})_H = 1/4\pi^2$, we find from (6.1.25):

$$B^i = 2\pi\left(0, \frac{d\Psi^{\mathrm{r}}_{\mathrm{pol}}}{dV}, \frac{d\Psi_{\mathrm{tor}}}{dV}\right) . \tag{6.8.5}$$

$\boldsymbol{B}$ itself keeps the same Clebsch form as before, since the Jacobian did not appear in the expression for v in (6.1.23):

$$\boldsymbol{B} = \nabla V \times \nabla\left(\frac{1}{2\pi}\frac{d\Psi_{\mathrm{tor}}}{dV}\theta_H - \frac{1}{2\pi}\frac{d\Psi^{\mathrm{r}}_{\mathrm{pol}}}{dV}\zeta_H\right) . \tag{6.8.6}$$

The contravariant current-density components follow from (6.4.13)

$$J^i = 2\pi\left(0, \frac{dI^{\mathrm{r}}_{\mathrm{pol}}}{dV}, \frac{dI_{\mathrm{tor}}}{dV}\right) , \tag{6.8.7}$$

while its Clebsch form, using (6.4.5) and (6.4.7) with $\tilde{\eta} = 0$, or (6.4.12), reads

$$\boldsymbol{J} = \nabla V \times \nabla\left(\frac{1}{2\pi}\frac{dI_{\mathrm{tor}}}{dV}\theta_H - \frac{1}{2\pi}\frac{dI^{\mathrm{r}}_{\mathrm{pol}}}{dV}\zeta_H\right) . \tag{6.8.8}$$

All other formulae and expressions discussed so far can be transformed to the Hamada set (V, θ_H, ζ_H) without any difficulties as long as we recall that $(\sqrt{g})_H = 1/4\pi^2$.

The expression for the Jacobian given in (6.6.4) allows us to compare the Hamada system with Boozer's system. With the Hamada choice of Jacobian, as given in (6.8.3), we have

$$(B)^2 - \boldsymbol{B}\cdot\nabla\tilde{\Phi} = \mu_0(\dot{\Psi}^{\mathrm{r}}_{\mathrm{pol}} I_{\mathrm{tor}} + \dot{\Psi}_{\mathrm{tor}} I^{\mathrm{d}}_{\mathrm{pol}}) \tag{6.8.9a}$$

or

$$B\frac{\partial\tilde{\Phi}}{\partial l} = (B)^2 - \mu_0(\dot{\Psi}^{\mathrm{r}}_{\mathrm{pol}} I_{\mathrm{tor}} + \dot{\Psi}_{\mathrm{tor}} I^{\mathrm{d}}_{\mathrm{pol}}) , \tag{6.8.9b}$$

with the dot representing here d/dV.

This equation says that θ_H and ζ_H are chosen such that the variation of $\tilde{\Phi}$ along a field line satisfies (6.8.9). Recall that Boozer chooses his θ_B and ζ_B such that $\tilde{\Phi} \equiv 0$ in his system. His Jacobian goes like $(B)^{-2}$ as shown by (6.6.5).

To actually find the new coordinates θ_H and ζ_H, we use (6.1.28)

$$\theta_H = \theta_f + \dot{\Psi}^{\mathrm{r}}_{\mathrm{pol}} G_H(\varrho, \theta_f, \zeta_f) \tag{6.8.10a}$$

$$\zeta_{\mathrm{H}} = \zeta_{\mathrm{f}} + \dot{\Psi}_{\mathrm{tor}} G_{\mathrm{H}}(\varrho, \theta_{\mathrm{f}}, \zeta_{\mathrm{f}}) \ . \tag{6.8.10b}$$

We must find the function G_{H} to ensure that the Jacobian $(\sqrt{g_{\mathrm{H}}})^{-1} = \nabla\varrho \cdot (\nabla\theta_{\mathrm{H}} \times \nabla\zeta_{\mathrm{H}})$ is a flux function. The procedure is analogous to the derivation of (6.6.13) for Boozer coordinates. We substitute the expressions for θ_{H} and ζ_{H} in the Hamada Jacobian $\sqrt{g_{\mathrm{H}}}$ and find a magnetic differential equation for $G_{\mathrm{H}}(\varrho, \theta_{\mathrm{f}}, \zeta_{\mathrm{f}})$

$$2\pi \boldsymbol{B} \cdot \nabla G_{\mathrm{H}} = \frac{1}{\sqrt{g_{\mathrm{H}}}} - \frac{1}{\sqrt{g_{\mathrm{f}}}} \ . \tag{6.8.11}$$

(The B^i used to arrive at (6.8.11) are those in terms of the old system $(\theta_{\mathrm{f}}, \zeta_{\mathrm{f}})$, i.e., given in (6.1.25).) As we discussed above (see also Solov'ev and Shafranov (1970)), this magnetic differential equation is subject to the usual solvability conditions. To actually solve for $G_{\mathrm{H}}(\varrho \equiv V, \theta_{\mathrm{f}}, \zeta_{\mathrm{f}})$, we write the $\boldsymbol{B} \cdot \nabla$ operator in terms of the old coordinates $(\theta_{\mathrm{f}}, \zeta_{\mathrm{f}})$ but with $\varrho \equiv V$. This last choice is not part of the transformation, it is merely to make things easy:

$$\frac{1}{\sqrt{g_{\mathrm{f}}}} \frac{d\Psi^{\mathrm{r}}_{\mathrm{pol}}}{dV} \frac{\partial G_{\mathrm{H}}}{\partial \theta_{\mathrm{f}}} + \frac{1}{\sqrt{g_{\mathrm{f}}}} \frac{d\Psi_{\mathrm{tor}}}{dV} \frac{\partial G_{\mathrm{H}}}{\partial \zeta_{\mathrm{f}}} = \frac{1}{(\sqrt{g})_{\mathrm{H}}} - \frac{1}{\sqrt{g_{\mathrm{f}}}}$$

or

$$\left\{ \iota \frac{\partial G_{\mathrm{H}}}{\partial \theta_{\mathrm{f}}} + \frac{\partial G_{\mathrm{H}}}{\partial \zeta_{\mathrm{f}}} \right\} = (4\pi^2 \sqrt{g_{\mathrm{f}}} - 1) \frac{dV}{d\Psi_{\mathrm{tor}}} \ . \tag{6.8.12}$$

Then expanding the expressions for G_{H} and $\sqrt{g_{\mathrm{f}}}$ in double Fourier series, and taking measures to deal with the resonant denominators of the form $(m\iota - n)$ allows us to find the Fourier coefficients for G_{H}. (See Sect. 4.10 on Magnetic Differential Equations, or the review by Solov'ev and Shafranov (1970) for details.) Once G_{H} is known, the new coordinates θ_{H} and ζ_{H} have been found.

Knowledge of the angle functions θ_{H} and ζ_{H} allows determination of the basis vectors and thus the coefficients of the metric tensor g_{ij}. Solov'ev (1968, 1975) has indicated an alternative route for obtaining the basis vectors of a Hamada system. For simplicity, we shall use the indexed notation:

$$(u^1, u^2, u^3) \equiv (V, \theta_{\mathrm{H}}, \zeta_{\mathrm{H}}) \ . \tag{6.8.13}$$

(Solov'ev (1968, 1975) uses index 3 to denote the radial variable V. Also, his Jacobian equals unity, whereas ours is $1/4\pi^2$; we have a *factor* of 2π in (6.8.5) and (6.8.7), whereas he *divides through* by 2π.)

The magnetic field $\boldsymbol{B}$ and the current density $\boldsymbol{J}$ are to be written in terms of Hamada coordinates as

$$\boldsymbol{B} = B^2 \boldsymbol{e}_2 + B^3 \boldsymbol{e}_3 = 2\pi \dot{\Psi}^{\mathrm{r}}_{\mathrm{pol}} \boldsymbol{e}_2 + 2\pi \dot{\Psi}_{\mathrm{tor}} \boldsymbol{e}_3 \tag{6.8.14}$$

$$\boldsymbol{J} = J^2 \boldsymbol{e}_2 + J^3 \boldsymbol{e}_3 = 2\pi \dot{I}^{\mathrm{r}}_{\mathrm{pol}} \boldsymbol{e}_2 + 2\pi \dot{I}_{\mathrm{tor}} \boldsymbol{e}_3 \ . \tag{6.8.15}$$

Again, the dot here stands for d/dV.

Recall that $B^1 = J^1 = 0$. The systems of equations, (6.8.14) and (6.8.15) can be solved for the unknowns $\boldsymbol{e}_2$ and $\boldsymbol{e}_3$ in terms of the known vector fields $\boldsymbol{B}$ and $\boldsymbol{J}$. With the help of the equilibrium condition, (6.7.12), where ϱ is identified with V, $\dot{I}^{\mathrm{r}}_{\mathrm{pol}}\dot{\Psi}_{\mathrm{tor}} - \dot{I}_{\mathrm{tor}}\dot{\Psi}^{\mathrm{r}}_{\mathrm{pol}} = \dot{p}$, we find:

$$\boldsymbol{e}_2 = \frac{1}{2\pi\dot{p}}(\dot{\Psi}_{\mathrm{tor}}\boldsymbol{J} - \dot{I}_{\mathrm{tor}}\boldsymbol{B}) \tag{6.8.16}$$

and

$$\boldsymbol{e}_3 = \frac{1}{2\pi\dot{p}}(\dot{I}^{\mathrm{r}}_{\mathrm{pol}}\boldsymbol{B} - \dot{\Psi}^{\mathrm{r}}_{\mathrm{pol}}\boldsymbol{J}) \ . \tag{6.8.17}$$

To find the first covariant-basis vector $\boldsymbol{e}_1$, we expand $\boldsymbol{e}^1 \equiv \nabla V$ along the covariant set $\boldsymbol{e}_i$ as follows:

$$\begin{aligned}\nabla V &= (\nabla V\cdot\boldsymbol{e}^1)\boldsymbol{e}_1 + (\nabla V\cdot\boldsymbol{e}^2)\boldsymbol{e}_2 + (\nabla V\cdot\boldsymbol{e}^3)\boldsymbol{e}_3 \\ &= |\nabla V|^2\boldsymbol{e}_1 + g^{12}\boldsymbol{e}_2 + g^{13}\boldsymbol{e}_3 \ .\end{aligned} \tag{6.8.18}$$

Here we left g^{11} as $|\nabla V|^2$. After solving this equation for $\boldsymbol{e}_1$, we obtain

$$\boldsymbol{e}_1 \equiv \boldsymbol{e}_V = \frac{\nabla V}{|\nabla V|^2} - \frac{g^{12}}{|\nabla V|^2}\boldsymbol{e}_2 - \frac{g^{13}}{|\nabla V|^2}\boldsymbol{e}_3 \ . \tag{6.8.19}$$

For clarity, we define two new symbols for the unknown coefficients of $\boldsymbol{e}_2$ and $\boldsymbol{e}_3$ in (6.8.19):

$$\omega = -\frac{g^{12}}{|\nabla V|^2} \quad \text{and} \quad \mu = -\frac{g^{13}}{|\nabla V|^2} \ ,$$

$$\boldsymbol{e}_1 = \frac{\nabla V}{|\nabla V|^2} + \omega\boldsymbol{e}_2 + \mu\boldsymbol{e}_3 \ . \tag{6.8.20}$$

After some straightforward but somewhat tedious vector manipulations (the main points of which are given below), one obtains two magnetic differential equations for ω and μ, respectively. They are a consequence of $\mu_0\boldsymbol{J} = \nabla\times\boldsymbol{B}$ and read:

$$\boldsymbol{B}\cdot\nabla\omega = -2\pi\ddot{\Psi}^{\mathrm{r}}_{\mathrm{pol}} - \frac{4\pi^2}{|\nabla V|^2}\boldsymbol{e}_3\cdot\left[\mu_0\boldsymbol{J} - \frac{2}{|\nabla V|^2}(\nabla V\times(\nabla V\cdot\nabla)\boldsymbol{B})\right] \tag{6.8.21}$$

$$\boldsymbol{B}\cdot\nabla\mu = -2\pi\ddot{\Psi}_{\mathrm{tor}} + \frac{4\pi^2}{|\nabla V|^2}\boldsymbol{e}_2\cdot\left[\mu_0\boldsymbol{J} - \frac{2}{|\nabla V|^2}(\nabla V\times(\nabla V\cdot\nabla)\boldsymbol{B})\right] \ . \tag{6.8.22}$$

When these equations have been solved, ω, μ and $\boldsymbol{e}_1$, and thus all three covariant basis vectors are known in terms of the known magnetic field and current density.

Proof of (6.8.21):[1]

We now indicate how the magnetic differential equations can be obtained. We focus on the first one, (6.8.21).

1) From the vector identity $\nabla(\boldsymbol{A}\cdot\boldsymbol{B}) = \boldsymbol{A}\times(\nabla\times\boldsymbol{B}) + \boldsymbol{B}\times(\nabla\times\boldsymbol{A}) + (\boldsymbol{A}\cdot\nabla)\boldsymbol{B} + (\boldsymbol{B}\cdot\nabla)\boldsymbol{A}$, with $\boldsymbol{A}\equiv\nabla V$, and $\boldsymbol{B}\equiv\boldsymbol{B}$, $\boldsymbol{B}\cdot\nabla V = 0$ and $\mu_0\boldsymbol{J} = \nabla\times\boldsymbol{B}$, we find that

$$\mu_0\boldsymbol{J}\times\nabla V = (\nabla V\cdot\nabla)\boldsymbol{B} + (\boldsymbol{B}\cdot\nabla)\nabla V\,. \tag{6.8.23}$$

2) The perpendicular component of a vector $\boldsymbol{A}$ with respect to a directional vector $\hat{\boldsymbol{a}}$ is $-\hat{\boldsymbol{a}}\times(\hat{\boldsymbol{a}}\times\boldsymbol{A})$. Since $\boldsymbol{J}\perp\nabla V$, it follows that $\nabla V\times$ (6.8.23) equals

$$\mu_0|\nabla V|^2\boldsymbol{J} = \nabla V\times(\nabla V\cdot\nabla)\boldsymbol{B} + \nabla V\times(\boldsymbol{B}\cdot\nabla)\nabla V\,. \tag{6.8.24}$$

To obtain terms similar to those of (6.8.21), we dot multiply (6.8.24) by $\boldsymbol{e}_3$:

$$\mu_0|\nabla V|^2\boldsymbol{e}_3\cdot\boldsymbol{J} = \boldsymbol{e}_3\cdot[\nabla V\times(\nabla V\cdot\nabla)\boldsymbol{B}] + \boldsymbol{e}_3\cdot[\nabla V\times(\boldsymbol{B}\cdot\nabla)\nabla V] \tag{6.8.25}$$

3) We concentrate on the last term of (6.8.25). Write the first ∇V of this term as $\boldsymbol{e}^1$ and replace the second ∇V by $|\nabla V|^2(\boldsymbol{e}_1 - \omega\boldsymbol{e}_2 - \mu\boldsymbol{e}_3)$ according to (6.8.18) or (6.8.20). The term proportional to $\boldsymbol{B}\cdot\nabla|\nabla V|^2$ vanishes because $\boldsymbol{e}^1\times\boldsymbol{e}^1/|\nabla V|^2\equiv 0$. Next, we expand $\boldsymbol{B}\cdot\nabla$ as $B^i\partial/\partial u^i$. The terms involving $\partial\boldsymbol{e}_k/\partial u^i$ can be expanded in contravariant components as $\partial\boldsymbol{e}_k/\partial u^i = \boldsymbol{e}_j(\boldsymbol{e}^j\cdot\partial\boldsymbol{e}_k/\partial u^i)$. Then writing $\boldsymbol{e}_j$ in terms of the $\boldsymbol{e}^m$, via the "back-and-forth" relations of (2.5.25) with a Jacobian equal to $1/4\pi^2$, and application of the *bac-cab* rule leads to $\boldsymbol{e}_3\cdot\boldsymbol{e}^1\times\boldsymbol{e}_j = (-g^{12}/4\pi^2, |\nabla V|^2/4\pi^2, 0)$ for $j = (1, 2, 3)$, respectively. These rules imply that the term $[-|\nabla V|^2(\boldsymbol{e}_3\cdot\boldsymbol{e}^1\times\boldsymbol{e}_3)\boldsymbol{B}\cdot\nabla\mu]$ vanishes, while $-|\nabla V|^2(\boldsymbol{e}_3\cdot\boldsymbol{e}^1\times\boldsymbol{e}_2)\boldsymbol{B}\cdot\nabla\omega = -|\nabla V|^4\boldsymbol{B}\cdot\nabla\omega/4\pi^2$ survives. After using the property of the Christoffel symbol, given in (2.6.22), $\boldsymbol{e}^j\cdot\partial\boldsymbol{e}_k/\partial u^i = -\boldsymbol{e}_k\cdot\partial\boldsymbol{e}^j/\partial u^i$, and (6.8.20), we have

$$\boldsymbol{e}_3\cdot[\nabla V\times(\boldsymbol{B}\cdot\nabla)\nabla V] = \frac{-|\nabla V|^4}{4\pi^2}\boldsymbol{B}\cdot\nabla\omega - B^i\frac{|\nabla V|^2}{4\pi^2}\nabla V\cdot\frac{\partial\boldsymbol{e}^2}{\partial u^i} + B^i\left(\frac{g^{12}}{4\pi^2}\right)\nabla V\cdot\frac{\partial\boldsymbol{e}^1}{\partial u^i}\,. \tag{6.8.26}$$

4) The last two terms of (6.8.26) can be manipulated into a term proportional to $\boldsymbol{e}_3\cdot\nabla V\times(\nabla V\cdot\nabla)\boldsymbol{B}$, leading to the factor 2 in (6.8.21) and a term proportional to $\ddot{\Psi}^{\mathrm{T}}_{\mathrm{pol}}$. The easiest way to prove this is to show that $\boldsymbol{e}_3\cdot\nabla V\times(\nabla V\cdot\nabla)\boldsymbol{B}$ leads to something similar to (6.8.26). Because $(\nabla V\cdot\nabla)\equiv g^{1k}\partial/\partial u^k$ and since $B^1\equiv 0$, we obtain by using the above mentioned expressions for $\boldsymbol{e}_3\cdot\boldsymbol{e}^1\times\boldsymbol{e}_j$:

$$\boldsymbol{e}_3\cdot[\nabla V\times(\nabla V\cdot\nabla)\boldsymbol{B}] = \frac{|\nabla V|^4}{2\pi}\ddot{\Psi}^{\mathrm{T}}_{\mathrm{pol}} - \frac{g^{12}}{4\pi^2}g^{1k}B^i\left(\boldsymbol{e}^1\cdot\frac{\partial\boldsymbol{e}_i}{\partial u^k}\right) + \frac{|\nabla V|^2}{4\pi^2}g^{1k}B^i\left(\boldsymbol{e}^2\cdot\frac{\partial\boldsymbol{e}_i}{\partial u^k}\right)\,. \tag{6.8.27}$$

We now replace $\boldsymbol{e}^j\cdot\partial\boldsymbol{e}_i/\partial u^k = \boldsymbol{e}^j\cdot\partial\boldsymbol{e}_k/\partial u^i$ by $-\boldsymbol{e}_k\cdot\partial\boldsymbol{e}^j/\partial u^i$ (for $j = 1, 2$ in the last two terms) and write g^{1k} as $\boldsymbol{e}^1\cdot\boldsymbol{e}^k$. After recognizing that we are then left with the covariant expansion of $\partial\boldsymbol{e}^j/\partial u^i$ as $\boldsymbol{e}^k\boldsymbol{e}_k\cdot\partial\boldsymbol{e}^j/\partial u^i$, we obtain for (6.8.27):

$$\boldsymbol{e}_3\cdot[\nabla V\times(\nabla V\cdot\nabla)\boldsymbol{B}] = \frac{|\nabla V|^4}{2\pi}\ddot{\Psi}^{\mathrm{T}}_{\mathrm{pol}} + \frac{g^{12}}{4\pi^2}B^i\left(\boldsymbol{e}^1\cdot\frac{\partial\boldsymbol{e}^1}{\partial u^i}\right) - \frac{|\nabla V|^2}{4\pi^2}B^i\left(\boldsymbol{e}^1\cdot\frac{\partial\boldsymbol{e}^2}{\partial u^i}\right)\,. \tag{6.8.28}$$

5) Comparing (6.8.28) with (6.8.26) gives

$$\boldsymbol{e}_3\cdot[\nabla V\times(\boldsymbol{B}\cdot\nabla)\nabla V] = -\frac{|\nabla V|^4}{4\pi^2}\boldsymbol{B}\cdot\nabla\omega + \boldsymbol{e}_3\cdot[\nabla V\times(\nabla V\cdot\nabla)\boldsymbol{B}] - \frac{|\nabla V|^4}{2\pi}\ddot{\Psi}^{\mathrm{T}}_{\mathrm{pol}}\,. \tag{6.8.29}$$

6) After substitution of (6.8.29) into (6.8.25), we recover (6.8.21).

[1] Can be skipped without loss of continuity.

7. Conversion from Clebsch Coordinates to Toroidal Flux Coordinates

In this Chapter, we discuss the conversion formulae from toroidal flux coordinates to Clebsch functions. First, we deal with the (ϱ, ν, l) system, and after that, we discuss the Boozer-Grad (ϱ, ν, χ) system.

7.1 The Generic Clebsch Coordinate System (ϱ, ν, l)

It is quite inconvenient to transform from angle flux coordinates to the Clebsch function ν and the "along the field line" coordinate l. The major problem has to do with the fact that the l coordinate curve does not close upon itself after one toroidal transit. For a qualitative discussion of the problems involved, see Sect. 4.6.1d.

In angle flux coordinates, the magnetic field is written as

$$\boldsymbol{B} = \boldsymbol{\nabla}\varrho \times \boldsymbol{\nabla}\left(\frac{\dot{\Psi}_{\text{tor}}}{2\pi}\theta_{\text{f}} - \frac{\dot{\Psi}^{\text{T}}_{\text{pol}}}{2\pi}\zeta_{\text{f}}\right) . \tag{7.1.1}$$

The coordinates are $(\varrho, \theta_{\text{f}}, \zeta_{\text{f}})$. In the Clebsch (ϱ, ν, l) representation, it is

$$\boldsymbol{B} = \boldsymbol{\nabla}\varrho \times \boldsymbol{\nabla}\nu = B\hat{\boldsymbol{l}} . \tag{7.1.2}$$

It is not difficult to see that

$$\varrho \equiv \varrho \tag{7.1.3}$$

$$\nu \equiv \frac{1}{2\pi}(\dot{\Psi}_{\text{tor}}\theta_{\text{f}} - \dot{\Psi}^{\text{T}}_{\text{pol}}\zeta_{\text{f}}) . \tag{7.1.4}$$

Finding l as a function of θ_{f} and ζ_{f} is not easy. The field lines are straight in flux coordinates, but the θ_{f} and ζ_{f} coordinate curves do not intersect at right angles. We must use the metric coefficients g_{ij} to relate l to θ_{f} and ζ_{f}, which results in an integral relationship. One way to see this is to compute the square of the differential arc length $(dl)^2$:

$$\begin{aligned}(dl)^2 &= g_{ij}du^i du^j \\ &= g_{\theta_{\text{f}}\theta_{\text{f}}}(d\theta_{\text{f}})^2 + 2g_{\theta_{\text{f}}\zeta_{\text{f}}}(d\theta_{\text{f}})(d\zeta_{\text{f}}) + g_{\zeta_{\text{f}}\zeta_{\text{f}}}(d\zeta_{\text{f}})^2 .\end{aligned} \tag{7.1.5}$$

The equation of a field line, expressed in flux coordinates, $d\theta_f = \iota d\zeta_f$, simplifies this a little. We can express $(dl)^2$ in terms of only one angle ζ_f,

$$(dl)^2 = (\iota^2 g_{\theta_f\theta_f} + 2\iota g_{\theta_f\zeta_f} + g_{\zeta_f\zeta_f})(d\zeta_f)^2$$

or

$$dl = \sqrt{\iota^2 g_{\theta_f\theta_f} + 2\iota g_{\theta_f\zeta_f} + g_{\zeta_f\zeta_f}}\, d\zeta_f \ . \tag{7.1.6}$$

Since the expression under the square root is in general a function of ζ_f, analytic integration is not generally possible. The best we can do is to write an integral relationship

$$l = \int_0^{\zeta_f} \sqrt{\iota^2 g_{\theta_f\theta_f} + 2\iota g_{\theta_f\zeta_f} + g_{\zeta_f\zeta_f}}\, d\zeta_f \ . \tag{7.1.7}$$

An alternative expression for $l \equiv l(\theta_f, \zeta_f)$ can be obtained by using the general equation of a field line as follows:

$$\frac{dl}{B} = \frac{d\zeta_f}{B^{\zeta_f}} \quad \text{or} \quad dl = \frac{B}{B^{\zeta_f}} d\zeta_f \ . \tag{7.1.8}$$

In flux coordinates $B^{\zeta_f} = \dot{\Psi}_{tor}/(2\pi\sqrt{g_f})$ so that dl is given by

$$dl = \frac{2\pi B\sqrt{g_f}}{\dot{\Psi}_{tor}} d\zeta_f \ . \tag{7.1.9}$$

$B \equiv |\boldsymbol{B}|$ as well as $\sqrt{g_f}$ are generally functions of θ_f and ζ_f so again an integral formula is the best we can do:

$$l = \frac{2\pi}{\dot{\Psi}_{tor}} \int_0^{\zeta_f} B\sqrt{g_f}\, d\zeta_f \ . \tag{7.1.10}$$

A second use of the equation of a field line, $d\theta_f = \iota d\zeta_f$ or $\theta_{f0} = \theta_f - \iota\zeta_f$ can remove the θ_f dependence from B and $\sqrt{g_f}$. Even in Hamada coordinates, the problem is not any simpler. Although the Jacobian can be taken out of the integral, $B \equiv B(\theta_f, \zeta_f)$ must remain under it.

7.2 Boozer-Grad Coordinates (ϱ, ν, χ)

The covariant and contravariant forms of $\boldsymbol{B}$ in the (ϱ, ν, χ) coordinate frame are, respectively,

$$\boldsymbol{B} = \nabla\chi + \lambda\nabla\varrho \ . \tag{7.2.1}$$

and

$$\boldsymbol{B} = \nabla\varrho \times \nabla\nu \ . \tag{7.2.2}$$

As discussed above, the covariant Boozer-Grad form reduces to a very simple form in vacuum: $\boldsymbol{B} = \nabla\chi$, where χ can *then* be thought of as a scalar magnetic potential. A second important feature of the "along the field line" label χ is the fact that it can very easily be expressed in terms of the flux coordinate angles θ_f and ζ_f. This is probably the major advantage of the variable χ over the length l.

The variables ϱ and v do not cause any difficulty. As in the previous section, they obey (7.1.3) and (7.1.4). Concerning the "along the field line" variable χ, we shall now argue that starting from the covariant form of $\boldsymbol{B}$, there is only one possible form that χ can have in terms of θ_f and ζ_f. (Here, we temporarily disregard the fact that we have already found χ in terms of ϱ, θ_f and ζ_f in (6.5.39), which, as one recalls, was obtained by finding the covariant components of $\boldsymbol{B}$ in flux coordinates and identifying the covariant expression for $\boldsymbol{B}$ with the Boozer-Grad form. What we are about to undertake now is a somewhat more direct approach.) The argument is closely analogous to that which we presented for the stream function v in (6.1.13).

The function χ in (7.2.1) need not be single valued (i.e., periodic) in θ_f and ζ_f. Even $\nabla\chi$ is exempted from the periodicity property, as long as the $\lambda\nabla\varrho$ term cancels off any non-single valuedness so that $\boldsymbol{B}$ is single valued. However, the term $\lambda\nabla\varrho$ can obviously only affect the terms in $\nabla\chi$ proportional to $\nabla\varrho$.

As was the case with v, any non-periodic function which is not a linear polynomial in θ_f or ζ_f must be rejected.

The "constant" in front of the linear term can be a function of ϱ. Although that term does produce a secular term in θ_f or ζ_f, it has $\nabla\varrho$ in its gradient which can be cancelled by the $\lambda\nabla\varrho$ term: $\nabla(C(\varrho)\theta_f) = C\nabla\theta_f + \theta_f\dot{C}\nabla\varrho$.

There is a possibility that a periodic function must be included, $\tilde{\Phi}(\varrho, \theta_f, \zeta_f)$. Thus χ must have the form

$$\chi = K(\varrho) + C(\varrho)\theta_f + D(\varrho)\zeta_f + \tilde{\Phi}(\varrho, \theta_f, \zeta_f) \ . \tag{7.2.3}$$

If we absorb the $K(\varrho)$ in the function $\tilde{\Phi}$, we obtain

$$\chi = C(\varrho)\theta_f + D(\varrho)\zeta_f + \tilde{\Phi}(\varrho, \theta_f, \zeta_f) \ . \tag{7.2.4}$$

With this χ, the covariant representation of $\boldsymbol{B}$ of (7.2.1) is

$$\boldsymbol{B} = \left(C + \frac{\partial\tilde{\Phi}}{\partial\theta_f}\right)\nabla\theta_f + \left(D + \frac{\partial\tilde{\Phi}}{\partial\zeta_f}\right)\nabla\zeta_f + \left(\lambda + \dot{C}\theta_f + \dot{D}\zeta_f + \frac{\partial\tilde{\Phi}}{\partial\varrho}\right)\nabla\varrho \ . \tag{7.2.5}$$

To find the functions $C(\varrho)$ and $D(\varrho)$, we use Ampère's law along a toroidal coordinate curve and along a poloidal coordinate curve,

$$\oint \boldsymbol{B}\cdot d\boldsymbol{l} = \mu_0 I_{\text{enclosed}} \ . \tag{7.2.6}$$

This has been explained in (6.5.22) and (6.5.23). We find that

$$C(\varrho) \equiv \frac{\mu_0}{2\pi} I_{\text{tor}}(\varrho) \tag{7.2.7}$$

$$D(\varrho) \equiv \frac{\mu_0}{2\pi}(I_{\rm pol}^{\rm plas} - I_{\rm pol}^{\rm r}(\varrho)) + \frac{\mu_0}{2\pi}\sum(I_{\rm coils}) = \frac{\mu_0}{2\pi} I_{\rm pol}^{\rm d}(\varrho) \ . \tag{7.2.8}$$

The symbols are all explained in Sect. 6.6.5. As expected, χ has the form already given in (6.5.39).

If we choose $\theta_{\rm f}$ and $\zeta_{\rm f}$ such that $\tilde{\Phi}$ vanishes in that system, which angles we now call $\theta_{\rm B}$ and $\zeta_{\rm B}$ (for Boozer), we recover (6.6.14c). Apart from the arbitrariness in ν and χ mentioned in (5.2.13) and (5.4.39), respectively, the functions ν and χ are completely determined.

It is clear that the expression for χ given in (7.2.4) is much simpler than that for l in (7.1.7) or (7.1.10). Thus, it appears that the (ϱ, ν, χ) system is preferable to (ϱ, ν, l). However, the flux coordinate angle system $(\theta_{\rm f}, \zeta_{\rm f})$ seems to have an advantage over both of these other systems because of the "ambiguity" in the constant ν surfaces as explained in Sect. 4.6.1d.

8. Establishment of the Flux-Coordinate Transformation; A Summary

The material in this section on finding the transformation from a particular coordinate system to flux coordinates has to a large extent already been discussed above, although the details were scattered throughout several sections. Here we summarize briefly the methods used and provide some practical references.

First, it should be understood that *use* of flux coordinates does not require knowledge of the transformation equations. In fact, the transformation itself is rarely undertaken. The whole idea of flux coordinates is based on computational convenience: they give a helpful decoupling of certain physics problems one wishes to focus on (e.g., transport across magnetic surfaces or along magnetic field lines) from complicated geometrical features of the device under investigation. To gain a theoretical understanding of a variety of physics issues this is sufficient. However, when we wish to obtain numbers which can be related to experimental data, quantities expressed in flux coordinates must be transformed.

A crucial assumption in the methods of establishing flux coordinates is that *the magnetic field and the current density are known everywhere* in a finite pressure plasma. Knowledge of $\boldsymbol{B}$ allows the construction of magnetic surfaces via analytic formulae in systems with symmetry and via approximate analytic calculations or via numerical field-line following in more general devices. The determination of the flux surfaces is part of the more general problem of finding the equilibrium state of the plasma. This will be discussed very briefly below. These flux surfaces then serve as our first coordinate surfaces. The fact that $\boldsymbol{B}$ is known at every point on the flux surface permits a choice of other suitable coordinates which make the field lines appear as straight lines. $\boldsymbol{J}$ lines also lie in flux surfaces and similarly, we can make the current-density lines straight (in addition to having straight magnetic-field lines or independently of the shape of the magnetic-field lines).

8.1 Toroidal Systems

In toroidal systems, we consider the components of $\boldsymbol{B}$ to be known as functions of the flux variable ϱ and two angular variables θ and ζ. For example, if $\boldsymbol{B}$ is known in terms of the Cartesian coordinates, a first transformation to an "ideal" toroidal-cylindrical or "elementary" toroidal coordinate system (r,θ_e,ζ_e) is easily performed. As soon as the flux surfaces are known the radial variable r can be replaced by a flux function $\varrho \equiv \varrho(\Psi)$. So far, we have $\boldsymbol{B} \equiv \boldsymbol{B}(\varrho, \theta_e, \zeta_e)$. If we keep the coordinate surfaces $\theta_e =$ constant and $\zeta_e =$ constant to start with, the contravariant components of $\boldsymbol{B}$, $\boldsymbol{B} \cdot \nabla\theta_e = B^{\theta_e}$ and $\boldsymbol{B} \cdot \nabla\zeta_e = B^{\zeta_e}$ are straightforwardly determined in terms of $(\varrho, \theta_e, \zeta_e)$.

In Sect. 6.1, we explained how from B^θ and B^ζ (in the above example, B^{θ_e} and B^{ζ_e}) the periodic part $\tilde{\nu}(\varrho, \theta, \zeta)$ of the streaming function $\nu(\varrho, \theta, \zeta)$ can be found (see underneath (6.1.13)). The original ζ ($\equiv \zeta_f$) coordinate, together with a deformed θ-coordinate, now called θ_f, and equal to $\theta + 2\pi\tilde{\nu}/\dot{\Psi}_{tor}$, make the magnetic field lines straight (see (6.1.22) to (6.1.24)).

For Boozer and Hamada coordinates, we go one step further. Having established the flux coordinates θ_f and ζ_f, we deform them further to get rid of the periodic part of the "magnetic scalar potential" in Boozer's case, or to make the $\boldsymbol{J}$ lines straight in the Hamada system, in addition to having straight $\boldsymbol{B}$ lines. (The Hamada system also has the advantage of a unity or constant Jacobian.) In both cases, $G(\varrho, \theta_f, \zeta_f)$ must be found and the new "angles" are then determined by (6.1.28). Equation (6.6.13) shows the solution for $G_B \equiv G_{Boozer}$, whereas (6.8.11) gives the differential equation that should be solved for $G_H \equiv G_{Hamada}$.

If a Clebsch coordinate system is desired, the streaming functions are the flux surfaces labeled by ϱ and the function $\nu =$ constant: $\boldsymbol{B} = \nabla\varrho \times \nabla\nu$. The third variable is either l or χ, where $\chi = \int B\, dl$ (see (5.4.26)). To find l from the available information, one could use the equation of a field line in the coordinate system in which $\boldsymbol{B}$ was originally specified.

A flux-coordinate system is completely specified (in terms of the variables of another system) when the transformation equations or the metric coefficients g_{ij} have been found.

Specific examples of establishing a flux-coordinate system have been treated in the literature. In one approach, Shafranov (1968) derives the metric coefficients starting from a Frenet-Serret coordinate system based upon the magnetic axis. The actual computation is performed via an expansion procedure in terms of the curvature of the magnetic axis. These axial coordinate systems are discussed in considerable detail in the review papers by Solov'ev and Shafranov (1970) and by Solov'ev (1975). The expansion procedure in terms of the curvature is described by Shafranov (1966).

In his 1982 paper, Boozer describes how to construct a magnetic coordinate system via field-line following and Fourier expansion. To limit the extent of the field line integration, a Gaussian window function is used. This seems to be a widely used technique in electronics when dealing with the fast Fourier transform. (The fast Fourier transform technique and the desirability of window functions are discussed in Bergland (1969). A variety of window functions has been reviewed in the article by Harris (1978).) For further details on field line integration for flux coordinate purposes, see Kuo-Petravic, Boozer, Rome and Fowler (1983), Kuo-Petravic (1984) and Beasley, Rome, Attenberger and Hirshman (1987). An improved field-line-following method, which requires integration along two field lines, has been devised by Rome (1989). Reiman and Pomphrey

(1989) have developed an algorithm which is a combination of iterative and trajectory-following methods. In the next section, we shall say more about Boozer's "Fourier field-line-following" method.

We shall now comment on the computation of the flux surfaces. This calculation is not a trivial exercise, even when the magnetic field is known everywhere. An analytic result can be obtained for symmetric devices (such as tokamaks or helically symmetric stellarators) or in a non-axisymmetric device an approximate solution may be found by using the method of averaging or an expansion technique near a closed field line. These methods are explained by Morozov and Solov'ev (1966). In addition, the review paper by Solov'ev and Shafranov (1970) covers a solution procedure for flux surfaces close to the magnetic axis by means of a perturbation technique analogous to the stellarator expansion introduced by Greene and Johnson (1961). For more general cases (e.g., where a suitable expansion parameter does not exist), one has to resort to numerical methods. This merely amounts to integrating the equations of a field line.

Recently, Kuo-Petravic and Boozer (1986) described a numerically economical method in which the Hamiltonian character of magnetic-field lines is utilized. According to the authors, this is an effective way to evaluate the quality of the magnetic surfaces. We shall discuss a few aspects of the magnetic field-line Hamiltonian in the next Chapter as this is a rather important concept.

Frequently, the determination of the flux surfaces is undertaken as part of a self-consistent equilibrium calculation. In devices with a symmetry direction, a substantial amount of information is summarized in the solution of an equation known as the Grad-Schlüter-Shafranov equation (after its discoverers, in alphabetical order). This is a second-order differential equation for the flux variable $\Psi \propto \Psi_{\mathrm{pol}}$ in terms of the flux functions $I \equiv I(\Psi) \equiv RB_{\mathrm{T}}$ and the pressure $p \equiv p(\Psi)$. It is obtained by substituting $\boldsymbol{J} = (\nabla \times \boldsymbol{B})/\mu_0$ into $\boldsymbol{J} \times \boldsymbol{B} = \nabla p = \dot{p}(\Psi)\nabla\Psi$, after which $\boldsymbol{B}$ is replaced by $I\nabla\zeta + (\nabla\zeta \times \nabla\Psi)$. Because of axisymmetry, $I \equiv I(\Psi)$ is a flux function and one obtains

$$R^2 \nabla\cdot(R^{-2}\nabla\Psi) = -I\frac{dI}{d\Psi} - R^2\mu_0\frac{dp}{d\Psi}\,. \tag{8.1.1}$$

The operator on the left may be written as:

$$\Delta^*\Psi \equiv \frac{\partial^2\Psi}{\partial R^2} - \frac{1}{R}\frac{\partial\Psi}{\partial R} + \frac{\partial^2\Psi}{\partial z^2} = R\frac{\partial}{\partial R}\left(\frac{1}{R}\frac{\partial\Psi}{\partial R}\right) + \frac{\partial^2\Psi}{\partial z^2}\,. \tag{8.1.2}$$

Here z is the vertical coordinate of the cylindrical system, (R, ζ_c, z). Δ^* is reminiscent of the notation for the Laplacian $\Delta \equiv \nabla^2$. Δ^* is a "magnetic differential" operator similar to the non-angular part of the Laplacian in cylindrical coordinates:

$$\left(\Delta - \frac{1}{R^2}\frac{\partial^2}{\partial\zeta_c^2}\right)_{\mathrm{cyl}} \equiv \left(\nabla^2 - \frac{1}{R^2}\frac{\partial^2}{\partial\zeta_c^2}\right)_{\mathrm{cyl}} = \frac{1}{R}\frac{\partial}{\partial R}\left(R\frac{\partial}{\partial R}\right) + \frac{\partial^2}{\partial z^2}\,. \tag{8.1.3}$$

The Grad-Schlüter-Shafranov equation was derived independently by Shafranov (1958) in 1957, by Lüst and Schlüter (1957) and by Grad and Rubin (1958). The derivation of the equation can also be found in the review papers by Shafranov (1966), Freidberg (1982), Zakharov and Shafranov (1986) and Hazeltine and Meiss (1985). See also, Callen and Dory (1972). For simple textbook treatments, we refer to Bateman (1978), Stacey (1981) and Freidberg (1987).

Since $\dot{p}$ and $\dot{I}$ are not necessarily linear functions of Ψ and since the boundary conditions can be nonlinear as well, the Grad-Schlüter-Shafranov equation is in general a second-order nonlinear differential equation. Since one must specify $p(\Psi)$ and $I(\Psi)$ before one knows what the flux surfaces are, there is some ambiguity in modelling a plasma. In most calculations of equilibria for experimental devices, the Grad-Schlüter-Shafranov equation is solved by numerical iteration. See e.g. Lackner (1976). A valuable variational method is that by Lao, Hirshman and Wieland (1981). (See also Lao, 1984). Certain idealized cases seem to be tractable analytically, as has been discussed by Solov'ev (1975) and Zakharov and Shafranov (1986).

The existence of a Grad-Schlüter-Shafranov equation is crucially dependent on the presence of a symmetry direction because this makes I in the expression for $\boldsymbol{B}$ a flux function. As a consequence, there also exists a G-S-S equation for a helically symmetric stellarator. This equation is derived in Zakharov and Shafranov (1986); see also, Pytte and Boozer (1981) and White and Chance (1984).

In devices without a symmetry direction such as toroidal stellarators, there is no Grad-Schlüter-Shafranov equation, and the equilibrium problem is even more complicated. To treat this problem, Greene and Johnson (1961) developed the so-called stellarator expansion. This procedure solves the equation $\boldsymbol{B} \cdot \nabla \Psi = 0$, where Ψ is expanded as $\sum_i \Psi^{(i)}$ such that higher order $\Psi^{(i)}$ are considered as perturbations with regard to the lowest order Ψ. To lowest order it models a uniform axial magnetic field. The first-order field is a superposition of multi-polar fields, and the second order consists of fields which take into account the effects of the curvature of the system and the currents in the plasma. This expansion procedure was suggested in an earlier paper by Johnson, Oberman, Kulsrud and Frieman (1958). The method has been further utilized and/or discussed by Anania, Johnson and Weimer (1983a), and by Anania and Johnson (1983b), by Johnson, et al., (1984) and by Johnson (1986). Dobrott and Frieman (1971) have computed the vacuum magnetic surfaces of an $l = 3$ stellarator using an "optimal stellarator expansion". This same expansion was used by Fielding and Hitchon (1980) to compute the equilibrium flux surfaces and fields and, in particular, the equilibrium value of the vertical field, in an $l = 3$ stellarator. Later, this was extended to arbitrary l-numbers (Hitchon and Fielding, 1981). Related procedures have been described by Solov'ev and Shafranov (1970) and Johnson (1986).

Besides analytic (mostly asymptotic) results which can provide physical insight, numerical equilibrium codes provide results which are asymptotic in the mesh size and attack more complicated cases. Reviews of numerical methods used to compute the plasma equilibrium are given in the papers of Zakharov and Shafranov (1986), and Johnson (1986). More details can be found in Methods in Computational Physics, Killeen (1976). In addition, we should mention the 3-D equilibrium codes by Chodura and Schlüter (1981) and by Bauer, Betancourt and Garabedian (1978 and 1984). See also Hirshman and Whitson (1983), Lao (1984), Schwenn (1984), Bhattacharjee et al. (1984) and Hirshman and Lee (1986), among others. Island formation and destruction of flux surfaces in three-dimensional equilibria have been investigated by Boozer (1984a, b) and Reiman and Boozer (1984).

Further discussion of toroidal equilibrium can be found in the papers by Grad (1967), Kruskal and Kulsrud (1958), Spies and Nelson (1974), and Boozer (1981).

8.2 Open-Ended Systems

The flux coordinates of concern in open-ended systems are of the Clebsch or the Boozer-Grad type. To obtain the transformation equations, the same equations as in toroidal systems must be solved. There is one crucial difference concerning the flux label and the magnetic surfaces. As pointed out above, in mirrors there are no flux surfaces which are traced out by a single field line. One must *define* a closed contour somewhere in the plasma, and the flux surface is that surface traced out by all the magnetic field lines intersecting the chosen contour. Conventionally, for the vacuum field, the curve is chosen to be a circle (concentric about the axis) located in the middle of the device at $z = 0$, when the device is bounded by $z = \pm L$. (In finite-β plasmas, the originally circular flux surfaces tend to acquire a "square" shape.) The vacuum flux surfaces for tandem mirrors in the "long-thin cylinder" or "paraxial" approximation have been computed by Furth and Rosenbluth (1964) and by Cordey and Watson (1969). See also the review paper by Baldwin (1977). More recent papers dealing with mirror equilibrium include those by Hall and McNamara (1975), Pearlstein, Kaiser and Newcomb (1981), Newcomb (1981), and Bulmer, et al., (1983).

9. Canonical Coordinates or "Generalized Magnetic Coordinates"

9.1 Flux Coordinates Versus Canonical Coordinates

As an extension to the subject of flux coordinates, we discuss concisely the more general topic of canonical coordinates for magnetic fields. The word canonical derives from the Hamiltonian character of the magnetic field, which will be demonstrated presently. Canonical coordinates for $\boldsymbol{B}$, and the canonical representation of $\boldsymbol{B}$ exist even if there are no flux surfaces. Flux coordinates, on the contrary, only exist in the presence of magnetic surfaces. The canonical coordinates are sometimes called "generalized magnetic coordinates". It must be understood, however, that this is *not* a synonym for flux coordinates. The reason for the name "generalized magnetic coordinates" is the superficial resemblance of the $\boldsymbol{B}$-field representation in canonical coordinates and flux coordinates.

9.2 On the Existence of Flux Surfaces, Revisited

In the section on flux surfaces, we pointed out that they only rigorously exist in devices with an ignorable coordinate. In such a case, the irrational field lines cover a two-dimensional *surface* ergodically. In more general systems, the magnetic-field lines do not all stay on a surface but in general ergodically fill a finite *volume*.

We are interested in devices (such as stellarators or real tokamaks) in which the field lines generate "surfaces" with very limited ergodic volumes. The magnetic field in such devices is considered as one that has the required symmetry, plus a perturbation field. One might expect that there would be no closed surfaces left at all due to coupling of the perturbation with the original field structure, giving rise to singularities or resonances. However, the KAM theorem (after Kolmogorov, Arnold and Moser – references are given below) guarantees the existence of certain invariant surfaces in some circumstances.

Loosely speaking, this theorem states that there are two regions of non-vanishing volume (technically, sets of non-zero measure), one of which is small compared to the other and shrinks to zero as the perturbation goes to zero. The (hopefully) larger region has the familiar structure of embedded tori (flux surfaces in the proper sense) covered with dense field-line trajectories. Interlaced between these "good" flux surfaces, there is a possibly dense set of surfaces which open up into ergodic regions or islands, where the field lines behave erratically and may move far from nearby confined trajectories. However, even if a given magnetic field line is ergodic within an entire annular volume, it is absolutely contained, provided only that there is a closed (KAM) surface on either side. Further, the KAM theorem guarantees that for small perturbations, the chaotic regions will be thin, albeit of finite volume.

In conclusion, magnetic surfaces in non-symmetric devices exist only in the sense that the perturbed invariant toroidal surfaces of KAM theory exist as a set of non-zero measure (Grad 1967; Balescu 1975; Whiteman 1977). The degree of flux-surface destruction and island formation depends on the vacuum field structure and on the plasma pressure (beta) via the diamagnetic and Pfirsch-Schlüter equilibrium currents (Boozer 1984 a, b; Reiman and Boozer 1984).

Stochastic motion of field lines and other KAM related issues have been studied by means of modern Hamiltonian mechanics. Although this is not the place to discuss these points in detail, it seems warranted to point out the Hamiltonian character of the magnetic-field lines and to show how the transformation to canonical coordinates is done in practice. This has been the subject of a series of papers by Boozer (1983, 1984c, 1984d, 1985) and by Kuo-Petravic and Boozer (1986). We shall base our discussion on their work. An alternative route for describing magnetic-field lines by means of *non*-canonical Hamiltonian mechanics has been advocated by Cary and Littlejohn (1983), and Littlejohn (1984). Hamiltonian mechanics in general is authoritatively discussed in the classic book by Goldstein (1981). Hamiltonian theory applied to stochasticity is treated in Lichtenberg and Lieberman (1983). Also recommended are the review paper by Whiteman (1977), the appendix in Balescu's (1975) book, and White's book (1989).

Before dealing with the canonical coordinates and the Hamiltonian, we must first reconsider flux coordinates as a preliminary.

9.3 Flux Coordinates

From our discussion of flux coordinates, we recall that the magnetic field can be written as

$$\boldsymbol{B} = \nabla\varrho \times \nabla\left(\frac{\dot{\Psi}_{\text{tor}}}{2\pi}\theta_{\text{f}} + \frac{\dot{\Psi}^{\text{d}}_{\text{pol}}}{2\pi}\zeta_{\text{f}}\right)$$

$$= \frac{\dot{\Psi}_{\text{tor}}}{2\pi}\nabla\varrho \times \nabla\theta_{\text{f}} + \frac{\dot{\Psi}^{\text{d}}_{\text{pol}}}{2\pi}\nabla\varrho \times \nabla\zeta_{\text{f}} \,. \tag{9.3.1}$$

Here, ϱ is a flux-surface label, and Ψ_{tor} and $\Psi^{\text{d}}_{\text{pol}}$ measure the toroidal flux enclosed by, and the poloidal flux outside, the surface labeled by ϱ, respectively. Both Ψ_{tor}

and $\Psi^{\mathrm{d}}_{\mathrm{pol}}$ depend only on ϱ. For simplicity of notation, we introduce two new flux functions:

$$\psi_{\mathrm{t}} \equiv \frac{\Psi_{\mathrm{tor}}}{2\pi} \tag{9.3.2a}$$

$$\psi_{\mathrm{p}} \equiv \frac{\Psi^{\mathrm{d}}_{\mathrm{pol}}}{2\pi} \; . \tag{9.3.2b}$$

In terms of these functions, we have

$$\boldsymbol{B} = \nabla\psi_{\mathrm{t}} \times \nabla\theta_{\mathrm{f}} + \nabla\psi_{\mathrm{p}} \times \nabla\zeta_{\mathrm{f}} \; . \tag{9.3.3}$$

The rotational transform $\iota\!\!\!-$, defined by $\iota\!\!\!- \equiv -\dot{\Psi}^{\mathrm{d}}_{\mathrm{pol}}/\dot{\Psi}_{\mathrm{tor}} \equiv -d\psi_{\mathrm{p}}/d\psi_{\mathrm{t}}$, allows us to write the above equation as

$$\boldsymbol{B} = \nabla\psi_{\mathrm{t}} \times \nabla\theta_{\mathrm{f}} - \iota\!\!\!- \nabla\psi_{\mathrm{t}} \times \nabla\zeta_{\mathrm{f}} \tag{9.3.4a}$$

$$= \nabla\psi_{\mathrm{t}} \times \nabla(\theta_{\mathrm{f}} - \iota\!\!\!- \zeta_{\mathrm{f}}) \; . \tag{9.3.4b}$$

This is the best known representation. The flux coordinates are ψ_{t}, θ_{f} and ζ_{f}. $\iota\!\!\!-$ is a flux function, $\iota\!\!\!- \equiv \iota\!\!\!-(\psi_{\mathrm{t}})$. Equation (9.3.4) is the contravariant representation of $\boldsymbol{B}$, not because it is written in terms of (cross products of) contravariant-basis vectors ∇u^i, but because it deals with the (somewhat hidden) contravariant components. The more transparent contravariant form is revealed if we make use of the $\boldsymbol{e}_i \leftrightarrow \boldsymbol{e}^j$ "back and forth transformation" relations, where $\boldsymbol{e}^j \equiv \nabla u^j$ and $\boldsymbol{e}_i \equiv \partial\boldsymbol{R}/\partial u^i$:

$$\boldsymbol{e}_i = \frac{\boldsymbol{e}^j \times \boldsymbol{e}^k}{\boldsymbol{e}^i \cdot \boldsymbol{e}^j \times \boldsymbol{e}^k} = \sqrt{g}\,\boldsymbol{e}^j \times \boldsymbol{e}^k \; . \tag{9.3.5}$$

Therefore, we can write instead of (9.3.4a):

$$\boldsymbol{B} = \frac{1}{\sqrt{g}}\left(\frac{\partial\boldsymbol{R}}{\partial\zeta_{\mathrm{f}}} + \iota\!\!\!-(\psi_{\mathrm{t}})\frac{\partial\boldsymbol{R}}{\partial\theta_{\mathrm{f}}}\right) \tag{9.3.6a}$$

$$= \frac{\iota\!\!\!-}{\sqrt{g}}\boldsymbol{e}_{\theta_{\mathrm{f}}} + \frac{1}{\sqrt{g}}\boldsymbol{e}_{\zeta_{\mathrm{f}}} \; . \tag{9.3.6b}$$

The contravariant components are thus $B^{\psi_{\mathrm{t}}} = 0$, $B^{\theta_{\mathrm{f}}} = \iota\!\!\!-/\sqrt{g}$ and $B^{\zeta_{\mathrm{f}}} = 1/\sqrt{g}$. $B^{\psi_{\mathrm{t}}}$ is zero because $\boldsymbol{B}$ is tangent to the flux surfaces, i.e., $\boldsymbol{B}\cdot\nabla\psi_{\mathrm{t}} \equiv 0$ and $\sqrt{g}$ is the Jacobian:

$$\sqrt{g} = \frac{1}{\nabla\psi_{\mathrm{t}}\cdot\nabla\theta_{\mathrm{f}} \times \nabla\zeta_{\mathrm{f}}} = \frac{\partial\boldsymbol{R}}{\partial\psi_{\mathrm{t}}}\cdot\frac{\partial\boldsymbol{R}}{\partial\theta_{\mathrm{f}}} \times \frac{\partial\boldsymbol{R}}{\partial\zeta_{\mathrm{f}}} = \frac{1}{\boldsymbol{B}\cdot\nabla\zeta_{\mathrm{f}}} \; . \tag{9.3.7}$$

(Note the distinction between this Jacobian in terms of the toroidal flux and that defined in (6.1.38) which is a function of poloidal flux.)

From (9.3.4), we see that $\boldsymbol{B}$ is uniquely determined if $\iota(\psi_t)$, ψ_t, θ_f and ζ_f are known as functions of the position vector $\boldsymbol{R}$ such that the gradients can be calculated. According to Boozer (1983, 1985), it is more useful to define the field by (9.3.6) as far as a practical evaluation is concerned. The magnetic field can be constructed if $\boldsymbol{R}(\psi_t, \theta_f, \zeta_f)$, i.e., $x(\psi_t, \theta_f, \zeta_f)$, $y(\psi_t, \theta_f, \zeta_f)$, and $z(\psi_t, \theta_f, \zeta_f)$, in addition to $\iota(\psi_t)$ are known functions. Here, we assume a finite value of the Jacobian $(\sqrt{g})^{-1} = \boldsymbol{B} \cdot \nabla\zeta_f$. In other words, the toroidal field is nonzero.

The establishment of the transformation $\boldsymbol{R}(\psi_t, \theta_f, \zeta_f)$ by means of numerical field line following and Fourier decomposition was described by Boozer (1982, 1984b), Kuo-Petravic, Boozer, Rome and Fowler (1983), Kuo-Petravic (1984), Kuo-Petravic and Boozer (1986) and Beasley, Rome, Attenberger and Hirshman (1987). (Improvements to this method were made by Rome (1989) and by Reiman and Pomphrey (1989).) Because of the double periodicity, the three Cartesian components, x, y and z of $\boldsymbol{R}$ can be written as a double Fourier series. We treat x, y and z simultaneously via the vector symbol $\boldsymbol{R}(\psi_t, \theta_f, \zeta_f) \equiv \boldsymbol{R}[x(\psi_t, \theta_f, \zeta_f), y(\psi_t, \theta_f, \zeta_f), z(\psi_t, \theta_f, \zeta_f)]$:

$$\boldsymbol{R}(\psi_t, \theta_f, \zeta_f) = \sum_{m,n} \boldsymbol{R}_{mn}(\psi_t) e^{i(m\theta_f - n\zeta_f)} . \tag{9.3.8}$$

Boozer's straight-$\boldsymbol{B}$ coordinate system (Boozer 1980, 1981) makes use of a function χ as a parameter to measure the "length" along a field line.

$$\chi = \int_0^l B\, dl' . \tag{9.3.9}$$

χ reduces to the magnetic scalar potential if the plasma is currentless (i.e., $\boldsymbol{B} = \nabla\chi$) and is equal to (6.6.14c):

$$\chi = \frac{\mu_0}{2\pi}(I_{tor}\theta_f + I^d_{pol}\zeta_f) . \tag{9.3.10}$$

I_{tor} and I^d_{pol} are flux functions. For flux coordinates where (9.3.10) holds, we earlier used a subscript B, i.e., $\theta_f \equiv \theta_B$ and $\zeta_f \equiv \zeta_B$.

Since the equation of a field line in flux coordinates is given by (4.9.15),

$$\theta_{f0} = \theta_f - \iota\zeta_f , \tag{9.3.11}$$

we can rewrite the exponent of the Fourier series in terms of θ_{f0} and χ:

$$\boldsymbol{R}(\psi_t, \theta_f, \zeta_f) = \sum_{m,n} \boldsymbol{R}_{mn}(\psi_t) \exp\left[i \frac{(mI^d_{pol} + nI_{tor})\theta_{f0} + (m\iota - n)2\pi\chi/\mu_0}{I_{tor}\iota + I^d_{pol}} \right] \tag{9.3.12}$$

To evaluate the Cartesian coefficients $x_{mn}(\psi_t)$, $y_{mn}(\psi_t)$ and $z_{mn}(\psi_t)$ of $\boldsymbol{R}_{mn}(\psi_t)$, we perform a field line integration, starting from an arbitrary point. We choose χ and θ_{f0} zero at the starting point. Since θ_{f0} remains constant along the field line trajectory, θ_{f0} remains zero everywhere. χ is determined by the integration via (9.3.9), and thus, this field line integration determines $\boldsymbol{R}(\psi_t,\chi) \equiv \boldsymbol{R}(\psi_t,\theta_{f0}=0,\chi)$. We then have for (9.3.12), where the L.H.S. is now known numerically from the field-line integration:

$$\boldsymbol{R}(\psi_t,\theta_f,\zeta_f) = \sum_{m,n} \boldsymbol{R}_{mn}(\psi_t)\exp\left\{\mathrm{i}\frac{(m\iota - n)2\pi\chi}{\mu_0(I_{tor}\iota + I^{d}_{pol})}\right\} . \tag{9.3.13}$$

This Fourier decomposition of $\boldsymbol{R}(\psi_t,\theta_f,\zeta_f) \equiv \boldsymbol{R}(\psi_t,\chi)$ has distinct peaks. The amplitudes of the peaks determine the $\boldsymbol{R}_{mn}(\psi_t)$, and their locations determine m, n and $\iota(\psi_t)$. (The practical numerical decomposition is done with the help of a suitable window function. For references, see Sect. 8.1.) In practice, it turns out that "about 5 poloidal and toroidal circuits of a field line must be followed to carry out the Fourier decomposition to reasonable accuracy" (Boozer 1982). This way, $\boldsymbol{R}(\psi_t,\theta_f,\zeta_f)$ and $\iota(\psi_t)$ are known, and thus, the flux coordinate representation of (9.3.6) can be formed. The method outlined above is only applicable in the Boozer coordinate system, where (9.3.10) holds and where $\zeta_f \equiv \zeta_B$ and $\theta_f \equiv \theta_B$. A method valid for other straight-$\boldsymbol{B}$ systems is explained below.

The above explanation is in principle only valid for irrational surfaces, i.e., surfaces that are traced out by a single field line. This is not the case for rational surfaces on which ι is the ratio of two integers. However, as long as the number of poloidal and toroidal transits is sufficiently large (e.g., larger than 5), the surface can be considered as being adequately covered for all practical purposes. When ι is the ratio of low integers, we invoke the fact that the real number system is dense and thus that any rational number (rational surface) can be approximated infinitesimally closely by an irrational number (irrational surface).

In a later paper, Kuo-Petravic and Boozer (1986) perform the field line integration and Fourier decomposition without mentioning the function χ (beware that in that paper, the symbol χ is used with a different meaning). There, they start out with a definite choice of the toroidal angle ζ_f, which for convenience is taken to be (minus) the azimuthal angle of the cylindrical system $(R, -\zeta_f, z)$. This is effectively the angle that we called ζ_e. In this case the field line integration is based on the equations

$$\frac{dx}{d\zeta_f} = \frac{B^x(x,y,z)}{\boldsymbol{B}\cdot\nabla\zeta_f}; \qquad \frac{dy}{d\zeta_f} = \frac{B^y(x,y,z)}{\boldsymbol{B}\cdot\nabla\zeta_f}; \qquad \frac{dz}{d\zeta_f} = \frac{B^z(x,y,z)}{\boldsymbol{B}\cdot\nabla\zeta_f}$$

or

$$\frac{d\boldsymbol{R}}{d\zeta_f} = \frac{\boldsymbol{B}(\boldsymbol{R})}{\boldsymbol{B}\cdot\nabla\zeta_f} . \tag{9.3.14}$$

This allows determination of the function $\boldsymbol{R}(\zeta_f)$ on a particular flux-surface. Again, $\boldsymbol{R}(\psi_t,\theta_f,\zeta_f)$ and $\iota(\psi_t)$ can be obtained via a Fourier decomposition

$$\boldsymbol{R}(\psi_t,\theta_f,\zeta_f) = \sum_{m,n} \boldsymbol{R}_{mn}(\psi_t)\mathrm{e}^{\mathrm{i}(m\theta_f - n\zeta_f)}$$

or

$$\boldsymbol{R}(\psi_t, \theta_f, \zeta_f) \equiv \boldsymbol{R}(\zeta_f) = \sum_{m,n} \boldsymbol{R}_{mn}(\psi_t) e^{i(m\iota - n)\zeta_f} . \tag{9.3.15}$$

The notation $\boldsymbol{R}(\zeta_f)$ indicates that we have chosen a fixed flux surface. As before, θ_{f0} has been chosen equal to zero. The expansion in (9.3.15) performed on a variety of flux surfaces determines $\boldsymbol{R}_{mn}$ and ι.

9.4 Canonical Coordinates; The Field-Line Hamiltonian

We are now ready to treat the magnetic-field representation in the more general case where field lines are not restricted to a flux surface but cover a finite volume ergodically. The types of systems that we are interested in for magnetic confinement and whose description is still more or less tractable analytically are those where the magnetic field $\boldsymbol{B}(\boldsymbol{R})$ does not differ very much from a field $\boldsymbol{B}_0(\boldsymbol{R})$ which has perfect surfaces. The difference between the two fields can then be considered as a perturbation to the "perfect" field.

When a field is represented by means of a "magnetic field line Hamiltonian", the field lines can be identified with the phase-space trajectories produced by the Hamiltonian. Mathematically, the issue of magnetic-surface quality then reduces to a problem in Hamiltonian mechanics, meaning that a whole wealth of methods, such as Canonical Transformations, Hamilton-Jacobi theory, Canonical Perturbation theory, mappings, etc. are at our disposal.

Hamilton's equations for magnetic fields were indicated by Kerst (1962). The Hamiltonian character of the magnetic field was implicitly used by Spitzer (1958) in the paper in which he presented the stellarator concept, and by Kruskal and Kulsrud (1958). More explicit references to the Hamiltonian relationship were made by Rosenbluth, Sagdeev and Taylor (1966), Filonenko, Sagdeev and Zaslavsky (1967), and by Grad (1967). The field-line problem in the context of Hamiltonian mechanics is discussed by Lichtenberg and Lieberman (1983), who provide more references. Other references to the Hamiltonian nature of field lines were given in Sect. 9.2.

The simplest proof that field line trajectories are derivable from a Hamiltonian was suggested by H. Weitzner (according to Boozer (1985)). A globally divergence free vector field can be represented by a single-valued vector potential $\boldsymbol{A}(\boldsymbol{R})$. If we consider a set of variables (ϱ, θ, ζ), where ϱ is a radius-like variable characterized by $\varrho = 0$ along the magnetic axis, we can write for the vector field $\boldsymbol{A}(\boldsymbol{R})$:

$$\boldsymbol{A}(\boldsymbol{R}) \equiv \boldsymbol{A}(\varrho, \theta, \zeta) = A_\varrho \nabla \varrho + A_\theta \nabla \theta + A_\zeta \nabla \zeta . \tag{9.4.1}$$

Here A_ϱ, A_θ, and A_ζ are single valued functions of (ϱ, θ, ζ). Note that ϱ is in general not a flux-surface label because flux surfaces may not exist.

We know that a magnetic vector potential is determined up to an additive term, which is the gradient of a scalar function, $\nabla G(\varrho, \theta, \zeta)$. Indeed, we have that $\boldsymbol{B} = \nabla \times \boldsymbol{A} = \nabla \times (\boldsymbol{A} + \nabla G)$. Therefore, we could choose a G such that $\partial G/\partial \varrho = A_\varrho$, $G = 0$ at $\varrho = 0$, $\partial G/\partial \theta = A_\theta - \Psi$ and $\partial G/\partial \zeta = A_\zeta - \Phi$. Ψ and Φ are two single-valued functions that will be interpreted below. With this choice for G, we have

$$\boldsymbol{A} = \Psi \nabla \theta + \Phi \nabla \zeta + \nabla G \ . \tag{9.4.2}$$

Upon taking the curl of this vector potential, we obtain for $\boldsymbol{B}$:

$$\boldsymbol{B} = \nabla \Psi \times \nabla \theta + \nabla \Phi \times \nabla \zeta \ . \tag{9.4.3}$$

This representation of $\boldsymbol{B}$ is known as the *canonical* form. An alternative proof (for which $\boldsymbol{B} \cdot \nabla \zeta$ must be non-zero) was given by Boozer (1985).

The expression in (9.4.3) resembles the flux-coordinate representation of (9.3.3), but the similarity is superficial; (9.3.3) is only valid for a $\boldsymbol{B}$ field with proper magnetic surfaces everywhere, whereas (9.4.3) is valid for any magnetic field in a toroidal device. The difference is that in the flux-surface representation, both ψ_p and ψ_t are functions of only the flux label ϱ. In other words, $\psi_p \equiv \psi_p(\psi_t)$ only. The functions Ψ and Φ of (9.4.3), on the contrary, are both functions of the three variables, ϱ, θ, and ζ: $\Phi \equiv \Phi(\varrho, \theta, \zeta)$, $\Psi \equiv \Psi(\varrho, \theta, \zeta)$. Because Ψ and Φ depend on all three coordinates ϱ, θ, ζ, it follows that θ and ζ are arbitrary (poloidal and toroidal) angles, whereas the θ_f and ζ_f in (9.3.3) must be flux coordinates for (9.3.3) to be valid.

From the knowledge of $\boldsymbol{B}(\boldsymbol{R})$ everywhere, it is possible to construct functions Ψ and Φ such that $\boldsymbol{B}$ is represented by (9.4.3). This will be discussed below. We first give a physical interpretation of Ψ and Φ after which we shall indicate the connection with Hamiltonian mechanics.

It is possible to infer the following statement from the formal similarity between (9.3.3) and (9.4.3), but a rigorous proof can also be constructed. The function $\Psi(\varrho, \theta, \zeta)$ measures the toroidal flux through a constant Ψ surface: the toroidal flux equals $2\pi\Psi$. Similarly, $2\pi\Phi$ is the poloidal flux outside a constant Φ surface (i.e., through a disk tangent to a constant Φ surface). The formal proof makes use of (9.4.3), the expression for a surface element in a constant u^i surface, $dS(i) = \sqrt{g}\, du^j\, du^k\, \nabla u^i$, the definition of (flux through surface u^i = constant) = $\int\int \boldsymbol{B} \cdot d\boldsymbol{S}(i)$, and the fact that $\Phi \equiv \Phi(\text{disk})$ with $d\Phi = -d\Phi(\text{ribbon})$. Note that the surfaces $\Psi(\varrho, \theta, \zeta) = \text{constant}$ and $\Phi(\varrho, \theta, \zeta) = \text{constant}$ *do not coincide, in general.* Also, in general, $\boldsymbol{B} \cdot \nabla \Psi \neq 0 \neq \boldsymbol{B} \cdot \nabla \Phi$. Only in a system where flux surfaces exist are Φ and Ψ equivalent. In that case, Ψ can be taken equal to ψ_t while $\Phi \equiv \psi_p$.

As mentioned above, (9.4.3) is called the canonical representation of $\boldsymbol{B}$ (Boozer 1985). The reason is that the four functions, Φ, Ψ, θ, ζ can be considered as canonical variables in a Hamiltonian context. Φ will be identified with the

"Hamiltonian", ζ with the canonical "time", Ψ with the "momentum", and θ with the "coordinate".

To see this, we first eliminate the "auxiliary" variable ϱ from the problem and consider Φ as a function of the coordinates, Ψ, θ, ζ; so $\Phi \equiv \Phi(\Psi, \theta, \zeta)$. In terms of Ψ, θ, ζ coordinates, the equation of a field line reads:

$$\frac{d\Psi}{\boldsymbol{B}\cdot\nabla\Psi} = \frac{d\theta}{\boldsymbol{B}\cdot\nabla\theta} = \frac{d\zeta}{\boldsymbol{B}\cdot\nabla\zeta} . \tag{9.4.4}$$

To find $\boldsymbol{B}\cdot\nabla\Psi$, we dot multiply (9.4.3) with $\nabla\Psi$ and obtain

$$\boldsymbol{B}\cdot\nabla\Psi = \nabla\Psi\cdot\nabla\Phi \times \nabla\zeta . \tag{9.4.5}$$

But with $\nabla\Phi = (\partial\Phi/\partial\Psi)\nabla\Psi + (\partial\Phi/\partial\theta)\nabla\theta + (\partial\Phi/\partial\zeta)\nabla\zeta$, this last equation becomes

$$\boldsymbol{B}\cdot\nabla\Psi = \frac{\partial\Phi}{\partial\theta}\nabla\Psi\cdot\nabla\theta \times \nabla\zeta . \tag{9.4.6}$$

Therefore, we have:

$$\frac{d\Psi}{d\zeta} = \frac{\boldsymbol{B}\cdot\nabla\Psi}{\boldsymbol{B}\cdot\nabla\zeta} = \frac{\partial\Phi}{\partial\theta}\frac{\nabla\Psi\cdot\nabla\theta \times \nabla\zeta}{\nabla\zeta\cdot\nabla\Psi \times \nabla\theta}$$

or

$$\frac{d\Psi}{d\zeta} = \frac{\partial\Phi}{\partial\theta} . \tag{9.4.7}$$

Analogously, from the second and third equations, we have

$$\frac{d\theta}{d\zeta} = -\frac{\partial\Phi}{\partial\Psi} . \tag{9.4.8}$$

Thus, after identification of ζ with t, of Φ with $-H$, of θ with the coordinate q, and of Ψ with the momentum p, (9.4.7) and (9.4.8) are Hamilton's equations of motion: $\dot{p} = -\partial H/\partial q$ and $\dot{q} = \partial H/\partial p$. To emphasize this result, $-\Phi$ is called the *field-line Hamiltonian.* Because they satisfy Hamilton's equations, (9.4.7) and (9.4.8), Ψ and θ are canonical variables or conjugate variables.

It is sometimes useful to consider the time as a coordinate and the negative Hamiltonian as its conjugate momentum. This then allows us to consider ζ and Φ as canonical variables as well; hence the name canonical representation for (9.4.3).

The underlying idea is that in the general case, our magnetic-field Hamiltonian is dependent on toroidal angle ζ or on "time". We are thus dealing with a "time" dependent Hamiltonian problem in one degree of freedom (characterized by one coordinate and one momentum). It is now possible by means of canonical transformation to "transform away" the time dependence by considering an "extended phase space", where Hamilton's equations are parameterized by a new "time" τ (Lichtenberg and Lieberman 1983):

$$\begin{aligned} q_1 &\equiv \theta & \qquad q_1 &\equiv \theta \\ q_2 &\equiv t & \text{or} \qquad q_2 &\equiv \zeta \\ p_1 &\equiv \Psi & \qquad p_1 &\equiv \Psi \\ p_2 &\equiv -H & \qquad p_2 &\equiv \Phi \ . \end{aligned}$$

The new Hamiltonian is time independent.

We are now dealing with an equivalent time-independent Hamiltonian in two degrees of freedom (2 coordinates and 2 momenta). This permits use of all the computational tools that have been devised for conservative systems. It also makes it clear that the field-line problem is characterized by a similar level of complexity as a two degree-of-freedom system, and this implies the existence of stochastic field-line behavior.

When flux surfaces exist, the magnetic field line Hamiltonian is only a function of Ψ: $-\Phi_o \equiv -\Phi_o(\Psi)$, since then $\Phi_o \equiv \psi_p(\varrho) \equiv \psi_p(\psi_t)$. This means that the Hamiltonian is in action-angle form and that the action Ψ (denoted J in Hamiltonian mechanics) is a conserved quantity, while the angle $\theta = \theta_f$ depends linearly on the toroidal angle $\zeta = \zeta_f$ (Hamiltonian time t). The latter relationship is exactly the equation of a field line in flux coordinates

$$\theta_f = \iota(\Psi)\zeta_f + \theta_{fo} \ . \tag{9.4.9}$$

The rotational transform ι plays the role of a constant "frequency" in the action-angle picture.

In Hamiltonian language, a magnetic field with perfect magnetic surfaces everywhere is *integrable* (Lichtenberg and Lieberman (1983); Goldstein (1981) uses the word "separable"). Since we are interested in magnetic field configurations that have "reasonably good" surfaces in the sense explained in Sect. 9.2., we also wish to consider *near-integrable* systems, where the canonical variables of the field-line Hamiltonian are chosen so that the Hamiltonian is a function of the canonical momentum ($\equiv$ action) alone, plus a small perturbation. Such Hamiltonians are said to be in "near action-angle form." In our field-line case considered above, it would look like:

$$\Phi(\Psi,\theta,\zeta) = \Phi_0(\Psi) + \Phi_1(\Psi,\theta,\zeta) \ . \tag{9.4.10}$$

The field-line Hamiltonian contains all the information on the existence of surfaces, islands and stochastic regions. The near-integrable Hamiltonian problem is treated by means of the canonical perturbation theory of Hamiltonian mechanics (see Goldstein (1981), and Lichtenberg and Lieberman (1983)).

9.5 Practical Evaluation of the Field-Line Hamiltonian

Critical to this evaluation of the Hamiltonian function $\Phi(\Psi, \theta, \zeta)$ is the assumption that the *given* magnetic field $\boldsymbol{B}(\boldsymbol{R})$ is close to one with perfect magnetic surfaces $\boldsymbol{B}_0(\boldsymbol{R})$. The method to be discussed was developed by Boozer (1983), and the numerical implementation of it has been considered by Kuo-Petravic and Boozer (1986).

By analogy to the case where perfect surfaces exist, we are interested in the transformation equation $\boldsymbol{R}(\psi_t, \theta_f, \zeta_f)$. Here (ϱ, θ, ζ) are set equal to $(\psi_t, \theta_f, \zeta_f)$, being the flux coordinates for the magnetic field related to the unperturbed Hamiltonian $\Phi_0(\Psi) \equiv \Phi_0(\psi_t)$. Note that the angles of the canonical set are arbitrary, thus it is perfectly legitimate to make this choice.

We can no longer evaluate $\boldsymbol{R}_{mn}(\psi_t)$ and $\iota\!\!\!-(\psi_t)$ for all values of ψ_t. However, it can still be done over much of the plasma volume, for many ψ_t values of interest. Smooth interpolation between these values then leads to $\boldsymbol{R}_{mn}(\psi_t)$ or $\boldsymbol{R}(\psi_t, \theta_f, \zeta_f)$ and $\iota\!\!\!-(\psi_t)$. Knowledge of these quantities allows the construction of the "gedanken" field with perfect surfaces $\boldsymbol{B}_0$, according to (9.3.6a), in which the left-hand side now reads $\boldsymbol{B}_0$. Keep in mind that $(\psi_t, \theta_f, \zeta_f)$ are *not* flux coordinates for $\boldsymbol{B}$, only for $\boldsymbol{B}_0$. Thus $\boldsymbol{B}_0 \cdot \nabla\psi_t \equiv 0$, but $\boldsymbol{B} \cdot \nabla\psi_t \neq 0$. According to Boozer (1983), $(\psi_t, \theta_f, \zeta_f)$ form an appropriate system for determining Ψ and Φ.

When we consider (9.4.3) with $\theta = \theta_f$, $\zeta = \zeta_f$, $\Psi \equiv \Psi(\psi_t, \theta_f, \zeta_f)$ and $\Phi \equiv \Phi(\psi_t, \theta_f, \zeta_f)$, and expand $\nabla\Psi$ and $\nabla\Phi$ along the basis vectors $\nabla\psi_t$, $\nabla\theta_f$ and $\nabla\zeta_f$ via the chain rule, we obtain

$$\boldsymbol{B} = \left(\frac{\partial\Phi}{\partial\theta_f} - \frac{\partial\Psi}{\partial\zeta_f}\right)(\nabla\theta_f \times \nabla\zeta_f) - \frac{\partial\Phi}{\partial\psi_t}(\nabla\zeta_f \times \nabla\psi_t) + \frac{\partial\Psi}{\partial\psi_t}(\nabla\psi_t \times \nabla\theta_f) \ . \quad (9.5.1)$$

The usual way to write $\boldsymbol{B}$ in terms of contravariant components is

$$\boldsymbol{B} = (\boldsymbol{B} \cdot \nabla\psi_t)\boldsymbol{e}_{\psi_t} + (\boldsymbol{B} \cdot \nabla\theta_f)\boldsymbol{e}_{\theta_f} + (\boldsymbol{B} \cdot \nabla\zeta_f)\boldsymbol{e}_{\zeta_f} \ . \quad (9.5.2)$$

If we now substitute the $\boldsymbol{e}_i \leftrightarrow \boldsymbol{e}^j$ "back and forth transformation", given in (9.3.5) into (9.5.2), and use its dual form,

$$\boldsymbol{e}^i = \nabla u^i = \frac{\boldsymbol{e}_j \times \boldsymbol{e}_k}{\boldsymbol{e}_i \cdot \boldsymbol{e}_j \times \boldsymbol{e}_k} = \frac{1}{\sqrt{g}}\left(\frac{\partial\boldsymbol{R}}{\partial u^j} \times \frac{\partial\boldsymbol{R}}{\partial u^k}\right) \quad (9.5.3)$$

in the $\boldsymbol{B} \cdot \nabla u^i$ expressions of (9.5.2), we obtain after comparing with (9.5.1):

$$\frac{\partial\Phi}{\partial\theta_f} - \frac{\partial\Psi}{\partial\zeta_f} = \boldsymbol{B} \cdot \left(\frac{\partial\boldsymbol{R}}{\partial\theta_f} \times \frac{\partial\boldsymbol{R}}{\partial\zeta_f}\right) , \quad (9.5.4a)$$

$$\frac{\partial\Phi}{\partial\psi_t} = \boldsymbol{B} \cdot \left(\frac{\partial\boldsymbol{R}}{\partial\psi_t} \times \frac{\partial\boldsymbol{R}}{\partial\zeta_f}\right) , \quad (9.5.4b)$$

and

$$\frac{\partial \Psi}{\partial \psi_t} = \boldsymbol{B}\cdot\left(\frac{\partial \boldsymbol{R}}{\partial \psi_t} \times \frac{\partial \boldsymbol{R}}{\partial \theta_f}\right) . \tag{9.5.4c}$$

The right-hand side of these equations is known, since $\boldsymbol{B}(\boldsymbol{R})$ is given and $\boldsymbol{R}(\psi_t, \theta_f, \zeta_f)$ is known after field-line integration and Fourier decomposition. In fact, the last two equations determine Ψ and Φ. The first one can be dropped as it does not provide any new information; it is consistent with the other two equations, since its derivation was based on the general form of $\boldsymbol{B}$ in (9.4.3), which in turn was based on (9.4.2) and $\boldsymbol{B} = \nabla \times \boldsymbol{A}$ and thus on $\nabla \cdot \boldsymbol{B} = 0$.

Integration of (9.5.4b) and (9.5.4c) allows determination of $\Phi(\psi_t, \theta_f, \zeta_f)$ and $\Psi(\psi_t, \theta_f, \zeta_f)$. In practice, one picks a number of different θ_f and ζ_f values and uses a fast Fourier transform. After elimination of the "auxiliary" radial variable ψ_t, one obtains the field-line Hamiltonian $-\Phi$ in terms of its variables $(\Psi, \theta_f, \zeta_f)$ in Fourier decomposed form.

The nature of θ_f and ζ_f, being flux coordinates for $\boldsymbol{B}_0$, is not important for the canonical representation of $\boldsymbol{B}$. Other coordinates, ϱ, θ, and ζ can do the job as well. This alternative and more general set was used by Kuo-Petravic and Boozer (1986). In terms of these general coordinates, the field-line Hamiltonian, in Fourier decomposed form, is

$$\Phi(\Psi, \theta, \zeta) = \Phi_o(\Psi) + \sum_{m,n} \Phi_{mn}(\Psi) \mathrm{e}^{\mathrm{i}(m\theta - n\zeta)} . \tag{9.5.5}$$

Φ_0 is the θ, ζ averaged part of Φ and is the Hamiltonian of $\boldsymbol{B}_0$. Since Ψ is known as a function of (ψ_t, θ, ζ) it is possible to construct the canonical transformation equations from Cartesian coordinates $\boldsymbol{R}(x, y, z)$ to (Ψ, θ, ζ) as $\boldsymbol{R}(\Psi, \theta, \zeta)$. All the information regarding the magnetic field is contained in the functions $\Phi(\Psi, \theta, \zeta)$ and $\boldsymbol{R}(\Psi, \theta, \zeta)$. A typical example based on the theory outlined above has been worked out by Kuo-Petravic and Boozer (1986).

Part III

Selected Topics

In the third and last part of this book, we deal with five subjects that are related to some issues we addressed in the earlier chapters, but which, in the interest of not interrupting the flow of reading, have been postponed until the end of this treatise. In some cases, the topics to be covered now are clarifications of material discussed earlier; in other cases, they should be seen as additional information, given for the sake of completeness. (The topics covered do not necessarily reflect a preference by the authors; the subjects presented themselves automatically during the course of writing.)

In Chap. 10, we treat "proper" toroidal coordinates. This coordinate system is to be considered as a three-dimensional version of the bipolar coordinate system. Although this system is not frequently used for toroidally-confined-plasma computations, we think it is important that the reader should be aware of its existence, and understand the fundamental distinction between this system and those treated in earlier chapters.

Chapter 11 deals with the dynamic equilibrium of an ideal tokamak plasma. Fully realizing that the notion of an infinitely conducting plasma only has meaning in a scaling sense (i.e., that the zero-resistivity plasma is an asymptotic case), we interpret $\sigma \to \infty$ literally, and consider the consequences for a tokamak "through" which the transformer is pumping magnetic flux. We show that both the plasma and the flux surfaces move radially inward due to the presence of a perpendicular electric field, confirming the frozen-flux theorem.

In Chap. 12, we clarify the relationship and the distinction between the specific flux volume $dV/d\Psi_{tor}$ and the "proper" length of a field line $\int dl/B$.

Chapter 13 treats the transformation properties of vector and tensor components. It clarifies the names covariant and contravariant. It also makes clear why the normally used Levi-Civita symbols and the Christoffel symbols do not qualify as tensors.

Finally, in Chap. 14, we give two alternative derivations of the divergence formula. Although the derivation given in Chap. 2 suffices from the "result" point of view, it is interesting to see how (more elaborate) approaches, used frequently in the mathematical-physics literature, lead to the same result.

10. "Proper" Toroidal Coordinates

10.1 Introduction

"Proper" toroidal coordinates are not frequently utilized for describing the physics of toroidally confined plasmas. However, they have been employed occasionally (Shafranov 1960 and Fielding and Hitchon 1980). Furthermore, these coordinates have their own intrinsic value, since many problems in mathematical physics in toroidal geometry have made use of them (see e.g. Morse and Feshbach 1953).

"Proper" toroidal coordinates are constructed by three-dimensionalizing the two-dimensional "bipolar" coordinate system by means of a rotation, just like spherical coordinates are the three-dimensional generalization of the two-dimensional polar system. We will first introduce the bipolar coordinate system in an intuitive way, explain its meaning, and simply write down the equations of the constant coordinate curves. Then we show the analogy between the 3-D spherical system and the 2-D polar system. Next, we will exploit this analogy to construct the "real toroidal" coordinate system. Finally, we derive the equations of the 2-D bipolar system, a task that we postpone until the end in order not to interrupt the logic.

10.2 Bipolar Coordinates: Intuitive Considerations

In two dimensions, we always work with curves, not with surfaces. The position of a point is determined by the intersection of two constant-coordinate curves $u^1 = c^1$ and $u^2 = c^2$, where c^1 and c^2 are constants. The two-dimensional bipolar coordinate system is based on two intersecting families of circles, $\eta =$ constant and $\xi =$ constant, as shown in Fig. 10.1. These curves resemble the loci of constant electrostatic potential and electric-field lines of two point charges of opposite signs located at points A and B. The coordinates ξ and η of a point P are defined as

$$\xi = \alpha_1 - \alpha_2 \qquad \text{and} \qquad \eta = \ln\left(\frac{d_2}{d_1}\right) \tag{10.2.1}$$

where d_1 and d_2 represent the distances from the fixed points A and B to the point P, respectively, and α_1 and α_2 are the angles between the x-axis and the

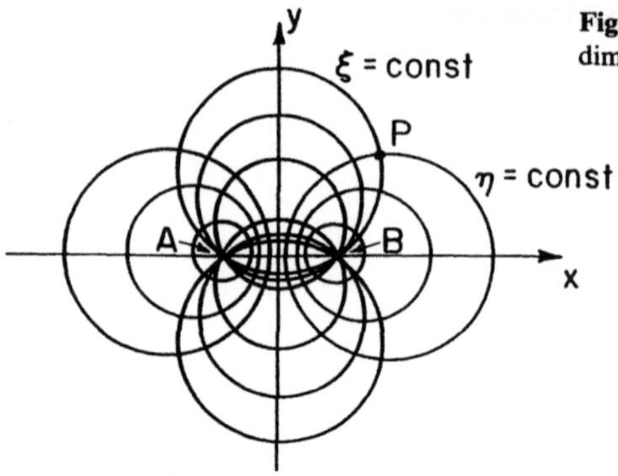

Fig. 10.1. Constant coordinate curves of the two-dimensional bipolar coordinate system

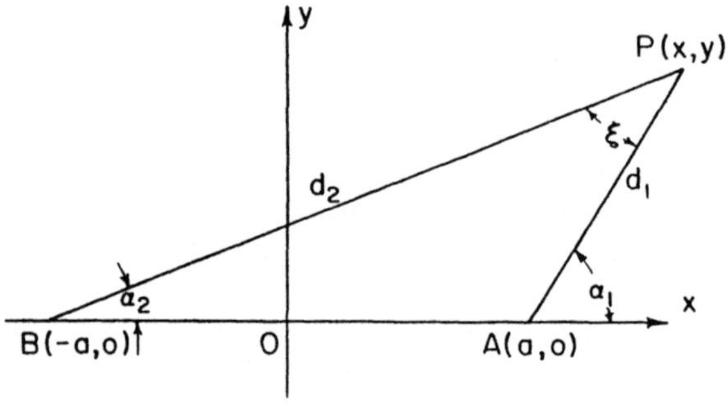

Fig. 10.2. Definition of the coordinates η and ξ. $\xi = \alpha_1 - \alpha_2$; $\eta = \ln(d_2/d_1)$

lines AP and BP. These quantities are shown in Fig. 10.2. The transformation formulae are

$$x = \frac{a \sinh \eta}{\cosh \eta - \cos \xi} \qquad 0 \leqslant \xi < 2\pi \tag{10.2.2a}$$

$$y = \frac{a \sin \xi}{\cosh \eta - \cos \xi} \qquad -\infty < \eta < +\infty \tag{10.2.2b}$$

and the constant coordinate curves are given by

$$x^2 + (y - a \cot \xi)^2 = a^2 \csc^2 \xi \tag{10.2.3}$$

$$(x - a \coth \eta)^2 + y^2 = a^2 \operatorname{csch}^2 \eta \tag{10.2.4}$$

where a is defined in Fig. 10.2. Both constant coordinate curves are circles. One family has its center on the y-axis (10.2.3), the other on the x-axis (10.2.4). Their radii are dependent on the values of ξ and η. Note that the family represented by (10.2.3) passes through the points A and B for all values of ξ.

10.3 The Relationship of 3-D Spherical Coordinates to 2-D Polar Coordinates

Now, we compare the transformation formulae of the common 2-D polar coordinate system of Fig. 10.3 with the 3-D spherical system of Fig. 10.4. The latter is obtained by rotating the 2-D system about its y-axis, after which we rename the axes: the 2-D y-axis becomes the z-axis of the 3-D system, and the 2-D x-axis is now identified with the axis denoted by an asterisk (*) in Fig. 10.4, which we now call the ϱ-axis. Thus, the 2-D x-y plane can be thought of as the z-ϱ plane in 3-D. Then the 3-D x- and y-axes are constructed such that they make a right angle with each other, and $\hat{x}$, $\hat{y}$, $\hat{z}$ form a right-handed basis. Note that $\varrho = \sqrt{x^2 + y^2}$ in the 3-D system. The location of the ϱ-axis is fixed by prescribing an angle ϕ such that $x = \varrho \cos \phi$ and $y = \varrho \sin \phi$. If we now observe that the angle Θ in 3-D is measured from the 3-D z-axis (our former 2-D y-axis), whereas angle θ in the polar system is measured from the 2-D x-axis (the ϱ-axis in 3-D), then we know that $\sin \theta$ in 2-D should be replaced by $\cos \Theta$ in 3-D and vice-versa.

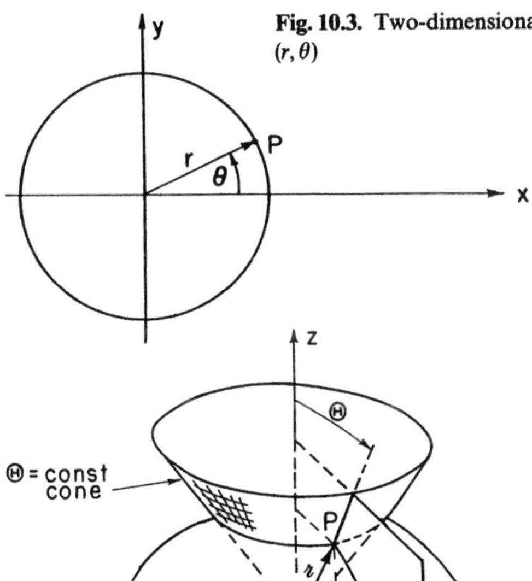

Fig. 10.3. Two-dimensional polar coordinate system. P has coordinates (r, θ)

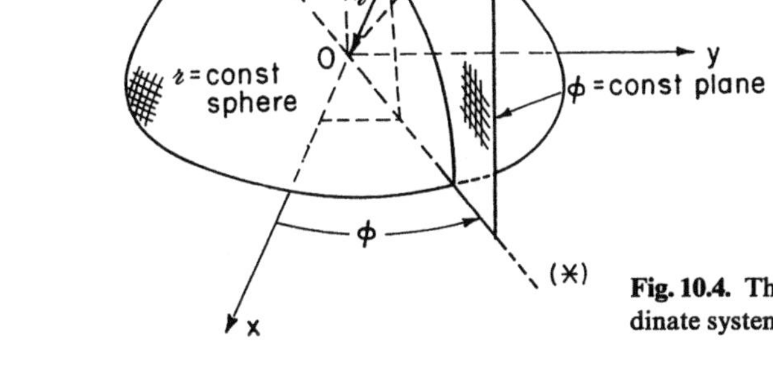

Fig. 10.4. Three-dimensional spherical coordinate system. P has coordinates (r, Θ, ϕ)

Thus, the 2-D equations $x = r\cos\theta$ and $y = r\sin\theta$, become in 3-D: $\varrho = \imath\sin\Theta$ $z = \imath\cos\Theta$ (r and $\imath$ are defined in Figs. 10.3 and 10.4, respectively; in the present context, $\imath \equiv r$). By using then the angle ϕ, we also obtain x and y in the 3-D system. Schematically, we have thus

2-D Polar				*3-D Spherical*
$x = r\cos\theta$	$\rightarrow$	$\varrho = \imath\sin\Theta$	$\rightarrow$	$x = \imath\sin\Theta\cos\phi$
$y = r\sin\theta$		$z = \imath\cos\Theta$		$y = \imath\sin\Theta\sin\phi$
		$x = \varrho\cos\phi$		$z = \imath\cos\Theta$.
		$y = \varrho\sin\phi$		

The constant coordinate curves are "transformed" in a similar way to constant coordinate surfaces. We see that the 2-D $x^2 + y^2 = r^2$ = constant circle becomes a sphere: $\varrho^2 + z^2 = \imath^2$ = constant or $x^2 + y^2 + z^2 = \imath^2$ = constant. The 2-D straight line θ = constant is transformed into a cone Θ = constant. The third constant coordinate surface is a new one, originating from the (rotation) angle ϕ = constant.

10.4 "Proper" Toroidal Coordinates as a 3-D Version of Bipolar Coordinates

Completely analogously, we can now establish the "*proper*" *toroidal coordinate system* by rotating the 2-D bipolar system of Fig. 10.1 about its y-axis, such that the ξ = constant circles go over into spheres, while the η = constant circles become tori. With similar replacements as above, i.e., x(2-D) $\rightarrow \varrho = \sqrt{x^2 + y^2}$ (3-D), y(2-D) $\rightarrow z$(3-D), a (rotation) angle ζ such that x(3-D) $= \varrho\cos\zeta$ and y(3-D) $=$ $\varrho\sin\zeta$, and with (the old bipolar variables) ξ and η now retaining their original meaning in contrast to the $\Theta \leftrightarrow \theta$ relationship, we obtain

$$x = \frac{a\sinh\eta}{\cosh\eta - \cos\xi}(\cos\zeta) \tag{10.4.1a}$$

and

$$y = \frac{a\sinh\eta}{\cosh\eta - \cos\xi}(\sin\zeta) \tag{10.4.1b}$$

from (10.2.2a). According to (10.2.2b), we have then for z:

$$z = \frac{a\sin\xi}{\cosh\eta - \cos\xi} \; . \tag{10.4.1c}$$

If we take the toroidal coordinates as (ξ, η, ζ) in that order, then a right-handed system implies that ζ should be measured from the x-axis towards the y-axis. This clarifies the above relationships between x, y, and ζ.

The equations for the constant coordinate surfaces are easily obtained from (10.2.3) and (10.2.4) by using our prescription above. Thus,

$$x^2 + y^2 + (z - a\cot\xi)^2 = a^2\csc^2\xi \; ; \qquad \xi = \text{constant} \; , \tag{10.4.2a}$$

is the equation of a family of spheres with center on the z-axis (at $z = a\cot\xi$) and radius ($a\csc\xi$). Also,

$$((x^2 + y^2)^{1/2} - a\coth\eta)^2 + z^2 = a^2\operatorname{csch}^2\eta \; ; \qquad \eta = \text{constant} \; , \tag{10.4.2b}$$

represents the equation of a toroidal surface, with the minor axis having a radius ($a\coth\eta$). The cross sections with the $z - \varrho$ plane (ζ = constant plane) are circles with radius ($a\operatorname{csch}\eta$). Finally, we have a plane ζ = constant (the $z - \varrho$ plane) with equation

$$y = x\tan\zeta \; ; \qquad \zeta = \text{const} \; . \tag{10.4.2c}$$

The coordinate system is an orthogonal one, and its scale factors h_i are

$$h_\xi = h_\eta = \frac{a}{\cosh\eta - \cos\xi} \tag{10.4.3a}$$

$$h_\zeta = \frac{|a\sinh\eta|}{\cosh\eta - \cos\xi} \; . \tag{10.4.3b}$$

10.5 The Bipolar Coordinate System: A Detailed Analysis

Our last task is to derive (10.2.2) for the bipolar system. (See e.g. Margenau and Murphy 1976.) We find it convenient to introduce a complex variable u, defined by $u = x + \mathrm{i}y$. Its complex conjugate is denoted by $u^* = x - \mathrm{i}y$. Since $x = (u + u^*)/2$ and $y = \mathrm{i}(u^* - u)/2$, we would like to find expressions for u and u^* as functions of the bipolar variables ξ and η defined in (10.2.1). From Fig. 10.2, we obtain that

$$u = a + d_1 \mathrm{e}^{\mathrm{i}\alpha_1} = -a + d_2 \mathrm{e}^{\mathrm{i}\alpha_2} \; . \tag{10.5.1}$$

If we write the first equation as $u - a$, the second as $u + a$, and divide the two by each other, we find

$$\frac{u + a}{u - a} = \frac{d_2}{d_1}\mathrm{e}^{-\mathrm{i}(\alpha_1 - \alpha_2)} = \mathrm{e}^{+\eta}\mathrm{e}^{-\mathrm{i}\xi} = \mathrm{e}^{-\mathrm{i}(\xi + \mathrm{i}\eta)} \; . \tag{10.5.2}$$

Here, we have used the definition equations of ξ and η given in (10.2.1). Then from (10.5.2), it follows that

$$u = a\frac{e^{-i(\xi+i\eta)}+1}{e^{-i(\xi+i\eta)}-1} = a\frac{e^{-iw}+1}{e^{-iw}-1} \tag{10.5.3}$$

with $w = \xi + i\eta$. From the Euler formula, $e^{iw} = \cos w + i\sin w$, and thus, $\tan w = \sin w/\cos w = i(e^{-iw} - e^{iw})/(e^{-iw} + e^{iw}) = i(1 - e^{2iw})/(1 + e^{2iw})$, so we can rewrite (10.5.3):

$$u = +ia\cot(w/2) \tag{10.5.4a}$$

$$u^* = -ia\cot(w^*/2)\ . \tag{10.5.4b}$$

Now, we substitute (10.5.4) into $x = (u + u^*)/2$, and use the identities $\cot p = \cos p/\sin p$, $w = \xi + i\eta$, $w^* = \xi - i\eta$, $\cos p - \cos q = -2\sin(p+q)/2\sin(p-q)/2$, and $\sin(p-q) = \sin p\cos q - \sin q\cos p$, in order to obtain:

$$x = ia\frac{\sin i\eta}{\cos\xi - \cos i\eta}\ .$$

Since $\sin(i\eta) = i\sinh\eta$ and $\cos(i\eta) = \cosh\eta$, this equation leads to (10.2.2a). The computation of (10.2.2b) is completely analogous.

Finally, we compute the constant coordinate curves given in (10.2.3) and (10.2.4). From the trigonometric identity $\tan p = i(1 - e^{2ip})/(1 + e^{2ip})$ given above, we derive easily that

$$2ip = \ln\left(\frac{i - \tan p}{i + \tan p}\right)\ . \tag{10.5.5}$$

If we now observe that $\tan\alpha_1 = y/(x-a)$ and $\tan\alpha_2 = y/(x+a)$, we obtain upon using (10.5.5) with $p \equiv \alpha_1$, and subsequently, $p \equiv \alpha_2$:

$$\begin{aligned}\xi = \alpha_1 - \alpha_2 &= \frac{i}{2}\left[\ln\left(\frac{i + y/(x-a)}{i - y/(x-a)}\right) - \ln\left(\frac{i + y/(x+a)}{i - y/(x+a)}\right)\right]\\ &= \frac{i}{2}\ln\frac{(ix - ia + y)(ix + ia - y)}{(ix - ia - y)(ix + ia + y)}\\ &= \frac{i}{2}\ln\frac{x^2 + y^2 - a^2 - 2iay}{x^2 + y^2 - a^2 + 2iay}\end{aligned} \tag{10.5.6}$$

After taking the anti-logarithm, and factoring, we find

$$x^2 + y^2 - a^2 + 2iay\left(\frac{1 + e^{2i\xi}}{1 - e^{2i\xi}}\right) = 0 \tag{10.5.7a}$$

or

$$x^2 + y^2 - a^2 - 2ay\cot\xi = 0\ . \tag{10.5.7b}$$

With $\csc^2\xi = 1 + \cot^2\xi$, we retrieve (10.2.3).

Analogously, with the definition of η, and the Pythagorean theorem, we easily obtain

$$e^{2\eta} = \frac{(d_2)^2}{(d_1)^2} = \frac{(x+a)^2 + y^2}{(x-a)^2 + y^2} = \frac{x^2 + y^2 + a^2 + 2ax}{x^2 + y^2 + a^2 - 2ax} \, . \tag{10.5.8}$$

After some rearranging, (10.2.4) is recovered.

11. The Dynamic Equilibrium of an Ideal Tokamak Plasma

11.1 Introduction

To acquire a good understanding of the relationship between the different flux measures in a tokamak (Ψ_{tor}, $\Psi^{\text{d}}_{\text{pol}}$, $\Psi^{\text{T}}_{\text{pol}}$), a careful investigation of the time evolution of the flux surfaces in an ideal plasma (with $\sigma \to \infty$) is called for. Plasma and flux surfaces will be shown to move in response to a time-varying transformer flux.

In this Chapter, we do not intend to "prove" relationships, but only to make certain statements plausible. We shall make use of certain ideas that are familiar to most readers, but which have not been discussed in detail in previous chapters, where we referred to this Chapter.

11.2 A Useful Identity in Tokamak Geometry

The representation of the magnetic field in terms of the contravariant-basis vectors ∇u^i is discussed in Sects. 6.1–3. In an axisymmetric tokamak, with symmetry angle ζ_o, the magnetic field can be written (see (6.3.10)) as:

$$\boldsymbol{B} = I\,\nabla\zeta_o + \nabla\zeta_o \times \nabla\Psi = \boldsymbol{B}_\text{T} + \boldsymbol{B}_\text{P}\ . \tag{11.2.1}$$

We assume that the toroidal field $\boldsymbol{B}_\text{T}$ and the poloidal field $\boldsymbol{B}_\text{P}$ are in the positive ζ_o and θ directions, respectively. Ψ is the radial variable and is related to $\Psi^{\text{d}}_{\text{pol}}$ by $\Psi = -\Psi^{\text{d}}_{\text{pol}}/2\pi$ (apart from an additive constant which is unimportant here but which will be mentioned at the end of this Chapter); $\nabla\Psi^{\text{d}}_{\text{pol}}$ points radially inward, whereas $\nabla\Psi$ points radially outward. I is a flux-surface quantity equal to $B_\text{T}R$ and $\nabla\zeta_o \cdot \nabla\Psi = 0$. As a consequence of this last property, it follows from (11.2.1) that $|\nabla\Psi|/R = B_\text{P}$ or $|\nabla\Psi| = RB_\text{P}$.

A very useful geometric relationship that we shall use below can be derived from the $\boldsymbol{B}$ representation in (11.2.1): we wish to decompose the component perpendicular to $\boldsymbol{B}$, which is tangent to the flux surface ($\propto \boldsymbol{B} \times \nabla\Psi$), into a toroidal component ($\propto \nabla\zeta_o$) and a parallel component ($\propto \boldsymbol{B}$). Take the cross product of $\nabla\Psi$ with (11.2.1):

$$\nabla\Psi \times \boldsymbol{B} = I(\nabla\Psi \times \nabla\zeta_o) + \nabla\Psi \times (\nabla\zeta_o \times \nabla\Psi)\ . \tag{11.2.2}$$

The expression between brackets of the first term equals $(I\,\nabla\zeta_o - \boldsymbol{B})$. To the last term we apply the *bac-cab* rule and take into account that $\nabla\zeta_o \cdot \nabla\Psi = 0$. We obtain

$$\begin{aligned}\nabla\Psi \times \boldsymbol{B} &= -I\boldsymbol{B} + I^2\nabla\zeta_o + \nabla\zeta_o|\nabla\Psi|^2\\ &= -I\boldsymbol{B} + \nabla\zeta_o(R^2B_T^2 + R^2B_P^2)\end{aligned}$$

or

$$\nabla\Psi \times \boldsymbol{B} = -I\boldsymbol{B} + B^2R^2\nabla\zeta_o \ . \tag{11.2.3}$$

The unit vector perpendicular to $\boldsymbol{B}$, but tangent to the flux surface is then

$$\frac{\nabla\Psi \times \hat{\boldsymbol{B}}}{|\nabla\Psi|} = -\frac{I\hat{\boldsymbol{B}}}{|\nabla\Psi|} + \frac{BR^2}{|\nabla\Psi|}\nabla\zeta_o \ . \tag{11.2.4}$$

Recalling that $|\nabla\Psi| = RB_P$,

$$\frac{\nabla\Psi \times \hat{\boldsymbol{B}}}{RB_P} = -\frac{B_T}{B_P}\hat{\boldsymbol{B}} + \frac{B}{B_P}\hat{\zeta}_o \ . \tag{11.2.5}$$

11.3 The Existence of an Electric Field in a Tokamak Plasma with Infinite Conductivity

A time-varying magnetic flux through the core of a transformer "generates" a non-electrostatic electric field $\boldsymbol{E}^{(A),\mathrm{tr}}$ which is tangent to any circle symmetric about the core of the transformer (see Fig. 11.1). The superscript (A) indicates that we are dealing with an $\boldsymbol{E}$ due to a $\partial A/\partial t$ (and not a $\nabla\Phi$). The value of $\boldsymbol{E}^{(A),\mathrm{tr}}$ at any position can be found from the law of Faraday-Lenz.

Next we consider a torus encircling the transformer core (Fig. 11.2). Suppose that the torus is an ideal conductor, $\sigma \to \infty$. There cannot exist an electric field in this ideal conductor, and Faraday's law is now obeyed via the generation of a surface current in response to a $d\Psi^{\mathrm{tr}}/dt$ such that $\Psi_{\mathrm{tot,disk}}$ remains constant. Said differently, the changing current density $\boldsymbol{J}$ generates a changing poloidal magnetic field whose flux change through the hole of the disk cancels $d\Psi^{\mathrm{tr}}/dt$ exactly. Still another way of looking at it is that the electric field $\boldsymbol{E}^{(A),\mathrm{tr}}$ due to the transformer is cancelled by an equally large, but reverse electric field $\boldsymbol{E}^{(A),\mathrm{react}}$ generated by the changing poloidal field so that at any moment $\boldsymbol{E}^{(A),\mathrm{tot}} \equiv 0$. In terms of loop voltages, $e = |E^{(A),\mathrm{tr}}2\pi R|$ equals $|d\Psi^{\mathrm{tr}}/dt|$, whereas, $|E^{(A),\mathrm{react}}2\pi R|$ is related to $|L\,di/dt|$, where L is the self-inductance of the torus ring, and i is the current. Hence, the current i changes proportionally to Ψ^{tr}.

The situation is different in a tokamak plasma. In Fig. 11.3, we consider a typical flux surface containing an ideal plasma at an instant $t = t_0$. At this time $t = t_0$, a spiraling magnetic field is present ($\boldsymbol{B}_0 = \boldsymbol{B}_{T_0} + \boldsymbol{B}_{P_0}$) as well as a non-zero

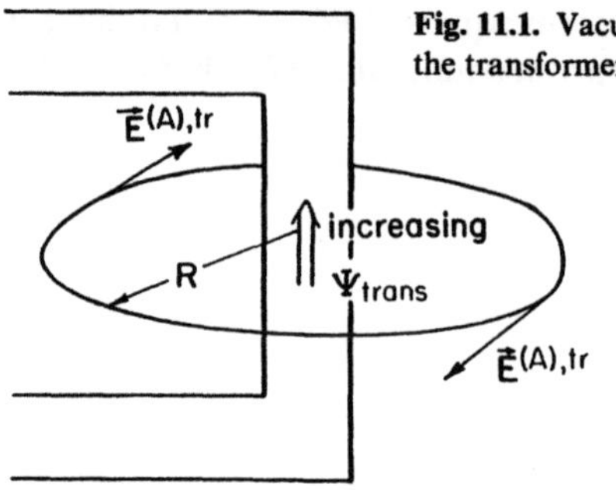

Fig. 11.1. Vacuum electric field $E^{(A),\mathrm{tr}}$, induced by a changing flux in the transformer: $d\Psi^{\mathrm{tr}}/dt$

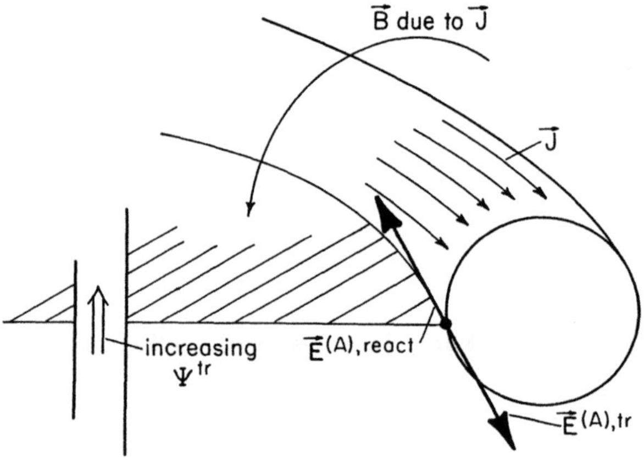

Fig. 11.2. The net electric field in an ideal conductor encircling a transformer is zero: $E^{(A),\mathrm{react}} = -E^{(A),\mathrm{tr}}$. See text

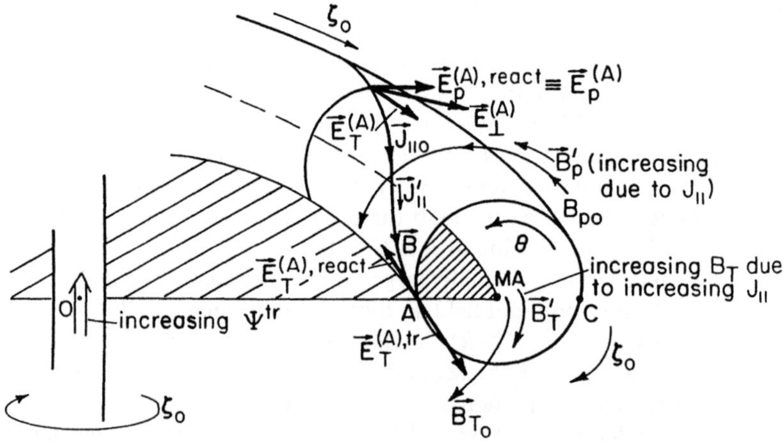

Fig. 11.3. Electric-field components, current-density components and magnetic-field components in an ideal plasma in response to an increasing transformer flux. See text

current density $J_{\|_0}$. We assume that the transformer flux has been increasing – say, linearly – from $t = 0$ to $t = t_0$, and will continue to do so for $t > t_0$. (The (perpendicular) diamagnetic current which is necessary for equilibrium does not influence our explanation here and is not mentioned henceforth.) The current density $J_{\|_0}$ is not only a surface current, but must exist inside the plasma to generate the poloidal magnetic field.

The requirement in an ideal plasma is that there *cannot* exist an *electric field parallel to* $\boldsymbol{B}$, since that would cause the current to blow up. In response to a linearly increasing Ψ^{tr}, the current density $J_\|$ will also increase linearly, somewhat similarly to the case considered in Fig. 11.2. In simplistic terms, as far as the current response is concerned, the tokamak case that we are dealing with now, could be compared with an ideal conductor in which the current is forced to flow helically; $\sigma_\| \to \infty$ along the helix, $\sigma_\perp \to 0$ perpendicular to the helix. (We shall show below, however, that $J_\|$ not only increases on the flux surface that we isolated, but also further inside the plasma.) The increase in $J_\|$ during a small time interval Δt is represented by $J_\|'$. The increasing current density $J_\| = J_{\|_0} + J_\|'$ has components in the toroidal and poloidal directions. Its toroidal component opposes the increase of flux through the hole (disk) of the toroidal surface (in much the same way as was the case in Fig. 11.2). The electric-field component "generated" by this changing toroidal component of $J_\|$ (and thus the changing poloidal field B_P) is labeled by $E_\text{T}^{(A),\text{react}}$. However, because the current density $\boldsymbol{J}$ is spiraling around, it will not be able to cancel the total electric-field component $E_\text{T}^{(A),\text{tr}}$. (The subscripts T specify that the $\boldsymbol{E}$-fields are in the toroidal direction.) So we might expect that a net toroidal component of $\boldsymbol{E}$ will survive. This is explained momentarily.

The poloidal component of the increasing $J_\|$ increases the toroidal magnetic field. The change in $\boldsymbol{B}_\text{T}$ during Δt is denoted by $\boldsymbol{B}_\text{T}'$. Application of Faraday's law along a poloidal loop shows that an increasing $\boldsymbol{B}_\text{T} = \boldsymbol{B}_{\text{T}_0} + \boldsymbol{B}_\text{T}'$ sets up a poloidal electric field. In Fig. 11.3, this field is indicated as $\boldsymbol{E}_\text{P}^{(A),\text{react}}$, since it follows from the "reaction current" $J_\|$. Since a poloidal $\boldsymbol{E}$-field has a parallel component, and no net parallel electric field can be tolerated, there must be another electric-field component that cancels this parallel component of $\boldsymbol{E}_\text{P}^{(A),\text{react}}$. As shown in Fig. 11.4, that field must be toroidal, so that the only surviving electric field is in the

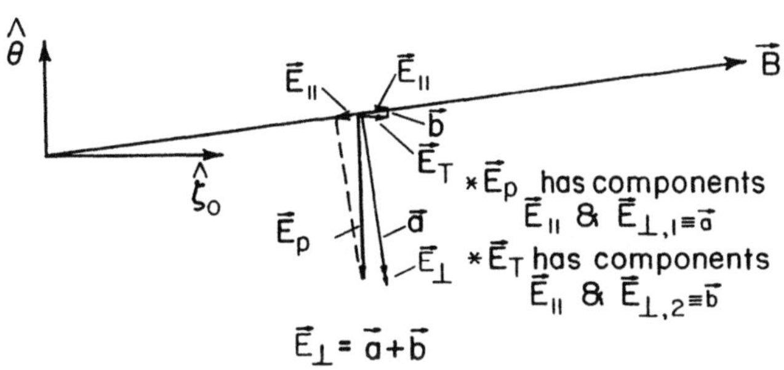

Fig. 11.4. Perpendicular and parallel components of $\boldsymbol{E}_\text{P}$ and $\boldsymbol{E}_\text{T}$; see text

perpendicular direction. This is exactly the surviving toroidal electric field we alluded to in the former paragraph (see also, Hinton and Hazeltine (1976), p. 252). In contrast thus to an ideal conductor (as in Fig. 11.2), there can be a toroidal electric field component because the magnetic field spirals (as long as the total $\boldsymbol{E}$ is in the perpendicular direction). The fields $E_T^{(A)}$ and $E_P^{(A)}$ are constant in time for a linearly increasing transformer flux. Although we did not say so, they obviously were already present from $t = 0$ to $t = t_0$.

The fact that a perpendicular $\boldsymbol{E}$ field can exist in an ideal plasma should be no surprise, since in a fully ionized plasma, the "perpendicular conductivity" defined as the proportionality constant between current density and electric field is zero. (The perpendicular conductivity found in the literature is the proportionality constant between the perpendicular current and the perpendicular friction force in analogy with the parallel one.) Rather than causing a current, a perpendicular electric field causes an $\boldsymbol{E} \times \boldsymbol{B}$ fluid motion (see Golant, Zhilinsky and Sakharov (1980) and Braginskii (1965)). Note that, even if we had assumed that $J_\parallel$ were only a surface current (in analogy with the ideal conductor), a field $E_\perp^{(A)}$ would still have existed *inside* the plasma.

11.4 Motion of the Plasma and the Flux Surfaces

11.4.1 Plasma Motion due to $\boldsymbol{E} \times \boldsymbol{B}$ Drift

When an electric field is present in a plasma, the latter will move with a fluid velocity $\boldsymbol{E} \times \boldsymbol{B}/B^2$. In the case we are considering here, we find that the plasma *pinches radially inward*. Letting $\boldsymbol{v}$ denote the plasma fluid velocity, we compute the radial "velocity" as follows:

$$\boldsymbol{v} \cdot \nabla\Psi = \nabla\Psi \cdot \frac{\boldsymbol{E} \times \boldsymbol{B}}{B^2} = \boldsymbol{E} \cdot \frac{\boldsymbol{B} \times \nabla\Psi}{B^2} . \tag{11.4.1}$$

By means of the geometric identity of (11.2.3), this gives

$$\boldsymbol{v} \cdot \nabla\Psi = \boldsymbol{E} \cdot \boldsymbol{B}\frac{I}{B^2} - \boldsymbol{E} \cdot \nabla\zeta_o R^2 , \tag{11.4.2}$$

or with

$$E_\parallel = \boldsymbol{E} \cdot \hat{\boldsymbol{B}} \text{ and } E_T = \boldsymbol{E} \cdot \boldsymbol{\zeta}_o = \boldsymbol{E} \cdot R\, \nabla\zeta_o,$$

$$\boldsymbol{v} \cdot \nabla\Psi = \frac{I}{B}E_\parallel - RE_T . \tag{11.4.3}$$

In our ideal plasma, the infinite conductivity requires that $E_\parallel$ vanishes, so we obtain

$$\boldsymbol{v} \cdot \nabla\Psi = -RE_T^{(A)} . \tag{11.4.4}$$

In the last expression, the label (A) has been added to the electric-field symbol.

$E_T^{(A)}$ is the *net* toroidal electric field, part of which is caused by the transformer, and a (counteracting) part is due to the reaction currents $dJ_{\parallel}/dt$. $\nabla\Psi$ is radially outward, and $E_T^{(A)}$ is a positive number (see Fig. 11.3); it thus follows that the plasma pinches *inward*.

11.4.2 Motion of Flux Surfaces Defined by the Poloidal Disk Flux $\Psi_{\rm pol}^{\rm d}$

Consider now the flux surfaces labeled by $\Psi_{\rm pol}^{\rm d}$, the *poloidal disk flux*. We define the flux surfaces by $\Psi_{\rm pol}^{\rm d}$ = constant. This implies that at that surface

$$\frac{d\Psi_{\rm pol}^{\rm d}}{dt}=0\ . \tag{11.4.5}$$

After expanding d/dt as a convective derivative, we get

$$\left.\frac{\partial\Psi_{\rm pol}^{\rm d}}{\partial t}\right|_{R}+\boldsymbol{u}_s\cdot\nabla\Psi_{\rm pol}^{\rm d}=0\ . \tag{11.4.6}$$

Here $\boldsymbol{u}_s$ is the velocity of a flux surface $\Psi_{\rm pol}^{\rm d}$ = constant. Note that only radial flux surface motion makes sense physically.

To find an expression for the partial derivative in (11.4.6), we start from the "radial" contravariant component of Faraday's law:

$$\nabla\Psi\cdot\frac{\partial\boldsymbol{B}}{\partial t}=-\nabla\Psi\cdot\nabla\times\boldsymbol{E}\ . \tag{11.4.7}$$

Since $\boldsymbol{B}$ is by definition tangent to a flux surface, we know that $\boldsymbol{B}\cdot\nabla\Psi\equiv 0$. It then follows from $\partial(\boldsymbol{B}\cdot\nabla\Psi)/\partial t=0$ that $\nabla\Psi\cdot\partial\boldsymbol{B}/\partial t=-\boldsymbol{B}\cdot\partial\nabla\Psi/\partial t$. Since we are dealing with smooth quantities, we can interchange the ∇ and $\partial/\partial t$ operators to obtain

$$\nabla\Psi\cdot\frac{\partial\boldsymbol{B}}{\partial t}=-\boldsymbol{B}\cdot\nabla\frac{\partial\Psi}{\partial t}\ . \tag{11.4.8}$$

For the right side of (11.4.7), we observe that

$$\nabla u^1\cdot(\nabla\times\boldsymbol{E})\equiv(\nabla\times\boldsymbol{E})^1=\frac{1}{\sqrt{g}}\left(\frac{\partial E_3}{\partial u^2}-\frac{\partial E_2}{\partial u^3}\right)\ .$$

Because $u^3=\zeta_o$ and $\partial/\partial\zeta_o\equiv 0$ as a consequence of axisymmetry, we see that only the E_{ζ_o} (covariant) component is of concern. Thus,

$$\begin{aligned}
-\nabla\Psi\cdot(\nabla\times\boldsymbol{E})&=-\nabla\Psi\cdot(\nabla\times E_{\zeta_o}\nabla\zeta_o)\\
&=-\nabla\Psi\cdot(\nabla E_{\zeta_o}\times\nabla\zeta_o)\\
&=-(\nabla\zeta_o\times\nabla\Psi)\cdot\nabla E_{\zeta_o}\\
&=-\boldsymbol{B}\cdot\nabla E_{\zeta_o}\\
&=-\boldsymbol{B}\cdot\nabla(RE_{\rm T})\ .
\end{aligned} \tag{11.4.9}$$

In this derivation we have used the fact that

$$\boldsymbol{B}_{\mathrm{T}} \cdot \boldsymbol{\nabla} E_{\zeta_o} = I\,\boldsymbol{\nabla}\zeta_o \cdot \boldsymbol{\nabla} E_{\zeta_o} = I\,\boldsymbol{\nabla}\zeta_o \cdot \left(\frac{\partial E_{\zeta_o}}{\partial \Psi} \boldsymbol{\nabla}\Psi + \frac{\partial E_{\zeta_o}}{\partial \theta} \boldsymbol{\nabla}\theta + \frac{\partial E_{\zeta_o}}{\partial \zeta_o} \boldsymbol{\nabla}\zeta_o \right) \equiv 0 \ ,$$

since $\partial/\partial\zeta_o = 0$ and $\boldsymbol{\nabla}\zeta_o \cdot \boldsymbol{\nabla}\Psi = 0 = \boldsymbol{\nabla}\zeta_o \cdot \boldsymbol{\nabla}\theta$, all because of axisymmetry. We employed $\boldsymbol{E} = E_{\mathrm{T}}\hat{\boldsymbol{\zeta}}_o = E_{\zeta_o}\boldsymbol{\nabla}\zeta_o$, where $\boldsymbol{\nabla}\zeta_o = \hat{\boldsymbol{\zeta}}_o/R$. Faraday's law is then, from (11.4.8) and (11.4.9),

$$\boldsymbol{B} \cdot \boldsymbol{\nabla}\left(-\frac{\partial \Psi}{\partial t} + RE_{\mathrm{T}} \right) = 0 \ . \tag{11.4.10}$$

This means that

$$\frac{\partial \Psi}{\partial t} = RE_{\mathrm{T}} + \text{flux function} \ . \tag{11.4.11}$$

With $\Psi \equiv -\Psi^{\mathrm{d}}_{\mathrm{pol}}/2\pi$, and with $E_{\mathrm{T}} \equiv E^{(A)}_{\mathrm{T}}$, the net electric field, we see that the flux function can be chosen to be zero. (This follows from the integral form of Ampère's law where a ζ_o loop is taken through point A of Fig. 11.3.) The time rate of change of the poloidal flux through a disk bounded by a fixed ζ_o loop is thus

$$\left.\frac{\partial \Psi^{\mathrm{d}}_{\mathrm{pol}}}{\partial t}\right|_{\substack{\text{fixed}\\ \text{disk}}} = -2\pi RE^{(A)}_{\mathrm{T}} \ . \tag{11.4.12}$$

After substitution of this result in (11.4.6), the latter can be solved for the radial flux-surface "velocity":

$$\boldsymbol{u}^{\mathrm{d}}_{\mathrm{s}} \cdot \boldsymbol{\nabla}\Psi^{\mathrm{d}}_{\mathrm{pol}} = -\left.\frac{\partial \Psi^{\mathrm{d}}_{\mathrm{pol}}}{\partial t}\right|_{D} = 2\pi RE^{(A)}_{\mathrm{T}} \ . \tag{11.4.13}$$

(In $\boldsymbol{u}^{\mathrm{d}}_{\mathrm{s}}$, s stands for surface and d for disk.) Taking into account that $\Psi \equiv -\Psi^{\mathrm{d}}_{\mathrm{pol}}/2\pi$, we observe from (11.4.4) and (11.4.13) that the plasma $\boldsymbol{E} \times \boldsymbol{B}$ inward pinch velocity is exactly equal to the radial velocity of the flux surfaces. This is a statement of the frozen flux theorem: plasma and flux surfaces are tied together.

An observer sitting on (and moving with) the plasma will find that the fluid-pinch motion generates a motional EMF. The electric field connected with this motional EMF is exactly the negative of the electric field that caused the $\boldsymbol{E} \times \boldsymbol{B}$ inward motion in the first place. This is as it should be: an observer moving with the plasma, or thus the plasma itself, does not see any electric field. The electric field is only present in the lab frame.

11.4.3 Motion of Flux Surfaces Defined by the Poloidal Ribbon Flux $\Psi^{\mathrm{r}}_{\mathrm{pol}}$

Suppose now that we wish to label a flux surface by the *poloidal ribbon flux* (see Fig. 11.3). To actually show that the flux surfaces labeled by $\Psi^{\mathrm{r}}_{\mathrm{pol}} = $ constant move with the same radially inward velocity as the plasma, we return to (11.4.10) or (11.4.11). When we now identify $\Psi = \Psi^{\mathrm{r}}_{\mathrm{pol}}/2\pi$, we must recognize that the ribbon flux does not know about the transformer; all it contains is the "reaction flux" due to an increase of B_p and $(J_\parallel)_T$. The integration constant that we called the "flux function" in (11.4.11) is the ubiquitous space independent transformer loop voltage (devided by 2π) $e(t)/2\pi = RE^{(A),\mathrm{tr}} = \oint \boldsymbol{E}\cdot d\boldsymbol{l}/2\pi = |d\Psi^{\mathrm{tr}}/dt|/2\pi$ that we encountered in Fig. 11.1 ($e(t)$ is spatially uniform, but $E^{(A),\mathrm{tr}} \propto 1/R$). Instead of (11.4.12), we obtain for the case of the ribbon flux:

$$\left.\frac{\partial \Psi^{\mathrm{r}}_{\mathrm{pol}}}{\partial t}\right|_{\substack{\text{fixed}\\ \text{ribbon}}} = +2\pi R E^{(A),\mathrm{react}}_{\mathrm{T}} + e(t)$$

$$= 2\pi R[+E^{(A),\mathrm{react}}_{\mathrm{T}} + E^{(A),\mathrm{tr}}_{\mathrm{T}}] \; . \tag{11.4.14}$$

The appropriate signs can be understood from (11.4.12): $\Psi = \Psi^{\mathrm{r}}_{\mathrm{pol}}/2\pi$ and $\boldsymbol{E}^{(A),\mathrm{react}}_{\mathrm{T}} = E^{(A),\mathrm{react}}_{\mathrm{T}}\hat{\zeta}_{\mathrm{o}} = -|E^{(A),\mathrm{react}}_{\mathrm{T}}|\hat{\zeta}_{\mathrm{o}}$; thus $E^{(A),\mathrm{react}}_{\mathrm{T}}$ is a negative quantity (see Fig. 11.3).

With the flux surfaces, $\Psi^{\mathrm{r}}_{\mathrm{pol}} = \text{constant}$ defined in a similar way to (11.4.5), we find for their radial "velocity"

$$\boldsymbol{u}^{\mathrm{r}}_{\mathrm{s}}\cdot \boldsymbol{\nabla}\Psi^{\mathrm{r}}_{\mathrm{pol}} = -\left.\frac{\partial \Psi^{\mathrm{r}}_{\mathrm{pol}}}{\partial t}\right|_{R} = -2\pi R[E^{(A),\mathrm{react}}_{\mathrm{T}} + E^{(A),\mathrm{tr}}_{\mathrm{T}}] \; . \tag{11.4.15}$$

From Fig. 11.3, it is clear that $E^{(A),\mathrm{tr}}_{\mathrm{T}} + E^{(A),\mathrm{react}}_{\mathrm{T}} = E^{(A),\mathrm{tr}}_{\mathrm{T}} - |E^{(A),\mathrm{react}}_{\mathrm{T}}| = E^{(A)}_{\mathrm{T}}$, where $E^{(A)}_{\mathrm{T}}$ is the magnitude of the net toroidal electric field ($E^{(A)}_{\mathrm{T}} > 0$). Thus

$$\boldsymbol{u}^{\mathrm{r}}_{\mathrm{s}}\cdot \boldsymbol{\nabla}\Psi^{\mathrm{r}}_{\mathrm{pol}} = -2\pi R E^{(A)}_{\mathrm{T}} \; . \tag{11.4.16}$$

A comparison with (11.4.4) and (11.4.13) shows that, indeed,

$$\boldsymbol{v}\cdot \boldsymbol{\nabla}\Psi = \boldsymbol{u}^{\mathrm{d}}_{\mathrm{s}}\cdot \boldsymbol{\nabla}\Psi = \boldsymbol{u}^{\mathrm{r}}_{\mathrm{s}}\cdot \boldsymbol{\nabla}\Psi \; . \tag{11.4.17}$$

Thus plasma and flux surfaces labeled by a constant poloidal (ribbon or disk) flux move together.

11.4.4 Motion of Flux Surfaces Defined by the Toroidal Flux Ψ_{tor}

Finally, we show that the *toroidal-flux label* Ψ_{tor} leads to a similar conclusion. We could invoke the general applicability of the frozen-flux theorem: the plasma and the flux surfaces (regardless of their label) move together. Simple manipulations with formulae, however, provide some physical insight: the toroidal-flux change should be connected with the *poloidal* electric field.

To find the inward $\boldsymbol{E}\times\boldsymbol{B}$ plasma velocity, we first derive a geometric relationship for the component perpendicular to $\boldsymbol{B}$ in terms of poloidal and toroidal components. From (11.2.2) we obtain:

$$\frac{\boldsymbol{\nabla}\Psi\times\hat{\boldsymbol{B}}}{|\boldsymbol{\nabla}\Psi|} = \frac{|\boldsymbol{\nabla}\Psi|}{B}\boldsymbol{\nabla}\zeta_{\mathrm{o}} - \frac{I}{B|\boldsymbol{\nabla}\Psi|}(\boldsymbol{\nabla}\zeta_{\mathrm{o}}\times\boldsymbol{\nabla}\Psi)$$

$$= \frac{|\boldsymbol{\nabla}\Psi|}{RB}\hat{\zeta}_{\mathrm{o}} - \frac{B_T}{B}\left(\hat{\zeta}_{\mathrm{o}}\times\frac{\boldsymbol{\nabla}\Psi}{|\boldsymbol{\nabla}\Psi|}\right) \; . \tag{11.4.18}$$

The only induced $\boldsymbol{E}$-field component present is in the perpendicular direction and points in the $+(\boldsymbol{\nabla}\Psi \times \hat{\boldsymbol{B}})$ direction. The $\boldsymbol{E} \times \boldsymbol{B}$ fluid flow is

$$\frac{\boldsymbol{E}^{(A)} \times \boldsymbol{B}}{B^2} = \frac{\boldsymbol{E}_\perp^{(A)} \times \boldsymbol{B}}{B^2} = E_\perp^{(A)} \frac{(\boldsymbol{\nabla}\Psi \times \hat{\boldsymbol{B}})}{|\boldsymbol{\nabla}\Psi|} \times \frac{\hat{\boldsymbol{B}}}{B} = -\frac{E_\perp^{(A)}}{B} \frac{\boldsymbol{\nabla}\Psi}{|\boldsymbol{\nabla}\Psi|} \;. \tag{11.4.19}$$

After multiplication of (11.4.18) with $E_\perp^{(A)}$, we find that the poloidal component of the electric field is given by

$$E_{\mathrm{P}}^{(A)} = -\frac{B_{\mathrm{T}}}{B} E_\perp^{(A)} \;. \tag{11.4.20}$$

With our conventions $E_\perp^{(A)}$ is positive (see also Fig. 11.4). The radial fluid velocity, in terms of the poloidal electric field is thus

$$\boldsymbol{v} \cdot \boldsymbol{\nabla}\Psi_{\mathrm{tor}} = E_{\mathrm{P}}^{(A)} \frac{|\boldsymbol{\nabla}\Psi_{\mathrm{tor}}|}{B_{\mathrm{T}}} \;. \tag{11.4.21}$$

In going from (11.4.18) and (11.4.19) to (11.4.21), we have converted Ψ into $\Psi_{\mathrm{tor}}/2\pi$. (We could have written down this result immediately from (11.4.4); multiply its RHS by $B_{\mathrm{P}}/B_{\mathrm{P}}$ and after replacing RB_{P} by $|\boldsymbol{\nabla}\Psi|$, set $E_{\mathrm{T}}^{(A)} = -(B_{\mathrm{P}}/B_{\mathrm{T}})E_{\mathrm{P}}^{(A)}$.) Note that $\boldsymbol{\nabla}\Psi$ in the relations (11.2.1), (11.2.3), (11.4.4) and (11.4.18), in fact, is the gradient of a poloidal flux (which is why we could set $|\boldsymbol{\nabla}\Psi| = RB_{\mathrm{P}}$). However, $\boldsymbol{\nabla}\Psi_{\mathrm{tor}} \propto \boldsymbol{\nabla}\Psi_{\mathrm{pol}}$, and the proportionality constant cancels in (11.4.21).

For the motion of the flux surfaces, we go back to (11.4.7), (11.4.8) and (11.4.9). Because $\boldsymbol{B} \cdot \boldsymbol{\nabla}\Psi_{\mathrm{tor}} \equiv 0$, (11.4.8) holds for $\Psi = \Psi_{\mathrm{tor}}/2\pi$. Noting that $\iota = |\boldsymbol{\nabla}\Psi_{\mathrm{pol}}^{\mathrm{r}}|/|\boldsymbol{\nabla}\Psi_{\mathrm{tor}}|$, the magnetic field $\boldsymbol{B}$, in terms of $\boldsymbol{\nabla}\Psi_{\mathrm{tor}}$ equals

$$\boldsymbol{B} = I\,\boldsymbol{\nabla}\zeta_{\mathrm{o}} + \frac{\iota}{2\pi}(\boldsymbol{\nabla}\zeta_{\mathrm{o}} \times \boldsymbol{\nabla}\Psi_{\mathrm{tor}}) \;. \tag{11.4.22}$$

If we go through the exercise of (11.4.9) for Ψ now replaced by $\Psi_{\mathrm{tor}}/2\pi$, and replace $(\boldsymbol{\nabla}\zeta_{\mathrm{o}} \times \boldsymbol{\nabla}\Psi_{\mathrm{tor}})$ by $2\pi(\boldsymbol{B} - I\,\boldsymbol{\nabla}\zeta_{\mathrm{o}})/\iota$, we obtain

$$-\boldsymbol{\nabla}\Psi_{\mathrm{tor}} \cdot (\boldsymbol{\nabla} \times \boldsymbol{E}) = -\frac{\boldsymbol{B}}{\iota} \cdot \boldsymbol{\nabla}(2\pi R E_{\mathrm{T}}) \;. \tag{11.4.23}$$

The combination of the adjusted (11.4.8) and (11.4.23) leads to

$$\boldsymbol{B} \cdot \boldsymbol{\nabla}\left(\frac{\partial \Psi_{\mathrm{tor}}}{\partial t} - 2\pi \frac{RE_{\mathrm{T}}}{\iota}\right) \equiv 0 \;. \tag{11.4.24}$$

Although the poloidal electric field component is completely generated by "reaction" currents, it is related to the net toroidal $\boldsymbol{E}$-field via $E_{\mathrm{P}}^{(A)} = -(B_{\mathrm{T}}/B_{\mathrm{P}})E_{\mathrm{T}}^{(A)}$. Equation (11.4.24) holds for E_{T} equal to both $E_{\mathrm{T}}^{(A)}$ and $E_{\mathrm{T}}^{(A),\mathrm{react}}$ because of the operator $\boldsymbol{B} \cdot \boldsymbol{\nabla}$. It gives, upon integration,

$$\frac{\partial \Psi_{\mathrm{tor}}}{\partial t} = -\frac{2\pi R B_{\mathrm{P}}}{B_{\mathrm{T}}\iota} E_{\mathrm{P}}^{(A)} \;. \tag{11.4.25}$$

For circular (idealized) flux surfaces, $\iota = 1/q = RB_P/rB_T$, where r is the minor radius. Thus with $RB_P/B_T\iota = r$, it is clear that (11.4.25) is in agreement with intuition.

The flux surface motion follows from $d\Psi_{tor}/dt \equiv 0$, expanded via the convective derivative,

$$U_s \cdot \nabla\Psi_{tor} = \frac{2\pi RB_P}{\iota B_T} E_P^{(A)} = \frac{|\nabla\Psi_{pol}^{d,r}|}{\iota} \frac{E_P^{(A)}}{B_T}$$

$$U_s \cdot \nabla\Psi_{tor} = |\nabla\Psi_{tor}| \frac{E_P^{(A)}}{B_T} . \tag{11.4.26}$$

Again, the flux-surface motion (11.4.26) equals the radial plasma motion given in (11.4.21). For mnemonic purposes, one should think of $|\nabla\Psi_{tor}|/B_T \sim 2\pi r$.

The physical picture is consistent. Plasma and flux surfaces, regardless of their labeling move radially inward together. By shrinking, the flux surfaces provide smaller areas for their (constant) contained fluxes. Consequently, both B_P and B_T increase (as we argued in Fig. 11.3). The increase of B_P within the plasma is the reason why the parallel current must also increase inside the plasma and not only on the surface. (Recall that a toroidal current on a toroidal surface of radius r_o influences only the magnetic field on the outside, for $r > r_o$, not that on the inside.)

11.5 Constancy of the Rotational-Transform Flux Function

From our discussion above, we learned that the magnetic fluxes within the flux surfaces remain constant and therefore that the concept of flux surfaces as flux tubes retains its physical meaning. This also implies that the poloidal and toroidal fluxes for an infinitesimal shell of flux volume remain constant. The toroidal and poloidal components of the magnetic field increase proportionally so that ι remains constant. Because of potential (notational) confusion when interpreting the literature, we shall now demonstrate the constancy of ι.

First, we need a relationship for the rotational transform in terms of the quantities appearing in the $\boldsymbol{B}$-field expression, (11.2.1). By definition, the rotational transform is

$$\iota = \frac{\delta\Psi_{pol}^{r}}{\delta\Psi_{tor}} = \frac{\nabla\Psi_{pol}^{r} \cdot d\boldsymbol{R}}{\nabla\Psi_{tor} \cdot d\boldsymbol{R}} . \tag{11.5.1}$$

Equation (11.5.1) stresses that ι is related to the spatial variations of Ψ_{pol}^{r} and Ψ_{tor}. The vector $d\boldsymbol{R}$ is an arbitrary differential position vector. The toroidal flux Ψ_{tor} equals

$$\begin{aligned}\Psi_{\text{tor}} &= \frac{1}{2\pi} \iiint\limits_V d^3R\, \boldsymbol{B} \cdot \nabla\zeta_o \\ &= \frac{1}{2\pi} \iiint\limits_V d^3R\, I |\nabla\zeta_o|^2 \\ &= \frac{1}{2\pi} \iiint\limits_V d^3R \frac{I}{R^2} \ .\end{aligned} \tag{11.5.2}$$

The differential toroidal flux $\delta\Psi_{\text{tor}}$ in terms of the poloidal ribbon flux label Ψ, i.e., $\delta\Psi = \delta\Psi_{\text{pol}}^r/2\pi$, is then

$$\begin{aligned}\delta\Psi_{\text{tor}} &= \frac{\delta\Psi}{2\pi} \int_0^{2\pi}\int_0^{2\pi} d\theta\, d\zeta_o \sqrt{g_o} \frac{I}{R^2} \\ &= \frac{\delta\Psi}{2\pi} I \int_0^{2\pi}\int_0^{2\pi} d\theta\, d\zeta_o \sqrt{g_o} \left(\frac{1}{R^2}\right) \ .\end{aligned} \tag{11.5.3}$$

Recall that $I \equiv I(\Psi)$. Since the flux-surface average of a quantity Φ is defined as

$$\langle\Phi\rangle \equiv \frac{\int_0^{2\pi}\int_0^{2\pi} d\theta\, d\zeta_o \sqrt{g_o}\,\Phi}{\int_0^{2\pi}\int_0^{2\pi} d\theta\, d\zeta_o \sqrt{g_o}} = \frac{\int_0^{2\pi}\int_0^{2\pi} d\theta\, d\zeta_o \sqrt{g_o}\,\Phi}{\delta V/\delta\Psi} \ , \tag{11.5.4}$$

we obtain for (11.5.3):

$$\frac{\delta\Psi_{\text{tor}}}{\delta\Psi} = \frac{1}{2\pi}\frac{\delta V}{\delta\Psi} I \langle R^{-2}\rangle \ . \tag{11.5.5}$$

With $q \equiv 1/\iota = \delta\Psi_{\text{tor}}/(2\pi\delta\Psi)$, we get for the safety factor:

$$q \equiv \frac{1}{\iota} = \frac{1}{4\pi^2}\frac{\delta V}{\delta\Psi}\langle IR^{-2}\rangle = \frac{I}{4\pi^2}\frac{\delta V}{\delta\Psi}\left\langle\frac{1}{R^2}\right\rangle \ . \tag{11.5.6}$$

The time rate of change of the safety factor q on a (moving) constant-Ψ flux surface can be related to its change at a fixed point $\boldsymbol{R}$ in space via the chain rule:

$$\left.\frac{\partial q}{\partial t}\right|_\Psi = \left.\frac{\partial q}{\partial t}\right|_R - \left.\frac{\partial q}{\partial\Psi}\right|_t \left.\frac{\partial\Psi}{\partial t}\right|_R = \left.\frac{\partial q}{\partial t}\right|_R - \left.\frac{\partial}{\partial\Psi}\right|_t \left[q \left.\frac{\partial\Psi}{\partial t}\right|_R\right] \ . \tag{11.5.7}$$

The last form follows from changing the order of differentiation in $\partial/\partial\Psi|_t\, \partial\Psi/\partial t|_R$ which makes this vanish. We now substitute the expression for q, given in (11.5.6) into the right side of (11.5.7). When we take into account that

$$\left.\frac{\partial}{\partial t}\right|_R \frac{\delta V}{\delta\Psi} = \frac{\delta}{\delta\Psi}\left.\frac{\partial V}{\partial t}\right|_R \equiv 0 \ , \tag{11.5.8}$$

because the volume of a flux surface through a fixed point $\boldsymbol{R}$ does not change, we obtain:

$$\left.\frac{\partial q}{\partial t}\right|_{\Psi} = \frac{1}{4\pi^2}\left\{\frac{\delta V}{\delta \Psi}\left\langle \left.\frac{\partial I R^{-2}}{\partial t}\right|_{\boldsymbol{R}}\right\rangle - \left.\frac{\partial}{\partial \Psi}\right|_{t}\left[\frac{\delta V}{\delta \Psi}\left\langle \frac{I}{R^2}\left.\frac{\partial \Psi}{\partial t}\right|_{\boldsymbol{R}}\right\rangle\right]\right\} \tag{11.5.9}$$

For the second term, we shall substitute (11.4.14). The first term can be obtained from the toroidal component of Faraday's law as follows:

$$\left\langle \nabla\zeta_o \cdot \left.\frac{\partial \boldsymbol{B}}{\partial t}\right|_{\boldsymbol{R}}\right\rangle = -\langle \nabla \times \boldsymbol{E} \cdot \nabla\zeta_o\rangle \;, \tag{11.5.10}$$

and thus

$$\left\langle \left.\frac{\partial}{\partial t}\right|_{\boldsymbol{R}} \nabla\zeta_o \cdot \boldsymbol{B}\right\rangle = -\langle \nabla\cdot(\boldsymbol{E} \times \nabla\zeta_o)\rangle \;. \tag{11.5.11}$$

In Sect. 4.9.3. we prove the relationship:

$$\langle \nabla \cdot \boldsymbol{W}\rangle = \frac{1}{\delta V/\delta \Psi}\left.\frac{\partial}{\partial \Psi}\right|_{t}\left(\frac{\delta V}{\delta \Psi}\langle \boldsymbol{W}\cdot \nabla\Psi\rangle\right) \;. \tag{11.5.12}$$

Therefore, (11.5.11) can be rewritten as:

$$\begin{aligned}\left\langle \left.\frac{\partial}{\partial t}\right|_{\boldsymbol{R}} (I R^{-2})\right\rangle &= -\frac{1}{\delta V/\delta \Psi}\left.\frac{\partial}{\partial \Psi}\right|_{t}\left[\frac{\delta V}{\delta \Psi}\langle \nabla\Psi\cdot \boldsymbol{E} \times \nabla\zeta_o\rangle\right] \\ &= -\frac{1}{\delta V/\delta \Psi}\left.\frac{\partial}{\partial \Psi}\right|_{t}\left[\frac{\delta V}{\delta \Psi}\langle \boldsymbol{E}\cdot \nabla\zeta_o \times \nabla\Psi\rangle\right]\end{aligned} \tag{11.5.13}$$

The quantity $\boldsymbol{E}\cdot \nabla\zeta_o \times \nabla\Psi$ is proportional to the magnitude of the poloidal electric field component (in Fig. 11.3). With $|\nabla\Psi| = RB_{\mathrm{P}}$ and $|\nabla\zeta_o| = 1/R$, we have that $\boldsymbol{E}\cdot \nabla\zeta_o \times \nabla\Psi = B_{\mathrm{P}}E_{\mathrm{P}}^{(A)} \equiv B_{\mathrm{P}}E_{\mathrm{P}}^{(A),\mathrm{react}}$. (Recall that $E_{\mathrm{P}}^{(A)} = -|E_{\mathrm{P}}^{(A)}|$.) After substitution in (11.5.9), we obtain

$$\begin{aligned}\left.\frac{\partial q}{\partial t}\right|_{\Psi} = \frac{1}{4\pi^2}&\left\{\left.\frac{\partial}{\partial \Psi}\right|_{t}\left[-\frac{\delta V}{\delta \Psi}\langle B_{\mathrm{P}}E_{\mathrm{P}}^{(A),\mathrm{react}}\rangle\right]\right. \\ &\left. -\left.\frac{\partial}{\partial \Psi}\right|_{t}\left[\frac{\delta V}{\delta \Psi}\left\langle \frac{I}{R}(E_{\mathrm{T}}^{(A),\mathrm{tr}} + E_{\mathrm{T}}^{(A),\mathrm{react}})\right\rangle\right]\right\} \;.\end{aligned} \tag{11.5.14}$$

Below (11.4.15) we saw that $E_{\mathrm{T}}^{(A),\mathrm{tr}} + E_{\mathrm{T}}^{(A),\mathrm{react}} = E_{\mathrm{T}}^{(A)}$ (see also Fig. 11.3), and below (11.4.21), we stated that $E_{\mathrm{T}}^{(A)} = -(B_{\mathrm{P}}/B_{\mathrm{T}})E_{\mathrm{P}}^{(A)}$. Taking into account that $I = RB_{\mathrm{T}}$ then leads to

$$\left.\frac{\partial q}{\partial t}\right|_{\Psi} = \frac{1}{4\pi^2}\left\{\left.\frac{\partial}{\partial \Psi}\right|_{t}\left[\frac{\delta V}{\delta \Psi}\langle -B_{\mathrm{P}}E_{\mathrm{P}}^{(A),\mathrm{react}}\rangle\right] + \left.\frac{\partial}{\partial \Psi}\right|_{t}\left[\frac{\delta V}{\delta \Psi}\langle B_{\mathrm{P}}E_{\mathrm{P}}^{(A)}\rangle\right]\right\}. \tag{11.5.15}$$

Or, since $E_{\mathrm{P}}^{(A)} \equiv E_{\mathrm{P}}^{(A),\mathrm{react}}$,

$$\left.\frac{\partial q}{\partial t}\right|_{\Psi} \equiv 0 \ . \qquad (11.5.16)$$

as argued intuitively above.

The reader might get a false impression when consulting the literature. In most cases, the above analysis is performed for a finite-conductivity plasma (see, e.g., Stacey (1981) or Hirshman and Sigmar (1981)). Instead of identifying $\boldsymbol{E}\cdot\boldsymbol{\nabla}\zeta_{\mathrm{o}}\times\boldsymbol{\nabla}\Psi$ in (11.5.13) with $B_{\mathrm{P}}E_{\mathrm{P}}^{(A)}$, we could have replaced $\boldsymbol{\nabla}\zeta_{\mathrm{o}}\times\boldsymbol{\nabla}\Psi$ by $(\boldsymbol{B}-I\boldsymbol{\nabla}\zeta_{\mathrm{o}})$. Then (11.5.14) would have been replaced by

$$\left.\frac{\partial q}{\partial t}\right|_{\Psi} = \frac{1}{4\pi^2}\left\{-\left.\frac{\partial}{\partial\Psi}\right|_t\left[\frac{\delta V}{\delta\Psi}\langle -E_{\parallel}^{(A),\mathrm{react}}B\rangle + \frac{\delta V}{\delta\Psi}\left\langle IR^{-2}\frac{e(t)}{2\pi}\right\rangle\right]\right\} \ . \qquad (11.5.17)$$

Now, for instance, Stacey (1981) does not clarify the term $E_{\parallel}^{(A)}$ further, although it is understood that it is our $E_{\parallel}^{(A),\mathrm{react}}$. One might get the impression that only his $\langle E_{\parallel}^{(A)}B\rangle$ term must be set to zero as $\sigma\to\infty$, while the "ubiquitous" $e(t)$ survives, erroneously leading to a nonzero $\partial q/\partial t|_{\Psi}$. When interpreting (11.5.17), we must realize that the *total* parallel electric field must vanish. Since we have that

$$\frac{e(t)}{2\pi} = RE_{\mathrm{T}}^{(A),\mathrm{tr}} = R\boldsymbol{E}_{\mathrm{T}}^{(A),\mathrm{tr}}\cdot\hat{\boldsymbol{\zeta}}_{\mathrm{o}} = R^2\boldsymbol{E}_{\mathrm{T}}^{(A),\mathrm{tr}}\cdot\boldsymbol{\nabla}\zeta_{\mathrm{o}}$$

$$= R^2\boldsymbol{E}_{\mathrm{T}}^{(A),\mathrm{tr}}\cdot(\boldsymbol{B}-\boldsymbol{\nabla}\zeta_{\mathrm{o}}\times\boldsymbol{\nabla}\Psi)/I = (R^2/I)E_{\parallel}^{(A),\mathrm{tr}}B \ ,$$

the parallel $\boldsymbol{E}$-field contribution of the transformer voltage must vanish as well in our ideal plasma.

11.6 Remarks on the Evolution of a Finite-Resistivity Plasma

When a *finite-resistivity plasma* ($\sigma \neq \infty$) is considered, the frozen-flux theorem no longer holds, and the plasma and flux surfaces "leak through" each other: the flux surfaces move inward faster than the plasma (the flux surfaces still obey (11.4.13), whereas the plasma follows (11.4.3)). Note, however, that in the tokamak neoclassical banana regime, the radial inward pinch velocity, now called the neoclassical or Ware pinch, approximately equals the radially inward velocity of the flux surfaces. Thus even in the presence of collisions and therefore resistivity, the plasma and flux surfaces can remain tied together. For details, see Hinton and Hazeltine (1976, Chap. 3). Also, in a finite resistivity plasma, the parallel current does not keep increasing but saturates at a value equal to $\sigma E_{\parallel}^{(A)}$. This can be understood qualitatively from an analogy with a circular conductor as in Fig. 11.2. When $\sigma\to\infty$, the current follows the flux change $|d\Psi/dt| = L|di/dt|$, where L is the self-inductance. When σ is finite, the current levels off to the above-mentioned value as $(1-\mathrm{e}^{-t/\tau})$ with the time constant $\tau = L/R$.

A more detailed discussion of the plasma dynamics in resistive plasmas is beyond the scope of this book. To do a proper quantitative analysis, the equilibrium and transport equations must be considered simultaneously, and the problem must be solved self-consistently. Interesting considerations concerning classical and neoclassical transport are discussed in Hinton and Hazeltine (1976) and Hirshman and Sigmar (1981); see also, the other references given in Chap. 4, above (4.7.7).

11.7 The Relationship Between Poloidal Disk and Ribbon Fluxes

Having considered the motion of an ideal plasma and flux surfaces embedded in it, we may say a few final words about the relationship between the poloidal disk and ribbon fluxes. In principle, the expression for the magnetic field given in (11.2.1), $\boldsymbol{B} = I\boldsymbol{\nabla}\zeta_0 + \boldsymbol{\nabla}\zeta_0 \times \boldsymbol{\nabla}\Psi$, makes most physical sense for Ψ proportional to the poloidal flux inside the plasma:

$$\Psi \equiv \frac{\Psi^{\mathrm{r}}_{\mathrm{pol}}}{2\pi} \ .$$

From Fig. 11.3, we see that the poloidal disk flux is related to the ribbon flux as follows:

$$\Psi^{\mathrm{d}}_{\mathrm{pol}} = \Psi^{\mathrm{total}}_{\mathrm{pol}} - \Psi^{\mathrm{r}}_{\mathrm{pol}} \ . \tag{11.7.1}$$

The total poloidal flux, represented by $\Psi^{\mathrm{total}}_{\mathrm{pol}}$ is the flux through a disk bounded by the magnetic axis. It contains the transformer flux and the flux generated by the plasma currents.

Above we showed that the time rate of change of the poloidal ribbon flux and the poloidal disk flux through the ribbon and disk, respectively, constructed through a *fixed* point A (on the line O-MA, where MA stands for magnetic axis) are exactly opposite. Compare (11.4.12) and (11.4.14). This implies that the total poloidal flux remains constant. From the *moving* flux surface picture, we can draw the same conclusions: poloidal disk and ribbon fluxes are each conserved through their enlarging and shrinking areas, respectively, again establishing that the poloidal flux is constant. The above points out that

$$\frac{\partial \Psi^{\mathrm{total}}_{\mathrm{pol}}}{\partial t} \equiv 0 \ . \tag{11.7.2}$$

Because $\Psi^{\mathrm{total}}_{\mathrm{pol}}$ is merely a constant, it was not mentioned previously. Indeed, all we are concerned with are *derivatives* of fluxes. It did not appear in the time derivatives because of (11.7.2), nor in the gradient expressions, since $\boldsymbol{\nabla}\Psi^{\mathrm{total}}_{\mathrm{pol}} = 0$. For this reason, we could identify $\Psi \equiv \Psi^{\mathrm{r}}_{\mathrm{pol}}/2\pi$ in (11.2.1) with $-\Psi^{\mathrm{d}}_{\mathrm{pol}}/2\pi$ and not as in (11.7.1). We must have $\boldsymbol{\nabla}\Psi = \boldsymbol{\nabla}\Psi^{\mathrm{r}}_{\mathrm{pol}}/2\pi = -\boldsymbol{\nabla}\Psi^{\mathrm{d}}_{\mathrm{pol}}/2\pi$, and so we may ignore the constant $\Psi^{\mathrm{total}}_{\mathrm{pol}}$.

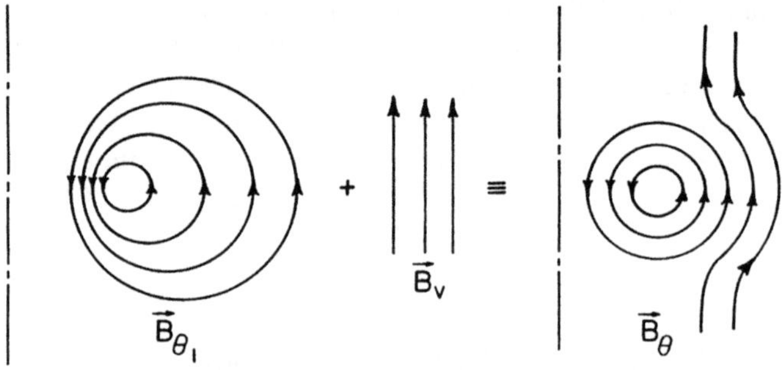

Fig. 11.5. The net poloidal magnetic field $\boldsymbol{B}_\theta$ is the sum of the vertical field $\boldsymbol{B}_v$ and the field $\boldsymbol{B}_{\theta_1}$

In our discussion above, the vertical magnetic-field component, which is needed for equilibrium with respect to horizontal motion, has not been mentioned. This is because the vertical field is merely part of the net poloidal field in the sense that it reduces the original poloidal field (due to the toroidal plasma current component) on the inside of the torus and strengthens it on the outside (see Fig. 11.5 for an idealized case). It is the *net* poloidal field including the vertical field that determines the position and the shape of the flux surfaces, which are "centered" – not necessarily concentrically – about the magnetic axis. This is in fact the position where the *net* vertical field vanishes.

The vertical field is varied in order to achieve a desired position of the plasma column. The required field has been calculated for tokamaks by Shafranov (1966), Shafranov and Zakharov (1972), and Greene, Johnson and Weimer (1971). See also Zakharov and Shafranov (1986). Formulae analogous to those of Greene, Johnson and Weimer (1971) were obtained for a stellarator by Fielding and Hitchon (1980). For elementary textbook discussions on the necessity of a vertical or "maintaining" field, see e.g., Bateman (1978); Golant, Zhilinsky and Sakharov (1980) or Stacey (1981). From the point of view of this discussion, the vertical field is not distinguishable from the other poloidal fluxes; the total poloidal field is still "frozen in".

In this Chapter we have idealized and neglected many influences concerning the vertical or "maintaining" field. For an overview of some of these potentially important effects, see the review papers by Mukhovatov and Shafranov (1971) and Zakharov and Shafranov (1986).

In most discussions concerning flux coordinates, this dynamic analysis is of secondary importance. One pretends that the flux surfaces are stationary and does not mention the time dependence. When we are interested in only the spatial dependences, time is merely a parameter, and we can establish the geometrical relationships at any time t.

12. The Relationship Between $\int dl/B$ and $dV/d\Psi_{\text{tor}}$

In many elementary discussions it is "proved" in an intuitive manner that the specific volume $\dot{V}(\Psi_{\text{tor}}) \equiv dV/d\Psi_{\text{tor}}$ equals the closed line integral $\oint dl/B$. That this is true only in an approximate sense, and that a distinction is to be made between rational and irrational surfaces is discussed in the authoritative review paper by Solov'ev and Shafranov (1970). For completeness, we reproduce their arguments in this Chapter albeit converted to our notation.

To evaluate $\int dl/B$, we make use of the equation of a field line in flux coordinates

$$\frac{dl}{B} = \frac{d\zeta_{\text{f}}}{\boldsymbol{B}\cdot\nabla\zeta_{\text{f}}} = \frac{d\zeta_{\text{f}}}{B^{\zeta_{\text{f}}}} = d\zeta_{\text{f}}\left(\frac{2\pi\sqrt{g_{\text{f}}}}{\dot{\Psi}_{\text{tor}}}\right) . \tag{12.1.1}$$

Recall that $\dot{\Psi}_{\text{tor}} = d\Psi_{\text{tor}}/d\varrho$. After multiplication in the numerator and the denominator by $dV/d\varrho$, we have

$$\int\frac{dl}{B} = \frac{dV}{d\Psi_{\text{tor}}}2\pi\frac{\int\sqrt{g_{\text{f}}}d\zeta_{\text{f}}}{dV/d\varrho} . \tag{12.1.2}$$

In Chap. 4, we explained that

$$\frac{dV}{d\varrho} = \int_0^{2\pi}\int_0^{2\pi}\sqrt{g_{\text{f}}}d\theta_{\text{f}}d\zeta_{\text{f}} . \tag{12.1.3}$$

If $\sqrt{g_{\text{f}}}$ is expanded in a double Fourier series in θ_{f} and ζ_{f} as

$$\sqrt{g_{\text{f}}} = \sum_{m,n} g^{\text{f}}_{mn}(\varrho)\mathrm{e}^{\mathrm{i}(m\theta_{\text{f}}-n\zeta_{\text{f}})} , \tag{12.1.4}$$

we obtain after substitution and integration in (12.1.3)

$$\frac{dV}{d\varrho} = 4\pi^2 g^{\text{f}}_{00}(\varrho) . \tag{12.1.5}$$

The integral of (12.1.2) becomes, upon Fourier expansion,

$$\begin{aligned}\int \sqrt{g_{\mathrm{f}}}\,d\zeta_{\mathrm{f}} &= \int \sum_{m,n} g^{\mathrm{f}}_{mn}(\varrho)\mathrm{e}^{\mathrm{i}(m\theta_{\mathrm{f}}-n\zeta_{\mathrm{f}})}d\zeta_{\mathrm{f}} \\ &= g^{\mathrm{f}}_{00}(\varrho)\zeta_{\mathrm{f}} + \int \sum_{m,n}{}' g^{\mathrm{f}}_{mn}(\varrho)\mathrm{e}^{\mathrm{i}(m\theta_{\mathrm{f}}-n\zeta_{\mathrm{f}})}d\zeta_{\mathrm{f}} \\ &= \frac{1}{4\pi^2}\frac{dV}{d\varrho}\zeta_{\mathrm{f}} + \int \sum_{m,n}{}' g^{\mathrm{f}}_{mn}(\varrho)\mathrm{e}^{\mathrm{i}(m\theta_{\mathrm{f}}-n\zeta_{\mathrm{f}})}d\zeta_{\mathrm{f}}\ . \end{aligned} \tag{12.1.6}$$

The prime on the summation sign indicates that the term $m = n = 0$ must be omitted. Combination of (12.1.2), (12.1.5) and (12.1.6) results in

$$\int \frac{dl}{B} = \frac{dV}{d\Psi_{\text{tor}}}\left\{\frac{1}{2\pi}\zeta_{\mathrm{f}} + \frac{1}{2\pi g^{\mathrm{f}}_{00}(\varrho)}\int \sum_{m,n}{}' g^{\mathrm{f}}_{mn}(\varrho)\mathrm{e}^{\mathrm{i}(m\theta_{\mathrm{f}}-n\zeta_{\mathrm{f}})}d\zeta_{\mathrm{f}}\right\}\ . \tag{12.1.7}$$

We now define what is called the "proper length" of a field line:

$$U(\varrho,\theta_{\mathrm{f}0}) \equiv \frac{1}{N}\int_0^{2\pi N}\frac{d\zeta_{\mathrm{f}}}{B^{\zeta_{\mathrm{f}}}}\ . \tag{12.1.8}$$

This integral is taken along a field line $\theta_{\mathrm{f}0} = \theta_{\mathrm{f}} - \iota\zeta_{\mathrm{f}}$, from $\zeta_{\mathrm{f}} = 0$, to $\zeta_{\mathrm{f}} = 2\pi N$. After performing the integral in (12.1.7), we have thus

$$U(\varrho,\theta_{\mathrm{f}0}) = \frac{dV}{d\Psi_{\text{tor}}}\left\{1 + \frac{1}{2\pi N g^{\mathrm{f}}_{00}(\varrho)}\sum_{m,n}{}' g^{\mathrm{f}}_{mn}(\varrho)\mathrm{e}^{\mathrm{i}m\theta_{\mathrm{f}0}}\left[\frac{\mathrm{e}^{\mathrm{i}(m\iota-n)2\pi N}-1}{\mathrm{i}(m\iota-n)}\right]\right\}\ . \tag{12.1.9}$$

When the field line lies on a rational surface, it closes upon itself after, say, n_{o} circuits of the minor axis and m_{o} toroidal circuits. For that field line,

$$\iota(\varrho) = \frac{n_{\mathrm{o}}}{m_{\mathrm{o}}}\ . \tag{12.1.10}$$

For this type of field line, $N \equiv m_{\mathrm{o}}$. Making use of the fact that

$$\lim_{x\to 0}\frac{\mathrm{e}^{\mathrm{i}ax}-1}{\mathrm{i}x} = a\ , \tag{12.1.11}$$

we have then with $a = 2\pi N$:

$$\begin{aligned}U(\varrho,\theta_{\mathrm{f}0}) &= \frac{1}{m_{\mathrm{o}}}\oint\frac{dl}{B} \\ &= \frac{dV}{d\Psi_{\text{tor}}}\left\{1 + \frac{1}{g^{\mathrm{f}}_{00}(\varrho)}\left[g_{m_{\mathrm{o}}n_{\mathrm{o}}}\mathrm{e}^{\mathrm{i}m_{\mathrm{o}}\theta_{\mathrm{f}0}} + g_{-m_{\mathrm{o}},-n_{\mathrm{o}}}\mathrm{e}^{-\mathrm{i}m_{\mathrm{o}}\theta_{\mathrm{f}0}}\right]\right\}\ .\end{aligned} \tag{12.1.12}$$

From this expression, it is clear that the integral $\oint dl/B$ taken along different field lines on the same rational magnetic surface will, in general, assume different values, depending on the value of the field-line label $\theta_{\mathrm{f}0}$. (Recall that $\theta_{\mathrm{f}0}$ is equivalent to v or β in the Clebsch representation). In fact, depending on the particular field-line label, U will oscillate about $dV/d\Psi_{\text{tor}}$. The fact that $\oint dl/B$ is not always constant on a rational surface has serious consequences as was mentioned above.

For the case of irrational surfaces, we observe that $\int dl/B$ integrated from $\zeta_{\text{f}} = 0$ to $\zeta_{\text{f}} = 2\pi$ should not be very different from the value of $U(\varrho, \theta_{\text{f}0})$ on the nearby rational surface characterized by $\iota = n_o/m_o$. For a large number of circuits, $N \to \infty$, the function $U(\varrho, \theta_{\text{f}0})$ is averaged, with the result:

$$\lim_{N\to\infty} \frac{1}{N} \int \frac{dl}{B} = \frac{dV}{d\Psi_{\text{tor}}} \; . \tag{12.1.13}$$

13. Transformation Properties of Vector and Tensor Components

In this Chapter we wish to investigate how the *components* of vectors and tensors change when another coordinate system is chosen. We want to find out what happens to the components under a general transformation of the form $u^{i'} \equiv u^{i'}(u^1, u^2, u^3)$ for $i = 1, 2, 3$. We find it convenient to use primes "*on*" *the indices* for quantities related to the new system; here, $u^{i'}$, rather than putting a prime or a bar on the u symbol, such as u'^i or $\bar{u}^i$. This is for notational purposes, since the indices then will match on the RHS and LHS of the equations. This will become clear below.

13.1 Transformation of the Basis Vectors

The first task is to find expressions for the basis vectors in the new coordinate system in terms of the basis vectors in the old one. Let there be two curvilinear coordinate systems (u^1, u^2, u^3) and $(u^{1'}, u^{2'}, u^{3'})$. Both systems are defined by transformation equations of the form

$$u^j = u^j(x, y, z) \tag{13.1.1a}$$

$$u^{i'} = u^{i'}(x, y, z) \ . \tag{13.1.1b}$$

There then exists a transformation directly from u^j to $u^{i'}$

$$\begin{aligned} u^{1'} &= u^{1'}(u^1, u^2, u^3) \\ u^{2'} &= u^{2'}(u^1, u^2, u^3) \\ u^{3'} &= u^{3'}(u^1, u^2, u^3) \end{aligned} \tag{13.1.2}$$

or

$$u^{i'} = u^{i'}(u^1, u^2, u^3) \ , \qquad \text{or} \qquad u^{i'} = u^{i'}(u^j) \tag{13.1.3}$$

and vice versa:

$$u^j = u^j(u^{1'}, u^{2'}, u^{3'}) \ , \qquad \text{or} \qquad u^j = u^j(u^{i'}) \ . \tag{13.1.4}$$

(The only limitation we impose on these transformations is that the Jacobian is nonzero.) We define the local basis vectors in each system as usual:

$$e_j = \frac{\partial R(u^1, u^2, u^3)}{\partial u^j} \tag{13.1.5}$$

and

$$e_{i'} = \frac{\partial R(u^{1'}, u^{2'}, u^{3'})}{\partial u^{i'}} \ . \tag{13.1.6}$$

Since we can think of the dependence of R on $u^{i'}$ (as in $R(u^{1'}, u^{2'}, u^{3'})$) as being established via the u^j, i.e., $R(u^j(u^{i'}))$, we can apply the chain rule of calculus to (13.1.6) as follows:

$$e_{i'} = \frac{\partial R(u^{i'})}{\partial u^{i'}} = \frac{\partial R(u^j(u^{i'}))}{\partial u^{i'}} = \frac{\partial R}{\partial u^j}\frac{\partial u^j}{\partial u^{i'}} \tag{13.1.7}$$

so that we have with definition (13.1.5)

$$e_{i'} = e_j \frac{\partial u^j}{\partial u^{i'}} \ , \tag{13.1.8}$$

where a summation over j is understood. Observe that the indices on both sides of the equation "line up" as usual. Equation (13.1.8) is called the "direct" transformation (from old to new). Application of the chain rule to the RHS of (13.1.5) leads to the coefficients for the "inverse" transformation

$$e_j = e_{i'} \frac{\partial u^{i'}}{\partial u^j} \ . \tag{13.1.9}$$

The summation is over i, not over i'(i' does not exist by itself); the prime is simply attached to i to denote transformed quantities. If we would sum over i', this would be a dummy index, and we could replace it by another, e.g., an unprimed index.

Analogous operations can be applied to the reciprocal-basis vectors $e^i = \nabla u^i$ and $e^{j'} = \nabla u^{j'}$. Via the chain rule, we obtain

$$e^{i'} = e^j \frac{\partial u^{i'}}{\partial u^j} \tag{13.1.10}$$

and

$$e^j = e^{i'} \frac{\partial u^j}{\partial u^{i'}} \ . \tag{13.1.11}$$

13.2 Transformation of Vector Components

In order to find the relationship between the contravariant components of a vector $A = A^j e_j = A^{i'} e_{i'}$ in different coordinate systems, we substitute (13.1.9) into $A^j e_j$ and equate it to $A^{i'} e_{i'}$. This gives

$$A^{i'} = A^j \frac{\partial u^{i'}}{\partial u^j} . \tag{13.2.1}$$

The inverse expression follows from substituting (13.1.8) into $A^{i'} \boldsymbol{e}_{i'}$ and equating it to $A^j \boldsymbol{e}_j$

$$A^j = A^{i'} \frac{\partial u^j}{\partial u^{i'}} . \tag{13.2.2}$$

A summation over repeated indices is implied, and the remaining index on the RHS is carried over to the LHS.

Objects that transform in the same fashion (having primed and non-primed indices in the same place) are said to transform *cogrediently* (Mathews and Walker 1970; Springer 1962; Behnke 1974; Menzel 1961). Objects that transform the other way around transform *contragrediently.* It is now *convention* to name the transformation properties after the similarities with those of the *tangent-basis vectors* $\boldsymbol{e}_i$. Objects that transform *co*grediently to the tangent-basis vectors (i.e., similarly to the $\boldsymbol{e}_i$) are called *covariant* objects. Those that transform *contra*grediently to the tangent-basis vectors (i.e., differently from the $\boldsymbol{e}_i$) are *contravariant* objects.

A comparison of (13.2.1) and (13.2.2) with (13.1.8) and (13.1.9) respectively, shows that the primes in the partial deriviates are in the "wrong" places. Therefore, the components A^i transform contragrediently to $\boldsymbol{e}_i$ and are therefore called *contravariant components.*

To find the transformation properties of the components of $\boldsymbol{A}$ which have a subscript, we start again from $\boldsymbol{A} = A_j \boldsymbol{e}^j$ with $A_j = \boldsymbol{A} \cdot \boldsymbol{e}_j$. We now dot multiply this expression for $\boldsymbol{A}$ by $\boldsymbol{e}_{i'}$ in order to obtain

$$A_{i'} = A_j \boldsymbol{e}^j \cdot \boldsymbol{e}_{i'} = A_j \boldsymbol{e}^j \cdot \boldsymbol{e}_k \frac{\partial u^k}{\partial u^{i'}} . \tag{13.2.3}$$

The scalar product of the basis vectors is a Kronecker delta δ_k^j so that

$$A_{i'} = A_j \frac{\partial u^j}{\partial u^{i'}} . \tag{13.2.4}$$

Similarly, the inverse transformation is

$$A_j = A_{i'} \frac{\partial u^{i'}}{\partial u^j} . \tag{13.2.5}$$

Equations (13.2.4) and (13.2.5) use primes in the same places as the tangent-basis vectors of (13.1.8) and (13.1.9), explaining the name *covariant* components for A_i.

These transformation properties can be used as a *definition* of "vectors". A "vector" could be defined as an object whose components transform in the way explained above. Depending on which transformation property applies, we talk about its co- or contravariant components. This definition is probably the most familiar to physicists and engineers.

13.3 Transformation of Tensor Components

The transformation rules given below are limited to second-order tensors. Generalization to higher order is straightforward. We shall only state the transformations because their derivation is completely analogous to that for vectors. Even better, we can simply write down the answer if we look carefully at the indices.

The *covariant* components F_{ij} of a tensor $F_{ij}e^i e^j$ transform as:

$$F_{i'j'} = F_{mn}\frac{\partial u^m}{\partial u^{i'}}\frac{\partial u^n}{\partial u^{j'}} \tag{13.3.1}$$

$$F_{ij} = F_{m'n'}\frac{\partial u^{m'}}{\partial u^i}\frac{\partial u^{n'}}{\partial u^j} \ . \tag{13.3.2}$$

For the *contravariant* components, we have

$$F^{i'j'} = F^{mn}\frac{\partial u^{i'}}{\partial u^m}\frac{\partial u^{j'}}{\partial u^n} \tag{13.3.3}$$

$$F^{ij} = F^{m'n'}\frac{\partial u^i}{\partial u^{m'}}\frac{\partial u^j}{\partial u^{n'}} \ . \tag{13.3.4}$$

There is no distinction between the transformation properties of the two sorts of *mixed* components. We just need mixed components of the same sort on both sides of the equation:

$$F^{i'}_{\bullet j'} = F^{i}_{\bullet j}\frac{\partial u^{i'}}{\partial u^i}\frac{\partial u^j}{\partial u^{j'}} \quad \text{or} \quad F_{j'}^{\bullet i'} = F_j^{\bullet i}\frac{\partial u^j}{\partial u^{j'}}\frac{\partial u^{i'}}{\partial u^i} \tag{13.3.5}$$

$$F^{i}_{\bullet j} = F^{i'}_{\bullet j'}\frac{\partial u^i}{\partial u^{i'}}\frac{\partial u^{j'}}{\partial u^j} \quad \text{or} \quad F_j^{\bullet i} = F_{j'}^{\bullet i'}\frac{\partial u^{j'}}{\partial u^j}\frac{\partial u^i}{\partial u^{i'}} \ . \tag{13.3.6}$$

13.4 Transformation of Components of Special Tensors and Symbols

13.4.1 The Kronecker Delta

We show now that the Kronecker delta written as δ^i_j represents the mixed components of the identity tensor. We have, according to (13.3.5),

$$\delta^{i'}_{j'} = \delta^m_n\frac{\partial u^{i'}}{\partial u^m}\frac{\partial u^n}{\partial u^{j'}} = \frac{\partial u^{i'}}{\partial u^m}\frac{\partial u^m}{\partial u^{j'}} = \frac{\partial u^{i'}}{\partial u^{j'}} = \delta^{i'}_{j'} \ . \tag{13.4.1}$$

Here we have used the chain rule. This shows that the Kronecker delta, having the values of either 1 or 0, remains unchanged in another coordinate system.

The symbols δ_{ij} and δ^{ij} being equal to 1 if the indices coincide and zero otherwise can likewise be shown *not* to qualify as components of a tensor. As shown in Chap. 3, the covariant and contravariant counterparts of δ^i_j are the metric coefficients g_{ij} and g^{ij}, all being components of one and the same "fundamental tensor". The transformation properties of the metric coefficients are shown momentarily.

13.4.2 Metric Coefficients

To show that the g_{ij} are legitimate components of a tensor, we make use of the property that the arc length is an invariant quantity not influenced by a change in coordinates. We must have that

$$(dl)^2 = g_{ij}du^i du^j = g_{m'n'}du^{m'}du^{n'} . \tag{13.4.2}$$

We apply the chain rule to the primed differentials as follows:

$$du^{m'} = \frac{\partial u^{m'}}{\partial u^i}du^i \qquad \text{and} \qquad du^{n'} = \frac{\partial u^{n'}}{\partial u^j}du^j . \tag{13.4.3}$$

This gives for (13.4.2)

$$(dl)^2 = g_{ij}du^i du^j = g_{m'n'}\frac{\partial u^{m'}}{\partial u^i}\frac{\partial u^{n'}}{\partial u^j}du^i du^j \tag{13.4.4}$$

or

$$g_{ij} = g_{m'n'}\frac{\partial u^{m'}}{\partial u^i}\frac{\partial u^{n'}}{\partial u^j} , \tag{13.4.5}$$

which is exactly the transformation law for covariant components.

13.4.3 Levi-Civita Symbols

To show that the Levi-Civita symbols ε_{ijk} and ε^{ijk}, which have a value $+1$, -1 or 0, are *not* tensor components, we assume for a moment that they are legitimate components and apply the tensor transformation rules. We will then find a contradiction. By way of example, we treat only ε_{ijk}. Its transformation properties would be

$$\varepsilon_{l'm'n'} = \varepsilon_{ijk}\frac{\partial u^i}{\partial u^{l'}}\frac{\partial u^j}{\partial u^{m'}}\frac{\partial u^k}{\partial u^{n'}} . \tag{13.4.6}$$

Since the RHS is summed over i, j and k, and because ε_{ijk} assumes only values $+1$, -1 and zero, there are nine terms. If we look carefully at that sum, we observe that it equals the Jacobian of the transformation for l, m and n equal to any even permutation of 1, 2, 3; it equals minus the Jacobian for l, m, n an odd permutation and is zero otherwise. Therefore, we have

$$\varepsilon_{l'm'n'} = \varepsilon_{lmn} \frac{\partial(u^1, u^2, u^3)}{\partial(u^{1'}, u^{2'}, u^{3'})} \tag{13.4.7}$$

in the notation of Chap. 2. There is a rule similar to the chain rule for Jacobians:

$$\frac{\partial(u^1, u^2, u^3)}{\partial(u^{1'}, u^{2'}, u^{3'})} = \frac{\partial(u^1, u^2, u^3)}{\partial(x, y, z)} \frac{\partial(x, y, z)}{\partial(u^{1'}, u^{2'}, u^{3'})} = \frac{\sqrt{g'}}{\sqrt{g}} . \tag{13.4.8}$$

Hence we obtain for (13.4.7)

$$\varepsilon_{l'm'n'} = \varepsilon_{lmn} \frac{\sqrt{g'}}{\sqrt{g}} . \tag{13.4.9}$$

This shows that for $\varepsilon_{lmn} = +1, -1$ or 0, the transformed component would be equal to $\pm\sqrt{g'}/\sqrt{g}$ and zero instead of $\pm 1, 0$. Therefore, the ε_{ijk} are not components of a tensor, as they do not transform properly. The argument for ε^{ijk} is analogous. Note that in (13.4.9), our rule of lining up indices no longer works. The reason is that we introduced a new symbol $\sqrt{g'}$ and we attached the prime to that.

The reader can check easily that the components of the "permutation tensor", $E_{ijk} = \varepsilon_{ijk}\sqrt{g}$ and $E^{ijk} = \varepsilon^{ijk}/\sqrt{g}$ do transform as required.

13.4.4 Christoffel Symbols

The Christoffel symbols were defined in Sect. 2.6.1c as

$$\begin{Bmatrix} j \\ i \; k \end{Bmatrix} = \frac{1}{2} g^{jn} \left[\frac{\partial g_{ni}}{\partial u^k} + \frac{\partial g_{nk}}{\partial u^i} - \frac{\partial g_{ik}}{\partial u^n} \right] \tag{13.4.10}$$

and

$$[j, ik] = \frac{1}{2} \left[\frac{\partial g_{ji}}{\partial u^k} + \frac{\partial g_{jk}}{\partial u^i} - \frac{\partial g_{ik}}{\partial u^j} \right] . \tag{13.4.11}$$

These symbols are *not* tensor components in general. We leave its proof as an exercise, because it is found in many textbooks on mathematical physics. We do give a few hints though as well as the answer. Compute $\partial g_{m'l'}/\partial u^{n'}$ where $g_{m'l'}$ is given by the inverse relationship of (13.4.5). The product rule of differentiation is applied, and the term involving $\partial g_{ji}/\partial u^{n'}$ is expanded via the chain rule as $(\partial g_{ji}/\partial u^k)(\partial u^k/\partial u^{n'})$. Doing this for all the terms in (13.4.11) leads to

$$[m', l'n'] = \frac{\partial u^j}{\partial u^{m'}} \frac{\partial u^i}{\partial u^{l'}} \frac{\partial u^k}{\partial u^{n'}} [j, ik] + \frac{\partial^2 u^i}{\partial u^{l'} \partial u^{n'}} \frac{\partial u^k}{\partial u^{m'}} g_{ik} . \tag{13.4.12}$$

For the other symbol, one also uses the transformed equation for the g^{jn}. This gives after use of the Kronecker delta:

$$\left\{ \begin{matrix} s' \\ r' \; t' \end{matrix} \right\} = \frac{\partial u^{s'}}{\partial u^j} \frac{\partial u^i}{\partial u^{r'}} \frac{\partial u^k}{\partial u^{t'}} \left\{ \begin{matrix} j \\ i \; k \end{matrix} \right\} + \frac{\partial^2 u^i}{\partial u^{r'} \partial u^{t'}} \frac{\partial u^{s'}}{\partial u^i} . \tag{13.4.13}$$

It is because of the second terms, which are, in general, nonzero that the symbols are not tensor components.

14. Alternative Derivations of the Divergence Formula

In this chapter, we discuss two alternative derivations for the divergence formula of (2.6.39). The one found most frequently in the applied mathematics literature involves Christoffel symbols. This derivation is rather remote from the usual plasma physics calculations, however. Nevertheless, for completeness and cohesiveness of our treatment, we include it as a last case. The method given first is the most natural.

14.1 A Straightforward Derivation of the Divergence Formula

Use the product rule of differentiation applied to $\boldsymbol{A} = A^i \boldsymbol{e}_i$ as follows

$$\begin{aligned} \operatorname{div} \boldsymbol{A} \equiv \boldsymbol{\nabla}\cdot\boldsymbol{A} = \boldsymbol{\nabla}\cdot(A^i\boldsymbol{e}_i) &= \boldsymbol{e}_i\cdot\boldsymbol{\nabla}A^i + A^i\boldsymbol{\nabla}\cdot\boldsymbol{e}_i \\ &= \boldsymbol{e}_i\cdot\boldsymbol{\nabla}u^k\frac{\partial A^i}{\partial u^k} + A^i\boldsymbol{\nabla}\cdot\boldsymbol{e}_i\ . \end{aligned} \tag{14.1.1}$$

Here, we have applied the definition of the $\boldsymbol{\nabla}$ operator to the $\boldsymbol{\nabla}A^i$ part. Thus we can write for the divergence

$$\operatorname{div} \boldsymbol{A} \equiv \boldsymbol{\nabla}\cdot\boldsymbol{A} = \frac{\partial A^k}{\partial u^k} + \{\boldsymbol{\nabla}\cdot\boldsymbol{e}_i\}A^i\ . \tag{14.1.2}$$

We now find the divergences of the basis vectors $\boldsymbol{e}_i$. Applying the definition of the $\boldsymbol{\nabla}$ operator gives:

$$\{\boldsymbol{\nabla}\cdot\boldsymbol{e}_i\} = \boldsymbol{e}^k\cdot\frac{\partial \boldsymbol{e}_i}{\partial u^k} = \boldsymbol{e}^1\cdot\frac{\partial \boldsymbol{e}_i}{\partial u^1} + \boldsymbol{e}^2\cdot\frac{\partial \boldsymbol{e}_i}{\partial u^2} + \boldsymbol{e}^3\cdot\frac{\partial \boldsymbol{e}_i}{\partial u^3}\ . \tag{14.1.3}$$

(We could have done this in (14.1.1) but we want to display (14.1.2) the way it stands for future reference.) After replacing the contravariant basis vectors $\boldsymbol{e}^1$, $\boldsymbol{e}^2$ and $\boldsymbol{e}^3$ by the cross products of the covariant-basis vectors, we have:

$$\begin{aligned} \{\boldsymbol{\nabla}\cdot\boldsymbol{e}_i\} &= \frac{1}{J}\left[\frac{\partial \boldsymbol{e}_i}{\partial u^1}\cdot\boldsymbol{e}_2\times\boldsymbol{e}_3 + \frac{\partial \boldsymbol{e}_i}{\partial u^2}\cdot\boldsymbol{e}_3\times\boldsymbol{e}_1 + \frac{\partial \boldsymbol{e}_i}{\partial u^3}\cdot\boldsymbol{e}_1\times\boldsymbol{e}_2\right] \\ &= \frac{1}{J}\left[\frac{\partial \boldsymbol{e}_1}{\partial u^i}\cdot\boldsymbol{e}_2\times\boldsymbol{e}_3 + \frac{\partial \boldsymbol{e}_2}{\partial u^i}\cdot\boldsymbol{e}_3\times\boldsymbol{e}_1 + \frac{\partial \boldsymbol{e}_3}{\partial u^i}\cdot\boldsymbol{e}_1\times\boldsymbol{e}_2\right]. \end{aligned}$$

In the first step, we made use of the commutativity of the dot product, while in the last line, we used (2.6.20). Using the product rule of differentiation in reverse, we have

$$\{\nabla\cdot e_i\} = \frac{1}{J}\frac{\partial}{\partial u^i}(e_1\cdot e_2\times e_3)\ , \tag{14.1.4}$$

which in turn, because of (2.5.22a) is

$$\begin{aligned}\{\nabla\cdot e_i\} &= \frac{1}{J}\frac{\partial J}{\partial u^i} = \frac{\partial}{\partial u^i}\ln J = \frac{\partial}{\partial u^i}\ln\sqrt{g} = \frac{1}{\sqrt{g}}\frac{\partial\sqrt{g}}{\partial u^i}\\ &= \frac{1}{2}\frac{\partial \ln g}{\partial u^i} = \frac{1}{2g}\frac{\partial g}{\partial u^i}\ .\end{aligned} \tag{14.1.5}$$

Substitution into the expression for $\nabla\cdot A$, (14.1.2), leads to the "standard" divergence expression

$$\operatorname{div} A \equiv \nabla\cdot A = \frac{\partial A^k}{\partial u^k} + \frac{A^i}{\sqrt{g}}\frac{\partial}{\partial u^i}(\sqrt{g}) \tag{14.1.6}$$

or because of the product-differentiation rule

$$\operatorname{div} A \equiv \nabla\cdot A = \frac{1}{\sqrt{g}}\frac{\partial}{\partial u^k}(\sqrt{g}A^k) = \frac{1}{J}\frac{\partial}{\partial u^k}(JA^k)\ , \tag{14.1.7}$$

since i and k are dummy indices. In the above derivations, we did not make explicit use of Christoffel symbols or covariant derivatives.

14.2 Divergence-Formula Derivation Employing Christoffel Symbols

Next, we give the derivation of (14.1.7) with Christoffel symbols and the covariant derivative, since it is this derivation that is found in most reference works. Rather than using the rule $\nabla\cdot$(product) as in (14.1.1), we apply the definition of the ∇ operator "acting on" the vector A, and subsequently invoke the covariant derivative formalism developed in Sect. 2.6. Thus,

$$\operatorname{div} A \equiv \nabla\cdot A = \nabla u^k\frac{\partial}{\partial u^k}\cdot A = \nabla u^k\cdot\frac{\partial A}{\partial u^k} = e^k\cdot e_j\left(\frac{\partial A}{\partial u^k}\right)^j = \left(\frac{\partial A}{\partial u^k}\right)^k\ , \tag{14.2.1}$$

where we have used the notation of (2.6.6). Because of (2.6.15) and (2.6.31), this becomes

$$\operatorname{div} A \equiv \nabla\cdot A = A^k_{;k} \tag{14.2.2}$$

or

$$\operatorname{div} \boldsymbol{A} \equiv \boldsymbol{\nabla} \cdot \boldsymbol{A} = \frac{\partial A^k}{\partial u^k} + \begin{Bmatrix} k \\ i\ k \end{Bmatrix} A^i \ . \tag{14.2.3}$$

A summation is implied over all repeated indices, i.e., k and i. If we compare this last result with (14.1.2), the curly brackets anticipated a similarity between $\boldsymbol{\nabla} \cdot \boldsymbol{e}_i$ and the Christoffel symbol in (14.2.3):

$$\boldsymbol{\nabla} \cdot \boldsymbol{e}_i \equiv \begin{Bmatrix} k \\ i\ k \end{Bmatrix} \ . \tag{14.2.4}$$

In (14.2.4) a summation over k is implied. Although we kept k fixed, above, the formulae can be used for k variable.

This identity also follows from (14.1.3) and the definition of the Christoffel symbol of the second kind, (2.6.25).

Because of the identity in (14.2.4), we can substitute the result for $\boldsymbol{\nabla} \cdot \boldsymbol{e}_i$ obtained in (14.1.5) into (14.2.3) to retrieve the expression for the divergence in (14.1.7). However, there is an alternative derivation that makes no explicit use of the basis vectors. This derivation has the advantage (for theoretical physics) that, once the divergence is defined, as in (14.2.2), no reference to basis vectors is needed, but only to metric coefficients. A determinant can be computed by means of the Laplace development. If we call $g = \det[g_{ij}]$, we have that

$$g = \sum_j g_{ij} G^{ij} \ , \qquad i = 1, 2 \text{ or } 3 \tag{14.2.5}$$

where G^{ij} is the cofactor of the element g_{ij}. The summation is only over j. Since G^{ij} does not contain the element g_{ij} explicitly, we can write the derivative of g with respect to g_{im} (i is the fixed index in (14.2.5), and m is one of the j's)

$$\frac{\partial g}{\partial g_{im}} = G^{im} \ . \tag{14.2.6}$$

Application of chain-rule differentiation to the expression $\partial g/\partial u^k$ gives

$$\frac{\partial g}{\partial u^k} = \sum_{i,m} \frac{\partial g}{\partial g_{im}} \frac{\partial g_{im}}{\partial u^k} = \sum_{i,m} G^{im} \frac{\partial g_{im}}{\partial u^k} = \sum_{i,m} g g^{im} \frac{\partial g_{im}}{\partial u^k} \ . \tag{14.2.7}$$

In the last equality, we have applied a relationship discussed underneath (2.5.14)

From the definition of the Christoffel symbol of the first kind, (2.6.26), we observe that

$$[m, ik] + [i, mk] = \frac{1}{2}\left[\frac{\partial g_{mi}}{\partial u^k} + \frac{\partial g_{mk}}{\partial u^i} - \frac{\partial g_{ik}}{\partial u^m}\right] + \frac{1}{2}\left[\frac{\partial g_{im}}{\partial u^k} + \frac{\partial g_{ik}}{\partial u^m} - \frac{\partial g_{mk}}{\partial u^i}\right]$$

or because of the symmetry of g_{im}:

$$[m, ik] + [i, mk] = \frac{\partial g_{im}}{\partial u^k} \ . \tag{14.2.8}$$

Then (14.2.7) becomes

$$\frac{\partial g}{\partial u^k} = \sum_{i,m} g g^{im}([m, ik] + [i, mk]) \ . \tag{14.2.9}$$

This can be rewritten with Christoffel symbols of the second kind according to (2.6.30), recalling the symmetry of g^{im},

$$\frac{\partial g}{\partial u^k} = \sum_{i,m} g\left(\left\{\begin{matrix} i \\ i\ k \end{matrix}\right\} + \left\{\begin{matrix} m \\ m\ k \end{matrix}\right\}\right) , \tag{14.2.10}$$

which, because i and m are dummy (summation) indices, becomes

$$\frac{\partial g}{\partial u^k} = g \sum_{i} 2\left\{\begin{matrix} i \\ i\ k \end{matrix}\right\} . \tag{14.2.11}$$

Application of (2.6.27), omitting the summation sign, and interchanging the dummy indices i and k, finally leads to

$$\left\{\begin{matrix} k \\ i\ k \end{matrix}\right\} = \frac{1}{2g}\frac{\partial g}{\partial u^i}$$

which is the same as (14.1.5).

References

Abraham, R., Marsden J.E. (1978) *Foundations of Mechanics*, 2nd edn. Benjamin/Cummings, Reading, MA

Akivis, M.A., Goldberg V.V. (1977) *An Introduction to Linear Algebra and Tensors.* Dover, New York

Anania, G., Johnson J.L. Weimer K.E. (1983) Phys. Fluids **26**, 2210

Anania, G., Johnson J.L. (1983) Phys. Fluids **26**, 3070

Arnold, V.I. (1978) *Mathematical Methods of Classical Mechanics.* Springer, New York

Baldwin, D.E. (1977) Reviews of Modern Physics **49**, 317

Balescu, R. (1975) *Equilibrium and Nonequilibrium Statistical Mechanics.* Wiley, New York

Balescu, R. (1988) *Transport Processes in Plasmas*, 2 vol. North-Holland, Amsterdam

Bateman, G. (1978) *MHD Instabilities.* MIT Press, Cambridge, MA

Bauer, F., Betancourt O., Garabedian P. (1978) *A Computational Method in Plasma Physics.* Springer, New York

Bauer, F., Betancourt O., Garabedian P. (1984) *Magneto Hydrodynamic Equilibrium and Stability in Stellarators.* Springer, New York

Beasley, C.O., Rome J.A., Attenberger S.E., Hirshman S.P. (1987) Oak Ridge Report, ORNL/TM-10222, Oak Ridge, TN 37831

Bedeaux, D. (1980) In: Cohen, E.G.D. (ed) *Fundamental Problems in Statistical Mechanics.* North-Holland, New York, Vol. 5, 313

Behnke, H., Bachmann F., Fladt K., Süss W. (eds) (1974) *Fundamentals of Mathematics.* MIT Press, Cambridge, MA

Bergland, G.D. (1969) IEEE Spectrum, **6**, 41 (July)

Bernstein, I.B. (1971) In: Simon A., Thompson W.B., (eds) *Advances in Plasma Physics.* Wiley, New York, Vol. 4, 311

Bernstein, I.B. (1974) Phys. Fluids **17**, 547

Betancourt, O., Garabedian P. (1985) Phys. Fluids **28**, 912

Bhattacharjee, A., Wiley J.C., Dewar R.L. (1984) Computational Phys. Commun. **31**, 213

Bird, R.B., Stewart W.E., Lightfoot E.N. (1960) *Transport Phenomena*, Wiley, New York

Blum, J., Le Foll J. (1984) Computer Physics Reports **1**, 465

Blum, J. (1989) *Numerical Simulation and Optimal Control in Plasma Physics.* Wiley (Gauthier-Villars), New York

Boozer, A.H. (1980) Phys. Fluids **23**, 904

Boozer, A.H. (1981) Phys. Fluids **24**, 1999

Boozer, A.H. (1982) Phys. Fluids **25**, 520

Boozer, A.H. (1983) Phys. Fluids **26**, 1288

Boozer, A.H. (1984) Princeton report, PPPL-2094, Princeton, NJ 08544

Boozer, A.H. (1984) Phys. Fluids **27**, 2055

Boozer, A.H. (1984) Phys. Fluids **27**, 2110

Boozer, A.H. (1984) Princeton report, PPPL-2154, Princeton, NJ 08544

Boozer, A.H. (1985) Princeton report, PPPL-2094R, Princeton, NJ 08544

Borisenko, A.I., Tarapov I.E. (1968) *Vector and Tensor Analysis with Applications.* Prentice-Hall, Englewood Cliffs, NJ, re-edited by Dover

Braginskii, S.I. (1965) In: Leontovich, M.A. (ed) *Reviews of Plasma Physics.* Consultants Bureau, New York, Vol. 1, 205

Buck, R.C. (1965) *Advanced Calculus*, 2nd edn. McGraw-Hill, New York
Bulmer, R.H., Kaiser T.B., Nevins W.M., Newcomb W.A., Pearlstein L.D., Strauss H.R. Wollman S., Wakatani M. (1983) *Plasma Physics and Controlled Nuclear Fusion Research* (Baltimore, 1982). IAEA, Vienna, Vol. 1, 531
Burke, W.L. (1985) *Applied Differential Geometry*. Cambridge University Press, Cambridge, England
Butkov, E. (1968) *Mathematical Physics*. Addison-Wesley, Reading, MA

Callen, J.D., Dory R.A. (1972) Phys. Fluids **15**, 1523
Cary, J.R., Littlejohn R.G. (1983) *Annals of Physics* **151**, 1
Catto, P.J., Hazeltine R.D. (1981) Phys. Fluids **24**, 1663
Catto, P.J., Myra J.R. (1986) Plasma Phys. and Contr. Fusion **28**, 959
Chapman, S., Cowling T.G. (1973) *The Mathematical Theory of Non-uniform Gases*, 3rd edn. Cambridge University Press, Cambridge, England
Chodura, R., Schlüter A. (1981) *J. Computational Physics* **41**, 681
Choquet-Bruhat, Y., DeWitt-Morette C., Dillard-Bleick M. (1982) *Analysis and Physics*, revised edn. North-Holland, New York
Cordey, J.G., Watson C.J.H. (1969) *Proc. of Conf. on Nuclear Fusion Reactors*. UKAEA, Culham Laboratory, England, 122
Cross, R.C. (1989) Am. J. Phys. **57**, 722

Davis, H.F., Snider A.D. (1975) *Introduction to Vector Analysis*, 3rd edn. Allyn and Bacon, Boston, MA
Deschamps, G.A., (1970) In: Roubine, E. (ed) *Mathematics Applied to Physics*. Springer Berlin, Heidelberg, Chap. 3, 111
Dobrott, D., Frieman E.A. (1971) Phys. Fluids **14**, 349
Drew, T.B. (1961) *Handbook of Vector and Polyadic Analysis*. Reinhold, New York

Ferziger, J.H., Kaper H.G. (1972) *Mathematical Theory of Transport Processes in Gases*. North-Holland, New York
Feynman, R.P., Leighton R.B., Sands M. (1963) *The Feynman Lectures on Physics*, Addison-Wesley, Reading, MA (3 vols)
Fielding, P.J., Hitchon W.N.G. (1980) J. Plasma Physics **24**, 453
Filonenko, N.N., Sagdeev R.Z., Zaslavsky G.M. (1967) Nuclear Fusion **7**, 253
Flanders, H. (1963) *Differential Forms with Applications to the Physical Sciences*. Academic, New York
Freidberg, J.P. (1982) Rev. Mod. Phys. **54**, 801
Freidberg, J.P. (1987) *Ideal Magnetohydrodynamics*, Plenum, New York
Furth, H.P., Rosenbluth M.N. (1964) Phys. Fluids **7**, 764

Gelfand, I.M. (1961) *Lectures on Linear Algebra*. Wiley, New York
Goetz, A. (1970) *Introduction to Differential Geometry*. Addison-Wesley, Reading, MA
Golant, V.E., Zhilinsky A.P., Sakharov I.E. (1980) *Fundamentals of Plasma Physics*. Wiley, New York
Goldstein, H. (1981) *Classical Mechanics*, 2nd edn. Addison-Wesley, Reading, MA
Grad, H., Rubin H. (1958) *Proc. of the Second United Nations Conference on the Peaceful Uses of Atomic Energy*, United Nations, Geneva, Vol. 31, 190
Grad, H. (1967) Phys. Fluids, **10**, 137
Grad, H., Hogan J. (1970) Phys. Rev. Lett. **24**, 1337
Grad, H., (1971) *Plasma Physics and Controlled Nuclear Fusion Research* (Madison, 1971). IAEA, Vienna, Vol. 3, 229
Grad, H. (1974) *Advances of Plasma Physics*. Wiley, New York, Vol. 5, 103
Grad, H., Hu P.N., Stevens D.C. (1975) *Proc. Nat. Acad. Sci. USA*, **72**, 3789
Greene, J.M., Johnson J.L. (1961) Phys. Fluids **4**, 875
Greene, J.M., Johnson J.L. (1962) Phys. Fluids, **5**, 510
Greene, J.M., Johnson J.L., Weimer K.E. (1971) *Phys. Fluids*, **14**, 671
Guggenheimer, H.W. (1977) *Differential Geometry*. Dover, New York

Hall, L.S., McNamara B. (1975) Phys. Fluids **18**, 552
Hamada, S. (1962) Nuclear Fusion **2**, 23
Harris, F.J. (1978) Proceedings of the IEEE, **66**, 51
Hazeltine, R.D., Meiss J.D. (1985) Physics Reports **121**, 1
Hinton, F.L., Hazeltine R.D. (1976) Reviews of Modern Physics **48**, 239
Hirshman, S.P., Jardin S.C. (1979) Phys. Fluids **22**, 731
Hirshman, S.P., Sigmar D.J. (1981) Nuclear Fusion **21**, 1079
Hirshman, S.P., (1982) Oak Ridge National Laboratory Report, ORNL/TM-8393, unpublished
Hirshman, S.P., Whitson J.C. (1983) Phys. Fluids **26**, 3553
Hirshman, S.P., Lee D.K. (1986) Computational Phys. Commun. **39**, 161
Hitchon, W.N.G., Fielding P.J. (1981) Nuclear Fusion **21**, 775
Hoffman, K., Kunze P. (1971) *Linear Algebra*. Prentice-Hall, Englewood Cliffs, NJ
Hoffmann, B. (1977) *About Vectors*. Dover, New York

Jackson, J.D. (1975) *Classical Electrodynamics*. 2nd edn. Wiley, New York
Jancel, R., Kahan T. (1966) *Electrodynamics of Plasmas*. Wiley, New York
Johnson, J.L., Oberman C.R., Kulsrud R.M., Frieman E.A. (1958) *Phys. Fluids*, **1**, 281
Johnson, J.L., Boozer A.H., Koniges A.E., Kuo-Petravic G., Manickam J., Monticello D.A., Mynick H., Park W., Reiman A., Rewoldt G. 1984, Princeton report, PPPL-2143, Princeton, NJ 08544
Johnson, J.L. (1986) Computational Phys. Rep. **4**, 37

Kerst, D.W. (1962) Plasma Physics (J. of Nuclear Energy, Part C) **4**, 253
Kieras, C.E. (1982) Ph.D. Thesis, University of Wisconsin, Physics, Appendix A
Killeen, J. (ed) (1976) *Methods in Computational Physics*, Vol. 16. Academic, New York
Kovrizhnykh, L.M. (1984) Nuclear Fusion **24**, 851
Krall, N.A., Trivelpiece A.W. (1973) *Principles of Plasma Physics*. McGraw-Hill, New York
Kruskal, M.D., Kulsrud R.M. (1958) Phys. Fluids **1**, 265
Kulsrud, R.M. (1983) In: Rosenbluth, M.N., Sagdeev, R.Z. (eds) *Handbook of Plasma Physics*. Vol. 1: *Basic Plasma Physics* I. North-Holland, New York, 115
Kuo-Petravic, G., Boozer A.H., Rome J.A., Fowler R.H. (1983) J. of Computational Physics **51**, 261
Kuo-Petravic, G. (1984) Computer Physics Communications **33**, 353
Kuo-Petravic, G., Boozer A.H. (1986) Princeton Report, PPPL-2320, Princeton, NJ 08544 (Also, in J. Computational Physics **73**, 107 (1987))

Lackner, K. (1976) Computational Phys. Rep. **12**, 33
Lamb, H. (1945) *Hydrodynamics*, 6th edn. Dover, New York
Lao, L.L., Hirshman S.P., Wieland R.M. (1981) *Phys. Fluids* **24**, 1431
Lao, L.L. (1984) Computational Phys. Rep. **31**, 201
Lee, D.K., Hedrick C.L., Callen J.D. (1986) Phys. Rev. A **33**, 2675
Lichnerowicz, A. (1962) *Elements of Tensor Calculus*. Wiley, New York
Lichtenberg, A.J., Lieberman M.A. (1983) *Regular and Stochastic Motion*. Springer, New York
Littlejohn, R.G. (1984) In: Horton C.W., Jr., Reichl L.E., (eds) *Statistical Physics and Chaos in Fusion Plasmas*. Wiley, New York, 43
Lüst, R., Schlüter A. (1957) Z. Naturforschung **12a**, 850 (in German)

Margenau, H., Murphy G.M. (1976) *The Mathematics of Physics and Chemistry*. Krieger, Huntington, NY
Mathews, J., Walker R.L. (1970) *Mathematical Methods of Physics*. Benjamin/Cummings, Menlo Park, CA
Menzel, D.H. (1961) *Mathematical Physics*. Dover, New York
Milne-Thomson, L.M. (1968) *Theoretical Hydrodynamics*, 5th edn. Macmillan, London
Misner, C.W., Thorne K.S., Wheeler J.A. (1973) *Gravitation*. W.H. Freeman, San Francisco, CA
Miyamoto, K. (1980) *Plasma Physics for Nuclear Fusion*. MIT Press, Cambridge, MA

Morozov, A.I., Solov'ev L.S. (1966) In: Leontovich, M.A. (ed) *Reviews of Plasma Physics.* Consultants Bureau, New York, Vol. 2, 1
Morse, P.M., Feshbach H. (1953) *Methods of Theoretical Physics.* McGraw-Hill, New York
Mukhovatov, V.S., Shafranov V.D. (1971) Nuclear Fusion **11**, 605
Myra, J.R., Catto P.J. (1982) Phys. Rev. Lett. **48**, 620
Myra, J.R., Catto P.J., Hazeltine R.D. (1982) Phys. Fluids **25**, 2028

Newcomb, W.A. (1958) Annals of Physics **3**, 347
Newcomb, W.A. (1959) Phys. Fluids **2**, 362
Newcomb, W.A. (1981) J. Plasma Physics **26**, 529

Panofsky, W.K.H., Phillips M. (1962) *Classical Electricity and Magnetism.* Addison-Wesley, Reading, MA
Pao, Y.-P. (1976) Phys. Fluids **19**, 1177
Pao, Y.-P. (1978) Phys. Fluids **21**, 1120
Pearlstein, L.D., Kaiser T.B., Newcomb W.A. (1981) Phys. Fluids. **24** 1326
Pereversev, G.V., Shafranov V.D., Zakharov L.E. (1978) In: *Theoretical and Computational Plasma Physics* (Proc. of the 3rd Kiev Conference on Plasma Physics Theory, Trieste, 1977) IAEA, Vienna, 469
Pytte, A., Boozer A.H. (1981) Phys. Fluids **24**, 88

Reiman, A.H., Boozer A.H. (1984) Phys. Fluids **27**, 2446
Reiman, A.H., Greenside H.S. (1988) J. Computational Phys. **75**, 423
Reiman, A.H, Pomphrey N. (1989) Princeton Report, PPPL-2652, to be published
Rome, J.A. (1989) J. Computational Phys. **82**, 348
Rosenbluth, M.N., Sagdeev R.Z., Taylor J.B. (1966) Nuclear Fusion **6**, 297

Schmidt, G. (1979) *Physics of High Temperature Plasmas.* Academic, New York
Schouten, J.A. (1951) *Tensor Analysis for Physicists.* Clarendon, Oxford
Schutz, B.F. (1980) *Geometrical Methods of Mathematical Physics.* Cambridge University Press, Cambridge, UK
Schwenn, U. (1984) Computational Phys. Commun. **31**, 167
Seshadri, S.R. (1973) *Fundamentals of Plasma Physics.* Elsevier, Amsterdam
Shafranov, V.D. (1958) Soviet Physics JETP **6**, 545 [J. Exptl. Phys. (USSR) **33**, 710 (1957)]
Shafranov, V.D. (1960) Soviet-Physics JETP **37**, 775 [J. Exptl. Phys. (USSR) **37**, 1088 (1959)]
Shafranov, V.D. (1966) In: Leontovich M.A. (ed) *Reviews of Plasma Physics.* Consultants Bureau, New York, Vol. 2, 103
Shafranov, V.D. (1968) Nuclear Fusion **8**, 253
Shafranov, V.D., Zakharov L.E. (1972) *Nuclear Fusion* **12**, 599
Shkarofsky, I.P., Johnston, T.W., Bachynski M.P. (1966) *The Particle Kinetics of Plasmas.* Addison-Wesley, Reading, MA
Solov'ev, L.S. (1968) Soviet Physics JETP **26**, 400 [*Zh. Eksp. Teor. Fiz.* **53**, 626 (1967)]
Solov'ev, L.S. (1968) Soviet Physics JETP **26**, 1167 [*Zh. Eksp. Teor. Fiz.* **53**, 2063 (1967)]
Solov'ev, L.S., Shafranov V.D. (1970) In: Leontovich, M.A. (ed) *Reviews of Plasma Physics.* Consultants Bureau, New York, Vol. 5, 1
Solov'ev, L.S. (1975) In: Leontovich, M.A. (ed) *Reviews of Plasma Physics.* Consultants Bureau, New York, Vol. 6, 239
Sowerby, L. (1974) *Vector Field Theory with Applications.* Longman, London
Spiegel, M.R. (1959) *Vector Analysis and an Introduction to Tensor Analysis.* McGraw-Hill, New York
Spies, G.O., Nelson D.B. (1974) Phys. Fluids **17**, 1879
Spies, G.O. (1978) Nuclear Fusion **18**, 1671
Spitzer, L., Jr. (1958) Phys. Fluids **1**, 253; also reprinted in IEEE PS **9**, 130 (1981)

Spivak, M. (1970 and 1975) *A Comprehensive Introduction to Differential Geometry.* Perish Publ., Boston, MA, Vols. 1–2, 1970, and Vols. 3–5, 1975
Springer, C.E. (1962) *Tensor and Vector Analysis.* Ronald, New York
Stacey, W.M., Jr. (1981) *Fusion Plasma Physics.* Wiley, New York
Stoker, J.J. (1969) *Differential Geometry.* Wiley, New York
Stratton, J.A. (1941) *Electromagnetic Theory.* McGraw-Hill, New York
Symon, K.R. (1971) *Mechanics,* 3rd edn. Addison-Wesley, Reading, MA

Van Bladel, J. (1964) *Electromagnetic Fields.* McGraw-Hill, New York

White, R.B., Boozer A.H., Goldston R., Hay R., Albert J., Karney C.F.F. (1983) *Plasma Physics and Controlled Nuclear Fusion Research* (Baltimore 1982), IAEA, Vienna, Vol. 3, 391
White, R.B., Chance M.S. (1984) Phys. Fluids **27**, 2455
White, R.B. (1989) *Theory of Tokamak Plasmas.* North-Holland, Amsterdam
Whiteman, K.J. (1977) Rep. Prog. Phys. **40**, 1033
Wimmel, H.K. (1982) European J. of Physics **3**, 223
Woodson, H.H., Melcher J.R. (1968) *Electromechanical Dynamics,* Part III. Wiley, New York
Wrede, R.C. (1963) *Introduction to Vector and Tensor Analysis.* Wiley, New York; re-edited by Dover

Zakharov, L.E., Shafranov V.D. (1981) In: Kadomtsev, B.B. (ed) *Plasma Physics.* MIR, Moscow, 13
Zakharov, L.E., Shafranov V.D. (1986) In: Leontovich, M.A. (ed) *Reviews of Plasma Physics.* Consultants Bureau, New York, Vol. 11, 153

Subject Index

FSC
www.fsc.org
MIX
Papier aus verantwortungsvollen Quellen
Paper from responsible sources
FSC® C105338

If you have any concerns about our products,
you can contact us on
ProductSafety@springernature.com

In case Publisher is established outside the EU,
the EU authorized representative is:
Springer Nature Customer Service Center GmbH
Europaplatz 3, 69115 Heidelberg, Germany

Printed by Libri Plureos GmbH
in Hamburg, Germany